An Introduction to
Electronic Materials
for Engineers

(2nd Edition)

An Introduction to
Electronic Materials
for Engineers

(2nd Edition)

Wei Gao & Zhangwei Li
University of Auckland, New Zealand

Nigel Sammes
Colorado School of Mines, USA

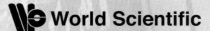
World Scientific

NEW JERSEY · LONDON · SINGAPORE · BEIJING · SHANGHAI · HONG KONG · TAIPEI · CHENNAI

Published by

World Scientific Publishing Co. Pte. Ltd.

5 Toh Tuck Link, Singapore 596224

USA office: 27 Warren Street, Suite 401-402, Hackensack, NJ 07601

UK office: 57 Shelton Street, Covent Garden, London WC2H 9HE

British Library Cataloguing-in-Publication Data
A catalogue record for this book is available from the British Library.

ISBN-13 978-981-4293-69-3
ISBN-10 981-4293-69-5

Printed in Singapore by Mainland Press Pte Ltd.

Preface

In 1998, the first edition of *Introduction to Electronic and Ionic Materials* was published by World Scientific. The book was assembled by Gao and Sammes from a series of lecture notes they had used for undergraduate and graduate courses to materials engineering majors and electrical/electronic engineering students. At that time, there really were no books of this kind published, particularly from properties and uses standpoint, rather than a fundamental physics viewpoint. The first edition was aimed at engineering and technology undergraduates who already have a background in physics and chemistry but only at the first year level. It provided a basic understanding of the electronic, magnetic, and other functional properties of materials, and uses of a wide range of electrically and ionically conducting materials. The book gave the students an overview of a range of electronic and ionic materials, and emphasized their applications in today's world.

We decided to write a second edition a couple of years ago, as the subject was expanding rapidly, particularly with the huge increase in interest in energy and the environment. There is also a tremendous amount of work being done today on nanomaterials and nano-technology, renewable energy materials and technologies. In fact, there are even majors on this subject now as such, we felt that an updated version of the book was in order, with more emphasis on nanomaterials and their applications, and materials for renewable energy. However, we have kept the basis of the original text in place: the book is aimed at the undergraduate level engineering students and technologists and not intended to be a solid state physics or chemistry book, hence the mathematics is still kept at a minimum. We have also included some exercise and problems at the end of each chapter, and have added and updated a major portion of the original material. The title has been altered as *Introduction to Electronic Materials*, to indicate it is an introductory text for a wide range of engineering readers.

We hope readers will enjoy this new edition. We would very much like to hear your feedback on where we can make further improvements. The field is expanding so rapidly, that we hope this edition will not become obsolete too quickly.

We would like to thank our colleagues at the University of Auckland, especially Drs. Michael Hodgson and Chongwen Zou, who also teach the course of "Electronic Materials and Their Applications". Gao and Li would like to thank their research group and the laboratories where research on electronic materials and transition metal oxide films are actively conducted. Sammes completed his text while on sabbatical at Karlsruhe Institute of Technology, in Germany, under a Humboldt Fellowship, and thus would like to thank the Alexander von-Humboldt-Stiftung for the opportunity to stay there.

Wei Gao
Zhengwei Li
Nigel Sammes

February 2011

Symbols and Units

Length	l	meter (m)
Area	A	m^2
Mass	m	kilogram (kg)
Density	d	kg/m^3 or g/cm^3
Time	t	second (s)
Temperature	T	degree Celsius ($^\circ$C) or Kelvin (K)
Force	F	newton (N, $kg \cdot m/s^2$)
Stress	σ	pascal (Pa, N/m^2)
Strain	ε	m/m
Young's modulus	E	Pa
Energy/work/quantity of heat	E, W	joule (J, N·m) or eV
Kinetic energy	E_K	joule (J)
Energy loss	E_{loss}	joule (J)
Power	N	watt (W, J/s)
Electric charge	q	coulomb (C, A·s)
Current flow	i	ampere (A)
Current density	j, J	A/m^2
Potential difference	V	volt (V)
Electric field	E	V/m
Electric resistance	R	ohm (Ω)
Electric resistivity	ρ	$\Omega \cdot m$
Electric conductivity	σ	$(\Omega \cdot m)^{-1}$
Temperature coefficient of resistivity	TCR, α	%/K
Electron drift velocity	v_d	m/s
Mobility of the electron	μ	$m^2/(V \cdot s)$
Hall voltage	V_H	V
Hall field strength	E_H	V/m
Hall constant	R_H	
Lorenz Number	$\mathbb{L}$	$W \cdot \Omega/K^2$
Fermi energy level	E_f	eV
Band gap energy	E_g	eV or J
Work function	$e\phi$	eV or J

Frequency	ν, f	hertz (Hz, s^{-1})
Wavelength	λ	m
Wave number	$\mathbf{K}$	m^{-1}
Bragg diffraction angle	θ	degree
Intrinsic carrier concentration	n_i	$/m^3$
Electron concentration	n_n	$/m^3$
Hole concentration	n_p	$/m^3$
Reverse saturation current	I_s	A/m^2
Diffusion coefficient	D	m^2/s
Activation energy	Q, E	kJ/mol, kcal/mol, eV/atom
Magnetic field	H	A/m
Magnetic induction (flux density)	B	tesla (T, $V \cdot s/m^2$)
Magnetic permeability	μ	H/m
Relative permeability	μ_r	$= \mu/\mu_0$
Magnetisation	M	A/m
Magnetic susceptibility	χ	$= M/H$
Curie temperature	T_C	K
Magnetostriction	λ_s	$= \Delta l/l$
Eddy current loss	W_e	W/m^3
Saturation induction	B_s	T
Remanent induction	B_r	T
Coercive force	H_c	A/m
Maximum energy product	$(BH)_{max}$	J/m^3
Capacitance	C	F
Permittivity	ε	F/m
Relative permittivity (dielectric constant)	ε_r, K	$= \varepsilon/\varepsilon_0$
Tangent of loss angle	$\tan \delta$	
Relative loss factor	F_L	
Dielectric strength (breakdown strength)	E_B	V/m
Polarisation	P	C/m^2
Concentration of charge centres	Z	$/m^3$
Power loss of a dielectric material	P_L	J/m^3
Light intensity	I	dB, lumen
Relaxation time	τ	s
Linear absorption coefficient of light	μ	

Index of refraction	n	
Reflectivity	R	%
Incident and refracted angles	α and β	degree
Heat capacity at constant volume	C_v	J/(mol·K)
Heat capacity at constant volume	C_v	J/(mol·K)
Linear coefficient of thermal expansion	α	m/(m·K)
Thermal conductivity	K	W/(m·K)
Thermoelectric (Seebeck) power	S	V/K
Superconducting transition temperature	T_C	K
Critical superconducting current	I_C	A
Critical superconducting current density	J_C	A/m^2 or A/cm^2
Critical magnetic field	H_C	A/m
Persistent current	I_0	A

Constants

Avogadro's number	N_0	6.023×10^{23} /mol
Bohr magneton	μ_B	9.27×10^{-24} J/T
Boltzmann's constant	K, K_B	8.62×10^{-5} eV/K or 1.38×10^{-23} J/K
Electron charge	e	1.6×10^{-19} C
Electron mass	m_e	9.1×10^{-31} kg
Faraday's constant	F	9.65×10^4 C/mol
Gravitational constant	g	9.8 m/s
Gas constant	R	8.314 J/(mol·K) or 1.987 cal/(mol·K)
Permeability of vacuum	μ_0	$4\pi \times 10^{-7}$ H/m
Planck's constant	h	6.63×10^{-34} J·s
Speed of light	c	3.0×10^8 m/s
Permittivity of vacuum	ε_0	8.85×10^{-12} F/m

Periodic Table of the Elements

1	2	3	4	5	6	7	8	9	10	11	12	13	14	15	16	17	18
hydrogen 1 **H** 1.0079																	helium 2 **He** 4.0026
lithium 3 **Li** 6.941	beryllium 4 **Be** 9.0122											boron 5 **B** 10.811	carbon 6 **C** 12.011	nitrogen 7 **N** 14.007	oxygen 8 **O** 15.999	fluorine 9 **F** 18.998	neon 10 **Ne** 20.180
sodium 11 **Na** 22.990	magnesium 12 **Mg** 24.305											aluminium 13 **Al** 26.982	silicon 14 **Si** 28.086	phosphorus 15 **P** 30.974	sulfur 16 **S** 32.065	chlorine 17 **Cl** 35.453	argon 18 **Ar** 39.948
potassium 19 **K** 39.098	calcium 20 **Ca** 40.078	scandium 21 **Sc** 44.956	titanium 22 **Ti** 47.867	vanadium 23 **V** 50.942	chromium 24 **Cr** 51.996	manganese 25 **Mn** 54.938	iron 26 **Fe** 55.845	cobalt 27 **Co** 58.933	nickel 28 **Ni** 58.693	copper 29 **Cu** 63.546	zinc 30 **Zn** 65.39	gallium 31 **Ga** 69.723	germanium 32 **Ge** 72.61	arsenic 33 **As** 74.922	selenium 34 **Se** 78.96	bromine 35 **Br** 79.904	krypton 36 **Kr** 83.80
rubidium 37 **Rb** 85.468	strontium 38 **Sr** 87.62	yttrium 39 **Y** 88.906	zirconium 40 **Zr** 91.224	niobium 41 **Nb** 92.906	molybdenum 42 **Mo** 95.94	technetium 43 **Tc** [98]	ruthenium 44 **Ru** 101.07	rhodium 45 **Rh** 102.91	palladium 46 **Pd** 106.42	silver 47 **Ag** 107.87	cadmium 48 **Cd** 112.41	indium 49 **In** 114.82	tin 50 **Sn** 118.71	antimony 51 **Sb** 121.76	tellurium 52 **Te** 127.60	iodine 53 **I** 126.90	xenon 54 **Xe** 131.29
caesium 55 **Cs** 132.91	barium 56 **Ba** 137.33	57-70 * lutetium 71 **Lu** 174.97	hafnium 72 **Hf** 178.49	tantalum 73 **Ta** 180.95	tungsten 74 **W** 183.84	rhenium 75 **Re** 186.21	osmium 76 **Os** 190.23	iridium 77 **Ir** 192.22	platinum 78 **Pt** 195.08	gold 79 **Au** 196.97	mercury 80 **Hg** 200.59	thallium 81 **Tl** 204.38	lead 82 **Pb** 207.2	bismuth 83 **Bi** 208.98	polonium 84 **Po** [209]	astatine 85 **At** [210]	radon 86 **Rn** [222]
francium 87 **Fr** [223]	radium 88 **Ra** [226]	89-102 ** lawrencium 103 **Lr** [262]	rutherfordium 104 **Rf** [261]	dubnium 105 **Db** [262]	seaborgium 106 **Sg** [266]	bohrium 107 **Bh** [264]	hassium 108 **Hs** [269]	meitnerium 109 **Mt** [268]	ununnilium 110 **Uun** [271]	unununium 111 **Uuu** [272]	ununbium 112 **Uub** [277]		ununquadium 114 **Uuq** [289]				

*Lanthanide series

lanthanum 57 **La** 138.91	cerium 58 **Ce** 140.12	praseodymium 59 **Pr** 140.91	neodymium 60 **Nd** 144.24	promethium 61 **Pm** [145]	samarium 62 **Sm** 150.36	europium 63 **Eu** 151.96	gadolinium 64 **Gd** 157.25	terbium 65 **Tb** 158.93	dysprosium 66 **Dy** 162.50	holmium 67 **Ho** 164.93	erbium 68 **Er** 167.26	thulium 69 **Tm** 168.93	ytterbium 70 **Yb** 173.04

**Actinide series

actinium 89 **Ac** [227]	thorium 90 **Th** 232.04	protactinium 91 **Pa** 231.04	uranium 92 **U** 238.03	neptunium 93 **Np** [237]	plutonium 94 **Pu** [244]	americium 95 **Am** [243]	curium 96 **Cm** [247]	berkelium 97 **Bk** [247]	californium 98 **Cf** [251]	einsteinium 99 **Es** [252]	fermium 100 **Fm** [257]	mendelevium 101 **Md** [258]	nobelium 102 **No** [259]

Contents

Chapter 1

Introduction

1.1 Definition and Development of Materials

What are materials? Materials are substances from which something is composed or made. Engineering materials are the materials we use to build our material world: appliances, bridges, buildings, communication facilities, devices, electricity systems, factories, furniture, instrumentation, irrigation systems, machines, pipelines, roads, transportation equipment, tools and various utilities.

Materials have been central to the growth, prosperity, security, and quality of life of human beings. Throughout history, the development of human civilisation has been closely tied to the materials that were produced and used in society. The history of human civilization, to a certain degree, could be regarded as a history of materials evolution. Actually, the civilisation levels have been named according to the materials used at that time. At the beginning of human civilisation, people used the materials that exist in nature, such as stone, sticks and clay, to fabricate tools and weapons. This is called the "Stone Age". With time people discovered techniques to produce metals that had properties superior to those natural materials. The industrial revolution and modern heavy industries were largely based on iron, bronze, steel and other metals, and therefore brought about the name of the "Metal Age" (Bronze Age, Iron Age and Industrial Revolution). It was not until relatively recent times that scientists started to study the relationships between the composition, structure and property of materials. The knowledge and understanding obtained through these studies enable us to design and create numerous materials with various properties that meet the needs of our modern society. Technology is rapidly changing and bringing us telephone, phonograph, wireless radio, motion picture, automobile, airplane and computer.

Material scientists and engineers now have a growing ability to tailor materials from the atomic scale upwards to obtain desired properties (bottom-up paradigm). A new materials age, the "Tailored (or Designed) Materials Age", has been used to describe the revolutionary changes in materials science and engineering (MSE), and their impact on our society. For instance, advanced composites have been developed to combine the properties of high stiffness, strength, toughness and low density to meet special structural requirements. Surface treatments including various coatings, thin films and surface modification techniques provide a combination of extremely high hardness, wear, corrosion and high temperature oxidation resistance on the top surface with a tough, shock-absorbing body. Artificial layered structures can offer limitless possibilities for creating new electronic and semiconductor devices. They can be produced by many methods including molecular beam epitaxy (MBE), chemical vapour deposition (CVD), vacuum evaporation, sputter deposition, ion beam deposition, and solid-phase epitaxy. We are now facing dramatic changes in the materials world, which give our industries and society endless development opportunities.

1.2 Classification and Properties of Materials

1.2.1 *Classification of materials*

The traditional way to classify materials is according to the nature of the materials:

Metals and alloys are inorganic materials, which are composed of one or more metallic elements and may also contain a small amount of non-metallic elements. Metals usually have a crystalline structure and are good thermal and electrical conductors. Many metals are strong and ductile at room temperature and maintain good strength at high and low temperatures.

Ceramics are inorganic materials, which consist of metallic and non-metallic elements chemically bonded together. Ceramics can be crystalline, non-crystalline or mixtures of both. They generally have high melting points and high chemical stabilities. They also have high hardness and high temperature strength, but tend to be brittle. They are generally more heat and corrosion resistant than metals or polymers and less dense than most metals and their alloys. Ceramics are usually poor electrical conductors. Ceramics can be further divided into two classes: traditional and advanced. Traditional ceramics include clay products, silicate glass and cement. Advanced ceramics consist of carbides (such as SiC and TiC), pure oxides (Al_2O_3 and SiO_2), nitrides (TiN, Si_3N_4), non-silicate glasses and many others.

Polymers are organic materials, which consist of carbon-containing long molecular chains or networks. Most polymers are non-crystalline but some consist of mixtures of crystalline and non-crystalline regions. They typically have low densities and are mechanically flexible. Their mechanical properties may vary considerably. Most polymers are poor electric conductors due to the nature of the atomic bonding, but conducting polymers have been developed in recent years.

Composites are mixtures of two or more different types of materials. Usually they consist of a matrix phase and a reinforcing phase. They are designed to obtain a combination of the best properties of each of the component materials.

There is also an increasing trend to classify engineering materials into another two categories: structural materials and functional materials (or electronic materials). Structural materials, as the name indicates, are materials used to build the structures, bodies and parts of everything. For instance in a car, the body, frame, wheels, seats, inside lining, engine and various mechanical transmission parts are all made of structural materials. Mechanical properties are the most important consideration for this type of application. The functional materials, on the other hand, are used to play special functions in equipment such as to conduct, insulate or store electricity, to generate or conduct light, to convert optical, mechanical or thermal signals into electric voltages, or to provide a strong magnetic field. The electronic devices in the control systems of a car are built with semiconductors, an important type of functional material.

1.2.2 *Properties of materials*

The properties of engineering materials can be classified into two main groups: physical and chemical properties.

Mechanical properties include Young's modulus, tensile and shear strengths, hardness, toughness, ductility, deformation and fracture behaviours, fatigue and creep strengths, and wear resistance etc. Other physical properties include electrical and electronic properties, dielectric, magnetic properties, optical and thermal properties etc.

Chemical properties of engineering materials generally include corrosion, oxidation, catalysis properties and chemical stabilities.

The present book concentrates not on the mechanical properties of materials, but on the other physical properties, such as electrical and electronic properties, magnetic properties, optical and thermal properties, although mechanical properties are always important in applications of functional materials. The present book emphasizes the performance and applications of the functional materials in our modern industries.

1.2.3 *Four elements of materials science and engineering*

Materials science is primarily concerned with the study of the basic knowledge about materials: the relationships between the composition/structure, properties and processing of materials. Materials engineering is mainly concerned with the use of this fundamental knowledge to design and to produce materials with the properties that meet the requirements of society. As subjects of study, materials science and materials engineering are very often closely related to each other. The subject "Materials Science and Engineering" combines both the basic knowledge and application, forms a bridge between the basic sciences (physics, chemistry and mathematics) and the various engineering disciplines, including electrical, mechanical, chemical, civil and aerospace engineering.

There are four essential elements in materials science and engineering: (1) processing/synthesis, (2) structure/composition, (3) properties and (4) performance/application (see Fig. 1.1). There is a good realisation among scientists and engineers that, to develop new materials and to provide materials efficiently for industry, all four elements should be considered. This gives materials science and engineering an interdisciplinary nature. It is common now and preferred in many cases, that people with different backgrounds, physics, chemistry, metallurgy, ceramics and electronics, etc, work together to solve materials problems and to make important contributions to our technological society.

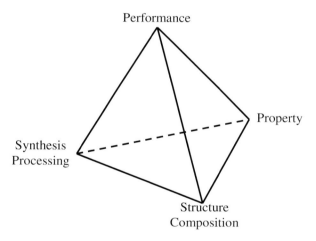

Fig. 1.1 Four elements of materials science and engineering (after National Research Council, "Materials Science and Engineering for the 1990s", National Academy Press, 1989).

1.3 Materials Used in Electrical and Electronic Industries

Electronic materials are the materials used in electrical and electronic industries. They are the substances for the fabrication of various electronic devices and integrated circuits, and can be used, for example, in circuit boards, communication cables, displays, packaging, optical fibers, and various control and monitoring devices.

According to their applications, electronic materials can be classified into two major groups: (1) structural electronic materials. They are relatively stable to pressure and weight, have good mechanical properties, and can be used as casings, frames, packaging, substrates and sealing materials; (2) functional electronic materials that can be used to realize chemical, electrical, magnetic, optical, and thermal functions. According to the basic composition of electronic materials, they can be divided into two groups as well: (1) inorganic electronic materials, typically include metallic and non-metallic (silicon, metal oxides, carbides, and nitrides); (2) organic electronic materials. They are mainly polymers (composed of C, H, O, N, Cl, and/or F) with covalent bonds and molecular links.

Functional electronic materials are often the critical part of equipment or instruments, which determines the overall performance and efficiency of the whole system. The subject of functional electronic materials has experienced rapid growth in the last thirty years. They are among the most active areas in modern science and engineering. The new developments of functional electronic materials such as semiconductors, superconductors and optoelectronic materials are making revolutionary changes in our modern society. Without these materials, our world of modern appliances, computers, machines, telecommunication systems, aircraft, etc, could not exist. For example, almost all the functions (engine, radio, air-conditioning and brake) in today's automobile rely on electronic control systems. The proportion of an automobile's value represented by electrical systems and electronics is growing rapidly. Currently, electrical and electronic components and software make up 20% of an automobile's value, on average, on a worldwide basis. By the year 2015, this proportion will grow to more than 30%, according to the Oliver Wyman study on auto electronics.

Electronic materials are one of the fastest growing and changing markets and the electronic industry is one of the most dynamic industries in the global economy. According to Report Buyer, global market for electronic products is expected to increase to $3.2 trillion by 2012, experiencing a compound annual growth rate of 9.5% over the next five years. The electronic industry is also a leader in the development and utilization of new materials. Highly engineered materials are vital to the progress of the electronic industry. Strong interrelationships between device design, materials science and engineering, and process chemistry determine the performance of a device. Semiconductors are the basic materials in the electronic industry. High-quality, single-crystal, defect-free silicon of 150 mm diameter is grown from the melt by a highly automated process. Oxygen levels for gathering impurities are controlled by computer. Epitaxial growth is widely used to form devices on Si wafers, and is also controlled by computer. The development of materials and processing in the semiconductor industry allows us to produce integrated circuits with a billion components on a single chip, and thus will further the revolution in information technologies that has reshaped our society.

The telecommunication industry is another example where functional electronic materials play an extremely important role. Most current devices in this industry are also based on single crystal silicon. The processing steps including masking, photolithography, etching, diffusion, ion implantation, metallisation, and oxidation, largely determine the performance and quality of the telecommunication devices. The shift from electronic to optical technology has required many new optical materials. Optical fibers have to be extremely transparent to transmit light signals over a long distance. The development of new process technologies has resulted in silica optical fibres with transmission losses 100 orders of magnitude lower ($\times 10^{-100}$) than ordinary optical glasses and approaching the theoretical minimum. Materials based on new systems including fluorides are being studied in an effort to further reduce the optical losses, although new light sources and detectors will be needed, because the transmission frequencies will move further into the infrared region.

1.4 Requirements and Future Developments of Electronic Materials

Advanced electronic materials are fundamental to the design and fabrication of electronic devices with higher performance. Thus the basic requirements for the development of electronic materials can be listed as below:

1. High purity and perfect crystal structures: this is critically important for growth of semiconducting materials. For example, purity of 6N is often required for GaAs and InP.
2. Advanced fabrication techniques: realization of smaller devices with higher performance requires manipulation at nano-scale or atomic levels. Advanced epitaxy techniques such as liquid or vapour phase epitaxy, molecular beam epitaxy (MBE) and metallorganic chemical vapour deposition (MOCVD) can control the growth of artificial structures at sub-micro or on atomic layer levels.
3. Large dimensions: this is particularly important for the fabrication of silicon single crystals (and their wafers). Si wafers of 400 mm diameter have been successfully produced; however, the demand for higher diameters is still increasing.
4. The service life of new electronic materials should be long and ideally controllable.
5. Development of electronic materials that can perform structural and functional properties.
6. Process for the fabrication of new electronic materials should have low contamination/pollution to environment and energy saving.

    To meet these requirements, new electronic material should be developed. This may include the production and utilization of:

1. Advanced electronic materials: nanostructured materials (such as quantum dots, nanowires, carbon nanotubes, and transition metal oxide thin films), advanced composite materials, high temperature superconducting materials, and bio-electronic materials for molecular devices, single electron devices, molecular computers, and biological computers with much smaller size, lower power consumption, and higher operating speed than the current computers.
2. Organic electronic materials: these may include organic electric conducting materials (for battery, microwave absorption, electrosatic control, data storage, static shielding, field effect transistor (FET), gas sensor); organic piezoelectrical materials (PVDF, PVF, PVC, PMLG, PC, PVDF/PZT composites for speakers, earphones, megaphones); organic optoelectronic materials (PVK, PBD, PPV for

LEDs and/or electroluminescence); and organic magnetic materials for data storage, electromagnetic interference shielding.
3. Thin film electronic materials for smaller width and higher integrated level.

Computer aided modelling and simulation will find more and more applications in design, growth and structural/functional manipulation of advanced electronic materials.

1.5 Typical Characterization Techniques for Electronic Materials

1.5.1 *Microstructural analysis*

The objective of microstructural characterisation is to provide a magnified image and to reveal features that are beyond the resolution of the human eye (~100 μm) such as phase, grain shape and size, grain boundary, precipitation, texture and defects.

1.5.1.1 *Scanning electron microscopy (SEM)*

SEM has an electron gun which provides a focused electron beam (in vacuum) into a fine probe that is rastered over the specimen surface. As the electrons hit and penetrate the surface, a number of interactions occur that can result in the emission of electrons or photons from (or through) the surface. A reasonable fraction of the electrons emitted can be collected by appropriate detectors to generate an image on screen; every point that the beam strikes on the sample is mapped directly onto a corresponding point on the screen. Principle images collected in SEM for analysis can include secondary electron images (SE), backscattered electron images (BSE), and elemental X-ray maps.

1.5.1.2 *Transmission electron microscopy (TEM)*

In TEM, a focused electron beam is incident on a thin (<200 nm) foil sample. The signal is obtained from both undeflected and deflected electrons that penetrate the sample thickness. A series of magnetic lenses at and below the sample position are delivering the signal to a detector.

1.5.1.3 *Scanning tunnelling microscopy (STM)*

When an atomically sharp tip (W or Pt-Ir) is within a few Å (0.1 nm) of the sample's surface, and a bias voltage is applied between the sample and the tip, quantum-mechanical tunnelling takes place across the gap. This tunnelling current depends exponentially on the separation between the tip and the sample, and linearly on the local density of states. The exponential dependence of the magnitude of tunnelling current upon separation means that, in most cases, a single atom on the tip will image the single nearest atom on the sample. When the tip is scanning the sample surface, if the tunnelling current is kept constant, the position of the tip can then reflect the arrangement of atoms on the sample surface. STM is a powerful tool for atomic scale imaging of morphology.

1.5.1.4 *Atomic force microscopy (AFM)*

In AFM, the tip in scanning tunnelling microscope is replaced by a force sensing cantilever to measure forces or to measure interactions between a sharp probing tip and sample surface. The force between tip and sample causes cantilever deflections which are monitored by a deflection sensor. While scanning the sample, a feedback-loop can keep the deflection constant (equiforce mode). AFM can be used for nanoscale measurement of surface morphology and forces between tip and sample surface. It can also scan the surface electrically or magnetically.

1.5.2 *Compositional analysis*

1.5.2.1 *Energy-dispersive X-ray spectroscopy (EDX or EDS)*

X-rays are produced as a result of the ionization of an atom by high-energy radiation wherein an inner shell electron is removed. To return the ionized atom to its ground state, an electron from a higher energy outer shell fills the vacant inner shell and, in the process, releases an amount of energy equal to the potential energy difference between the two shells. This excess energy, which is unique for every atomic transition, will be emitted by the atom either as an X-ray photon or will be self absorbed and emitted as an Auger electron. The photon or electron can ionize a Si atom in detector through a photoelectric effect which achieves signal generation by charge carriers and partition of incoming X-rays into proper energy channels.

EDS has several advantages: fast data collection, high detector's efficiency (both analytical and geometrical), ease of use, portability, and relative ease of interfacing to existing equipment. However its disadvantages are obvious: poor energy resolution of peaks, a relatively low peak-to-background ratio in electron-beam instruments, and a limit on the input signal rate because of pulse processing requirements. EDS is finding applications in electronic industry for quality control and test analysis of various semiconductors, metals, and polymers.

1.5.2.2 *X-ray photoelectron spectroscopy (XPS)*

X-rays of sufficiently short wavelength can ionize an atom, producing an ejected free electron. The kinetic energy KE of the photoelectron depends on the energy of the photon hv expressed by the Einstein photoelectric law: $KE = hv - BE$. BE is the binding energy of the particular electron to the atom concerned. Since hv is known, a measurement of KE determines BE and then identify atom of concerned. In addition to quantitative compositional analyses, XPS is also finding wide application in characterisation of the valence state of the target elements.

1.5.2.3 *Electron probe X-ray microanalysis (EPMA)*

EPMA is a spatially resolved and quantitative elemental analysis technique based on the generation of characteristic X-rays by a focused beam of energetic electrons. It is usually used to measure the concentrations of elements (beryllium to the actinides) at levels as low as 100 ppm and to determine lateral distributions by mapping.

1.5.2.4 *Rutherford backscattering (RBS)*

RBS is an elemental analysis technique that takes advantage of the energy loss of a high energy penetrating particle when it collides with an atom and is 'backscattered' back through the surface of the material.

Its detection limits are typically ranging from a few ppm for heavy elements to a few percent for light elements, while its depth resolution is on the order of 20–30 nm and can be as low as 2–3 nm near the sample surface.

RBS analysis is commonly used for quantitative depth profiling of semiconductor thin films, multilayered structures, polymers and high-T_c superconductors; areal concentration measurements; and crystal quality and impurity lattice site analysis in semiconductor materials.

1.5.3 *Structural analysis*

1.5.3.1 *X-ray diffraction (XRD)*

XRD is a powerful technique used to uniquely identify the crystalline phases present in materials and to measure the structural properties (strain state, grain size, epitaxy, phase composition, preferred orientation, and defect structure) of these phases; also used to determine the thickness of thin films and multilayers, and atomic arrangements in amorphous materials (including polymers) and at interfaces.

1.5.3.2 *Low-energy electron diffraction (LEED)*

LEED is a technique for investigating the crystallography of surfaces and overlayers or films adsorbed on surfaces. LEED is generally performed with electron energies of 10–1000 eV. The limited penetration of electrons in this energy range provides the sensitivity to the surface. Diffraction of electrons occurs because of the periodic arrangement of atoms in the surface.

LEED is mainly used to determine the atomic structure of surfaces, surface structural disorder, and to some extent, surface morphology, as well as changes in structure with time, temperature, and externally controlled conditions like deposition or chemical reaction.

1.6 Nature and Purpose of This Book

Engineers in all disciplines should have some basic and applied knowledge of materials so they can do their work more effectively when using materials. This is a textbook for an introductory course for electrical and electronic engineering and other engineering/technology students. The emphasis of this book is put on the applications of materials in electric, electronic and telecommunication industries. The basic understanding of the properties, performance of materials and their relations with processing and microstructure provide the tools for this purpose. The main purposes of this course for engineering students are:

1. To appreciate the important roles of materials played in modern technology;
2. To understand the basic principles of electronic properties of materials; and
3. To familiarise yourself with the various groups of materials that commonly used in electrical and electronic industries.

With the above goals in mind, we will focus on the understanding of a few basic principles of electronic properties of materials first and use them to discuss a wide range of properties including electrical and electronic properties, magnetic properties, optical and thermal properties. After the examining the properties, typical materials in this group will be discussed together with brief processing techniques that provide the materials with these properties and affect these properties. Applications of these materials follow and are emphasised and enhanced with case studies, wherever this is appropriate.

It is hoped that this book will provide engineering students with the basic concept and some working knowledge of materials, and this book may be used as a general reference for reviewing electronic materials in the future when they need. This book has also designed as a first reference book at the introduction level for engineers and technologists whose work involves in designing, improving, maintaining, repairing and/or using electrical and electronic devices, instruments and equipment. Finally, we believe that basic materials knowledge is necessary for people working and living in our modern technological society.

Questions:

1. Give two examples each to explain what are structural and functional materials? List the properties you know for all functional materials.

2. What are electronic materials? What are the basic requirements of modern society to electronic materials?

3. What is the future direction for the development of advanced electronic materials?

4. The correlation between composition, structure and property is the core of materials science and engineering. How do you understand the importance of this in research and development of electronic materials?

5. Briefly describe the operation principle and major application of typical characterisation techniques used in electronic materials research.

Chapter 2

Classical Theory of Electrical Conduction and Conducting Materials

In this chapter, the free electron conduction theory is described first, then it is used to explain the conduction properties of materials. Finally we will introduce materials which are used for electrical conduction in electrical and electronic industries.

In classical electron conduction theory, an electron is treated as a very small particle with certain mass and electric charge:

$$\text{Electron mass } m_e = 9.1 \times 10^{-31} \text{ kg,}$$
$$\text{Electron charge } e = -1.6 \times 10^{-19} \text{ C.}$$

Because electrons behave like particles in this theory, they are assumed to obey Newton's Laws of motion. We will apply this theory to describe the electron conduction behaviour in conductors in secs. 2.3 –2.5.

2.1 Resistivity, TCR (Temperature Coefficient of Resistivity) and Matthiessen's Rule

2.1.1 *Resistivity*

The electrical resistance, R, of a material is defined as:

$$R \propto l/A \quad \text{or} \tag{2.1}$$
$$R = \rho \cdot l/A, \tag{2.2}$$

where l is the length, A is the area of the cross section of the conductor and ρ is called electrical resistivity. Equation (2.2) indicates that ρ is the resistance of a material in unit length and unit cross section area.

2.1.2 *Matthiessen's rule and TCR*

For pure metals, resistivity ρ is the sum of two items: a residual part ρ_r and a thermal part ρ_t (see Fig. 2.1). This is called Matthiessen's rule.

$$\rho_{(total)} = \rho_r + \rho_t, \tag{2.3}$$

$$\rho = \rho_r \times (1 + \rho_t/\rho_r), \tag{2.4}$$

$$\rho_t/\rho_r = f(T), \tag{2.5}$$

$$\rho = \rho_r \times [1 + f(T)], \tag{2.6}$$

For most metals and alloys, ρ is approximately proportional to temperature T, and can be written as (see Fig. 2.2):

$$\rho = \rho_o \times (1 + \alpha \cdot \Delta T), \tag{2.7}$$

where α is called the temperature coefficient of resistivity (TCR).

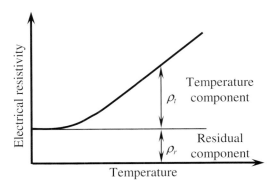

Fig. 2.1 Resistivity vs. temperature for a typical metal, Matthiessen's Rule.

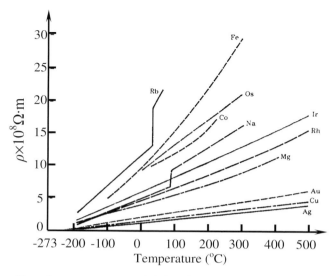

Fig. 2.2 The effect of temperature on resistivity of selected metals (after Zwikker, "Physical Properties of Solid Materials," Pergamon, 1954, p247).

2.2 Traditional Classification of Metals, Insulators and Semiconductors

Table 2.1 lists the resistivity and TCR for various materials. Materials can be classified into three groups according to their electrical conduction properties: conductors, insulators and semiconductors. The general conduction properties of materials can be described as follows:

(1) Compared with other physical properties of materials (e.g. density or Young's modulus), electrical resistivity varies over a much greater range ($\sim 10^{24}$). Metals and alloys have resistivity over the range of 10^{-8}–10^{-6} $\Omega \cdot m$. Insulators have resistivity over the range of 10^{8}–0^{16} $\Omega \cdot m$. The resistivities of semiconductors are typically ranging from 10^{-2} to 10^{8} $\Omega \cdot m$.

(2) For pure metals (99.9%), $\rho = 10^{-8}$–10^{-7} $\Omega \cdot m$. A higher purity (99.999%) only makes a small difference.

(3) TCR values for various pure metals are similar, in the range of $\sim 0.004/K$, regardless of their resistivity.

(4) Alloying elements increase the resistivity of metals but reduce TCR. TCR is positive, as for pure metals.

(5) Non-metals have much higher resistivity. Purity does not have much effect on the resistivity.

(6) Typical semiconductors have a high and negative TCR (see Fig. 2.3).

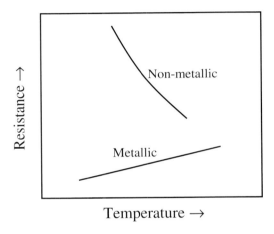

Fig. 2.3 A schematic to show the typical 'metallic' and 'non-metallic' conduction behaviours.

Electrical resistance measurement methods:

For a conductor, its electrical resistance (R) is easy to measure with standard "two-point" or "four-point" methods (see Fig. 10.4). Resistance can be calculated directly from the drop in voltage. However, it is difficult to measure the resistance of an insulator accurately with the above methods, because a very high voltage is needed. The measurements will be strongly influenced by a number of factors including contacting points, surface finishing, environmental conditions (such as moisture level) and defects in the material to be measured.

Table 2.1 Resistivities and TCRs of various materials at 293 K.

Classification	Material	Resistivity ($\Omega \cdot m$)	TCR (%/K)
Conductor	Silver	1.6×10^{-8}	+0.41
	Copper	1.7×10^{-8}	+0.43
	Aluminium	2.7×10^{-8}	+0.43
Heating element	Sodium	5.0×10^{-8}	+0.4
	Tungsten	5.7×10^{-8}	+0.45
	Iron	9.7×10^{-8}	+0.5
	Platinum	10.5×10^{-8}	+0.39
	Tantalum	13.5×10^{-8}	+0.38
	Manganin (87Cu13Mn)	38×10^{-8}	+0.001
	Constantan (57Cu43Ni)	49×10^{-8}	+0.002
	Nichrome (80Ni20Cr)	112×10^{-8}	+0.0085
	SiC, commercial	$1\text{-}2 \times 10^{-6}$	-0.15
	Graphite, commercial	1×10^{-5}	-0.07
Semiconductor	InAs, very pure	3×10^{-3}	-1.7
	Tellurium, very pure	4×10^{-3}	-2
	Germanium, diode grade	1×10^{-3}	+0.4
	Germanium, very pure	5×10^{-1}	-4
	Silicon, transistor grade	1×10^{-1}	+0.8
	Silicon, very pure	1×10^{-3}	-7
	Anthracene	3	-10
Insulator	Selenium, amorphous	$c.\ 1 \times 10^{10}$	-15
	Silica	$c.\ 1 \times 10^{13}$	negative
	Alumina	$c.\ 1 \times 10^{14}$	negative
	Sulphur	$c.\ 1 \times 10^{15}$	negative
	PTFE	$c.\ 1 \times 10^{16}$	negative

Source: Electronic Materials by L.A.A. Warnes, MaCmillan Education Ltd. 1990.

2.3 Drude's Free Electron Theory

2.3.1 *Drude's free electron theory*

Paul. K. Drude (1863–1906) developed a theory of electron conduction in 1900. In his theory, electrons are taken to be particles that move through the metal lattice freely, obeying Newton's Laws of motion and Maxwell–Boltzmann statistics. When an electric field E is applied to a metal, the force acting on an electron is

$$F = -e \cdot E, \tag{2.8}$$

According to the Newton's Law:

$$a = F/m = -e \cdot E/m = 1.75 \times 10^{11} \times E, \tag{2.9}$$

where a is the acceleration, m is the mass of an electron, $m = 9.1 \times 10^{-31}$ kg, $e = 1.6 \times 10^{-19}$ C. In a time τ, the electron will obtain a velocity of v_d:

$$v_d = a \cdot \tau, \tag{2.10}$$

v_d is called drift velocity.

From Eqs. (2.9) and (2.10)

$$v_d = -e \cdot E \cdot \tau/m. \tag{2.11}$$

This can be rewritten as

$$v_d = -\mu \cdot E, \tag{2.12}$$

$$\mu = e \cdot \tau/m, \tag{2.12a}$$

where μ is called the mobility of the electron, or drift velocity in unit electrical field.

Considering a unit volume of a conductor, which contains n free electrons, the total charge will be $-ne$, the charge crossing unit area in unit time must be $-nev_d$, which is the current density J.

$$J = -n \cdot e \cdot v_d. \tag{2.13}$$

Combining Eqs. (2.13) and (2.11), we have

$$J = n \cdot e^2 \cdot E \cdot \tau/m, \tag{2.14}$$

or $$J = \sigma \cdot E, \tag{2.15}$$

where $\sigma = 1/\rho$ is the conductivity of the metal, and has a unit of $\Omega^{-1}\cdot m^{-1}$. Therefore, we have

$$\sigma = n\cdot e^2 \cdot \tau/m, \qquad (2.16)$$

or
$$\sigma = n\cdot e\cdot \mu. \qquad (2.17)$$

We see σ is proportional to μ.

The resistivity was defined as

$$R = \rho\cdot l/A, \qquad (2.2)$$

also $E = V/l$ and from Eq. (2.15)   (2.18)

$$J = V/(\rho\cdot l). \qquad (2.19)$$

From $\rho\cdot l = R\cdot A$, so Eq. (2.19) can be written as

$$J = V/(R\cdot A), \qquad (2.20)$$

and
$$J\cdot A = I = V/R. \qquad (2.21)$$

This is Ohm's law. However, $J = \sigma\cdot E$ (Eq. (2.15)) is more often used in solving electromagnetic problems.

2.3.2 *Mean free path*

The average distance travelled by an electron between collisions is called mean free path.

$$l_m = \tau\cdot v, \qquad (2.22)$$

where τ is the time between two collisions and v is the average velocity of the electron, consisting of $v_{th} + v_d$.

Because $v_{th} \gg v_d$, Eq. (2.22) can be written as

$$l_m = \tau \cdot v_{th}. \qquad (2.23)$$

τ can be found from Eqs. (2.11) and (2.12), $\mu = e/\tau \cdot m$, and $\tau = \mu \cdot m/e$

$$l_m = \mu \cdot m \cdot v_{th}/e. \qquad (2.24)$$

Typically for a metal, the average time between two collisions τ is ranging from 10^{-14} to 10^{-15} s, and electron concentration n is about $10^{28}/m^3$. Using Eq. (2.16), we have:

$$\sigma = n \cdot e^2 \cdot \tau/m = 0.3\text{–}3 \times 10^6 /(\Omega \cdot m). \qquad (2.16)$$

2.3.3 *Example: calculation of mobility and drift velocity*

Cu is the most important conducting metal. There is one valance electron per Cu atom. Cu has a *fcc* crystal structure with four atoms in a unit cell, a lattice parameter of 0.360 nm and resistivity of 1.7×10^{-8} $\Omega \cdot m$. Assuming free electrons = valance electrons (valence = +1)

(1) Calculate the mobility, μ, of an electron in Cu.
(2) A typical house wire 3 m long has a resistance $R = 0.03$ Ω and carries a current of 15 A. Calculate the drift velocity, v_d, of an electron in the wire.
(3) Calculate the time τ between two collisions of electrons.
(4) Assuming the thermal velocity of an electron is ~105 m/s, calculate the mean free path, l_m, of an electron in Cu.

Answers:

(1) According to $\sigma = n \cdot e \cdot \mu$

$$\mu = \sigma/(n \cdot e) = 1/(\rho \cdot n \cdot e).$$

To calculate μ, we need σ and n. ρ can be measured experimentally.

$$\rho_{Cu} = 1.7 \times 10^{-8} \ \Omega \cdot m.$$

Density of free electrons, n, can be calculated from the atomic structure
Because here we assume valence electron = free electron,

The atomic density of Cu = $4/(3.6 \times 10^{-10})^3 = 8.6 \times 10^{28}$ atoms/m^3

The free electron density $n = 1 \times 8.6 \times 10^{28}$/m^3, therefore,

$$\mu = 1/(1.7 \times 10^{-8} \times 8.6 \times 10^{28} \times 1.6 \times 10^{-19}) = 4.3 \times 10^{-3} \ m^2/V \cdot s$$

(2) To calculate v_d from μ ($v_d = \mu \cdot E$), we need an electrical field E, which is the voltage drop over a unit distance. The voltage drop V in the wire is:

$$V = I \cdot R = 0.45 \ V$$

$$E = V/l = 0.45/3 = 0.15 \ V/m$$

$$v_d = \mu \cdot E = 0.6 \times 10^{-3} \ m/s = 0.6 \ mm/s$$

(3) $v_d = -e \cdot E \cdot \tau/m$

$$\tau = -v_d \cdot m/(e \cdot E) = 2.3 \times 10^{-14} \ s$$

(4) Mean free path of an electron:

$$l_m = \tau \times v_{th} = 2.3 \times 10^{-14} \times 10^5 = 2.3 \ nm \ (23 \ \text{Å})$$

Compared with the ~0.2 nm separating atoms in a Cu lattice, electrons travel ~10 times the average distance of the atoms before colliding with one.

2.4 Hall Effect

To show the motion behaviour of an electron, Edwin H. Hall (1855–1938) conducted the following experiment in 1879. As shown in Fig.

2.4, electrons in motion in a magnetic field are subjected to the Lorenz force, F.

$$F = -e \cdot v \cdot B, \qquad (2.25)$$

where B is the magnetic induction and v is the instantaneous velocity of electron. F, v and B are vectors.

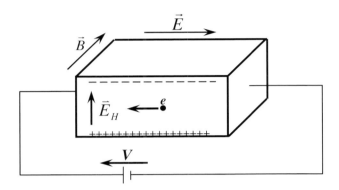

Fig. 2.4 A schematic of Hall effect.

The direction of F is given by the right hand corkscrew rule. The Lorenz force moves the electrons upwards as shown in Fig. 2.4, there they will accumulate and produce a field E_H, called the Hall field. Hall field acts on the electrons as to oppose the Lorenz force. The magnitudes of the two forces are equal:

$$e \cdot E_H = e \cdot v \cdot B. \qquad (2.26)$$

The magnitude of the vector product $v \times B$ must be $v \cdot B$ because they are perpendicular. Actually, E, E_H and B are mutually perpendicular.

The v in Eq. (2.26) is the electron instantaneous velocity. The average value of the Lorenz force should be calculated from average v.

Because $v = v_{th} + v_d$, and $v_{th} \approx 0$ as the thermal velocities are random in direction,

$$E_H = v_d \cdot B. \tag{2.27}$$

Replace v_d with $\mu \cdot E$,

$$E_H = \mu \cdot E \cdot B. \tag{2.28}$$

This means that the Hall field is the product of electrical field, magnetic induction and the mobility of the electrons in this material.

Since $\qquad\qquad v_d = -J/n \cdot e, \qquad\qquad (2.13)$

using Eqs. (2.27) and (2.13), we also have

$$E_H = (-1/n \cdot e)J \cdot B, \tag{2.29}$$

or $\qquad\qquad E_H = R_H \cdot J \cdot B. \qquad\qquad (2.30)$

R_H is the Hall constant of the material.

From Eq. (2.29), $\qquad R_H = -1/n \cdot e$ or $R_H \cdot n \cdot e = -1 \qquad (2.31)$

R_H can also be obtained by measuring E_H, J and B experimentally:

$$R_H = E_H/(J \cdot B). \tag{2.32}$$

The results can be used to verify Drude's free electron theory. Some experimental results are listed in Table 2.2. Some metals (e.g., alkali metals) obey the $R_H \cdot n \cdot e = -1$ relation, it is quite different for others. Some metals even have the opposite sign ($R_H \cdot n \cdot e > 0$), indicating that free electrons were not the only charge carriers in electrical conduction. The explanation is that there were missing electrons, called holes, which behaved like positively charged electrons.

Table 2.2 $R_H$ne values for some metals.

Metal	$R_H$ne
Li	-1.15
Na	-1.05
K	-1.08
Rb	-1.05
Cs	-1.04
Cu	-0.68
Ag	-0.8
Au	-0.69
Pd	-0.73
Pt	-0.21
Cd	+0.5
W	+1.2
Be	+5.0

Source: "Electronic Materials" by L.A.A. Warnes, Macmillan Education Ltd. 1990.

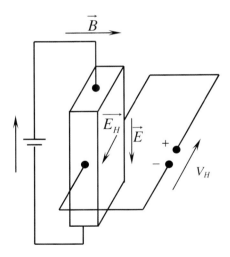

Fig. 2.5 A schematic of Hall probe.

The Hall probe (Fig. 2.5) has been used to measure magnetic field. For a given electrical field E and magnetic induction B, the Hall field is proportional to the mobility (μ) of the charge carriers. Therefore, materials of high μ are used for Hall probes. For example, InSb has an electron mobility of 8 m^2/V·s, ~2000 times the mobility of electron in Cu.

2.5 Wiedemann–Franz Law

Good electrical conductors are often good thermal conductors, implying a similar mechanism for the conduction of electric current and heat.

Assuming only electrons are responsible for both the electrical and thermal conduction in metals, and free electrons behave like an ideal gas. At a constant temperature, the ratio of the electrical and thermal conductivities should be a constant for metal conductors. This is called the Wiedemann–Franz law.

For an ideal gas, thermal conductivity K can be treated as

$$K = 1/3 \times C_v{\cdot}l_m{\cdot}v_{th}, \qquad (2.33)$$

where C_v = specific heat at constant volume
$\quad\quad l_m$ = mean free path
$\quad\quad v_{th}$ = thermal speed of electrons
C_v can be taken as $3/2nK_B$ for a monatomic ideal gas, so now

$$K = 1/2 \times n{\cdot}K_B{\cdot}l_m{\cdot}v_{th}, \qquad (2.34)$$

where n = number of free electrons, and K_B is Boltzmann's constant = 1.38×10^{-23} J/K.

Electrical conductivity is given by Eq. (2.14)

$$\sigma = n{\cdot}e^2{\cdot}\tau/m, \qquad (2.14)$$

τ redefined as $2\tau = l_m/v_{th}$,

$$K/\sigma = m{\cdot}K_B{\cdot}v_{th}^2/e^2. \qquad (2.35)$$

Using $\qquad\qquad v_{th} = \sqrt{\dfrac{3K_B T}{m}}, \qquad (2.36)$

Eq. (2.35) can be written as

$$K/\sigma = 3K_B^2 \cdot T/e^2, \tag{2.37}$$

or
$$K/(\sigma \cdot T) = 3K_B^2/e^2, \tag{2.38}$$

$3K_B^2/e^2$ is a constant, therefore, the left side of Eq. (2.38) is also a constant.

$$3K_B^2/e^2 = L = 2.23 \times 10^{-8} \text{ W} \cdot \Omega/\text{K}^2. \tag{2.38a}$$

Table 2.3 Lorenz numbers of some metals and alloys at 293 K.

Metal or Alloy	Lorenz number ($L \times 10^{-8}$ WΩ/K^2)
Aluminium	2.18
Cadmium	2.26
Copper	2.30
Indium	2.4
Lead	2.49
Lithium	2.48
Magnesium	2.38
Molybdenum	2.41
Nickel	2.15
Niobium	2.29
Palladium	2.60
Potassium	2.16
Rhodium	2.27
Tantalum	2.40
Tin	2.47
Uranium	2.77
Stainless steel (18/8)	3.33
Phosphor bronze (1.25%)	2.46
Yellow brass	2.50
Constantan (55Cu45Ni)	3.56

Source: Electronic Materials by L.A.A. Warnes, Macmillan Education Ltd. 1990.

L is called Lorenz number, which can also be measured by experiments. Some results are listed in Table 2.3. Metals with good conductivity generally obey Wiedemann-Franz law. However, some alloys with relatively low electrical conductivities have a larger L than calculated with Wiedemann–Franz law. This is because that the heat is not conducted solely by free electrons. The high vibration of atoms in the lattice contributes to thermal conductivity, increases thermal conductivity, and therefore increases L. However, non-metals have different thermal conducting mechanisms and do not follow the Wiedemann–Franz law.

2.6 Resistivity of Alloys, Nordheim's Rule

2.6.1 *Nordheim's rule*

According to the Matthiessen's rule, electrical resistivity is the sum of two items: a residual part ρ_r and a thermal part ρ_t. Impurity raises the residual resistivity ρ_r and the increase of ρ_r on a single impurity can be:

$$\rho_r(x) = A \cdot x \cdot (1 - x). \tag{2.39}$$

Eq. (2.39) is called Nordheim's rule, where x is the concentration (atomic fraction) of the impurity, and A is a constant which is called "solution resistivity coefficient". A depends on the base metal and the impurity (Figs. 2.6 and 2.7). The material factors that influence the constant A include the atomic sizes, crystal structures and electro-negativities.

Eq. (2.39) can also be written as:

$$\rho_r(x) = Ax - Ax^2. \tag{2.40}$$

Eq. (2.40) indicates a parabolic relation between resistivity and impurity concentration.

When $x \ll 1$, $1 - x \approx 1$, then we have

$$\rho_r \approx A \cdot x. \tag{2.41}$$

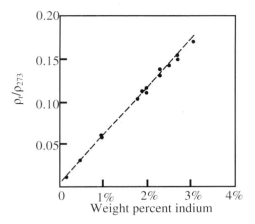

Fig. 2.6 The residual resistivity of Sn-In alloy (After A. Pippard, Roy. Soc. Philos T. A, **248**, 1955, pp97-129).

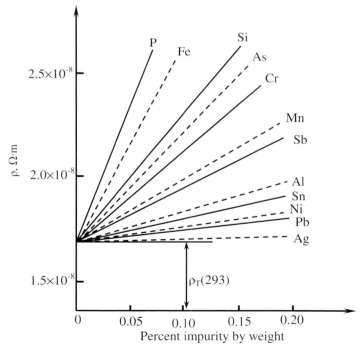

Fig. 2.7 The effect of small additions of various elements (After F. Pawlek *et al*, Z. Metallk., **47**, 1956, pp357-63).

This indicates a linear relation between ρ and x in a dilute solid solution alloy. Figure 2.7 shows the effect of small additions of various elements in Cu on ρ. From Fig. 2.6, we see:

1. The linear increase of ρ with x is caused by ρ_r, ρ_{th} remains constant. Therefore, the total $\rho = \rho_o + \rho_r = \rho_o + A \cdot x \cdot (1-x)$; and
2. The effects are quite different for different elements. For instance, silver (Ag) has the slightest effect because of the similarity it has with Cu. Phosphorus (P) has the strongest effect. The elements have strong effect on ρ often have strong effects on the mechanical strengthening properties.

2.6.2 *Resistivity ratio*

Resistivity ratio, R_r, is defined as the ratio of ρ at room temperature to that at 4 K:

$$R_r = \rho_{room} / \rho_{4K}. \tag{2.41a}$$

At 4 K, $\rho \approx \rho_r$; at room temperature, $\rho_{room} = \rho_r + \rho_t$. Therefore, Eq. (2.41a) can be written as:

$$R_r = (\rho_r + \rho_t)/(\rho_r). \tag{2.41b}$$

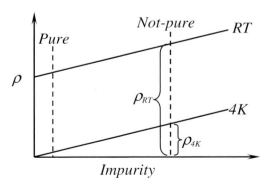

Fig. 2.8 Resistivity ratio at room temperature and 4K (Source: Electronic Materials by L.A.A. Warnes, Macmillan Education Ltd. 1990).

The ratio $(\rho_r + \rho_t)/(\rho_r)$ can be very large when ρ_r is very small, which is the case when the metal is pure (see Fig. 2.8 and Table 2.4). Therefore, this ratio can be used to estimate or compare the purity of metals.

Table 2.4 Resistivity ratios for high purity metals.

Metal	$\rho_{293}/\rho_{4.2}$
Tungsten	90,000
Rhenium	45,000
Aluminium	40,000
Molybdenum	14,000
Tantalum	7,000
Gold	2,000
Niobium	2,000
Platinum	2,000
Vanadium	300
Zirconium	200

Using special techniques such as zone refining and long periods of vacuum disgasing, impurities can be reduced to a very low level. The resistivity ratio can reach 100,000. On the other hand, metals with commercial purity usually have a resistivity ratio below 100. For some alloys, the ratio can be as low as 1. The resistivity ratio is also different for different metals.

2.6.3 *Resistivity from strain damage*

Cold work (mechanical deformation) increases resistivity. Figure 2.9(a) shows the effect of cold work on the resistivity of two Al alloys. The increase arises from the dislocations and other structural defects caused by the mechanical deformation. These defects deflect electrons from their wavelike movement, shorten the mean free path and therefore increase ρ in a similar way as the effect of alloying additions.

Radiation also introduces structural defects and increases electrical resistivity. It has notable effects in materials that are used in space technology and nuclear reactors.

Annealing can be used to remove most of the structural defects and restore the original resistivity. Annealing can also remove the hardening effect from the lattice distortion, and improve ductility of the metals (for further winding or mechanical working). Figure 2.9(b) shows the combinations of the effects of alloying elements, cold work and annealing.

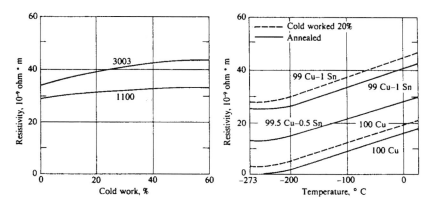

Fig. 2.9 Factors affecting resistivity of (a) Cold work of Al. 1100 Al is commercially pure and 3003 Al contains 1.2% Mg. (b) The resistivity is the sum of the contributions from temperature, alloying elements, and deformation.

2.7 Resistivity of Alloys and Multiphase Solids

According to the phase equilibrium, a binary alloy system can be in the forms of: (a) solid solution, (b) multiphase mixture, and (c) intermetallic compounds. Complete solid solution of two metals can only occur when the following conditions are satisfied:

1. The two metals have a similar crystal structure;
2. The atomic volumes of both elements do not differ by more than 15%;
3. The two elements have the same valence; and
4. They have similar electrochemical properties.

These are the Hume–Rothery Rules. For example, Cu-Ni, Au-Ag and Ag-Pd can form complete solid solution alloys.

2.7.1 *Solid solution alloys*

Nordheim's rule can be used for this type of alloys to build up parabolic curves for each component, up to 50 at.% (see Fig. 2.10).

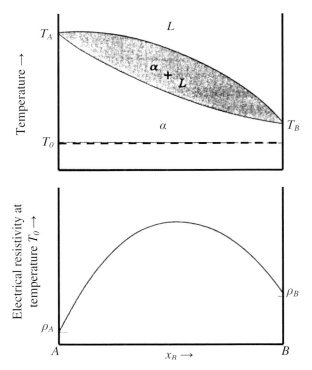

Fig. 2.10 Resistivity versus composition in a binary solid solution alloy system.

The increase of the resistivity can be written as:

$$\rho_r(x) = A \cdot x - A \cdot x^2. \tag{2.40}$$

The total resistivity should be:

$$\rho = \rho_A + \rho_r(x) = \rho_A + A \cdot x - A \cdot x^2, \tag{2.40a}$$

where $x = x_B$ is the atomic fraction of B.

Note that the maximum resistivity is not necessary at the position of 50 at.%. TCR usually decreases as ρ increases.

2.7.2 *Multiphase mixture*

In a two-phase mixture alloy, the situation is more complicated. Assume an alloy has a randomly distributed $\alpha + \beta$ phase mixture and V_α and V_β are the volume fractions of α and β phases. The length of the rod is l and the area of the cross section is A. Cut the alloy rod into N parallel fibres and calculate the resistance of each fibre, as shown in Fig. 2.11:

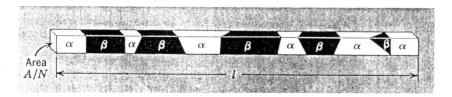

Fig. 2.11 A section from a two-phase alloy.

$$R_f = (\rho_\alpha \cdot V_\alpha \cdot l)/(A/N) + (\rho_\beta \cdot V_\beta \cdot l)/(A/N) = N \cdot l(\rho_\alpha \cdot V_\alpha + \rho_\beta \cdot V_\beta)/A. \quad (2.42)$$

Because the rod consists of N fibres in parallel, we have

$$1/R_{rod} = 1/R_1 + 1/R_2 + ... = A/(\rho_\alpha \cdot V_\alpha + \rho_\beta \cdot V_\beta). \quad (2.43)$$

Thus $\quad\quad\quad\quad R_{rod} = (\rho_\alpha \cdot V_\alpha + \rho_\beta \cdot V_\beta)l/A$, or

$$\rho = \rho_\alpha \cdot V_\alpha + \rho_\beta \cdot V_\beta. \quad (2.44)$$

Therefore, the electrical resistivity of a two-phase material is a linear function of the volume fractions of the two phases, as shown in Fig. 2.12.

If the densities of the two phases are similar, mass (weight) fractions can be used in Eq. (2.44).

$$\rho \approx \rho_\alpha wt.\% \cdot \alpha + \rho_\beta wt.\% \cdot \beta. \quad (2.45)$$

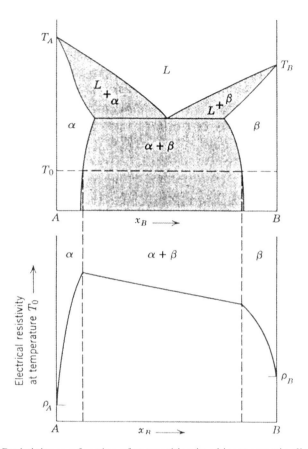

Fig. 2.12 Resistivity as a function of composition in a binary eutectic alloy system.

2.7.3 *Intermetallic compounds*

When the electronegativities of two metals are quite different, they will combine chemically to form a type of "intermetallic compounds" (such as Fe-Al, Ni-Fe, Zn-Mg etc.). The resistivity and TCR may vary in a complicated way depending on the nature of the intermetallic compounds. From Table 2.5, it can be seen that the resistivity of the intermetallic compound might be quite different from that of its individual components.

Table 2.5 Resistivity of intermetallic compounds.

	$MgCu_2$	Mg_2Cu	Mg_2Al_3	Mn_2Al_3	$FeAl_3$
First component	4.35×10^{-8}	4.35×10^{-8}	4.35×10^{-8}	4.41×10^{-8}	9.09×10^{-8}
Second component	1.56×10^{-8}	1.56×10^{-8}	2.85×10^{-8}	2.85×10^{-8}	2.85×10^{-8}
Compound	5.24×10^{-8}	11.9×10^{-8}	38.0×10^{-8}	500×10^{-8}	141×10^{-8}
	$NiAl_3$	Ag_3Al	Ag_3Al_2	$AgMg_3$	Cu_3As
First component	28.5×10^{-8}	1.47×10^{-8}	1.47×10^{-8}	1.47×10^{-8}	1.56×10^{-8}
Second component	2.85×10^{-8}	2.85×10^{-8}	2.85×10^{-8}	4.35×10^{-8}	35.1×10^{-8}
Compound	28.8×10^{-8}	36.4×10^{-8}	26.0×10^{-8}	16.2×10^{-8}	58.8×10^{-8}

2.8 Materials for Electricity Transmission

Materials used in electricity conduction can be divided into two groups according to their applications: The first group is for electricity transmission, the second group is for resistors and heating elements.

2.8.1 *Requirements for electricity transmission materials*

The materials with high conductivity are used as wires and cables for electricity transmission and distribution and all kinds of winding and connection in electric machines, apparatus and devices. The fundamental requirements for these applications are:

1. The lowest possible resistivity;
2. Low TCR;
3. Adequate mechanical properties; mainly high tensile strength and good flexibility;
4. Rollability and drawability for manufacturing processes;
5. Good weldability and solderability;
6. Adequate corrosion resistance; and
7. Reasonable cost.

2.8.2 *Typical electricity transmission materials*

2.8.2.1 *Copper*

Copper (Cu) is the most extensively used conducting metal. It has an electrical conductivity $\sigma = 6.0 \times 10^7$ /$\Omega \cdot$m. High conductivity requires high purity. 99.90–99.99% Cu can be produced by electrolytic refining.

As for the mechanical properties, Cu is soft, ductile, easy to be manufactured by rolling and drawing. Wires of all size, down to a few μm in diameter, can be drawn without much difficulty. Cu also has very good weldability and solderability.

Table 2.6 Relative strengths and conductivity of Cu alloys.

Materials	Composition wt.%	Relative conductivity	Tensile strength (N/mm^2)	General application
Cathode Cu	99.9% Cu	100	200	Power transmission
Cu-Cr	0.5-1 Cr	82	340	High strength conductors
Cu-Ag	~1	98	~200	Soldering material
Cu-Ni-P	0.85Ni + 0.15P	58	620	High strength
Cu-Be	1.8Be + 0.2Ni	30	1235	High strength
Cu-Cd	1.0Cd	90	278	Wear resistant

2.8.2.2 *Copper-based alloys*

Cu is not strong enough when it is pure. Therefore, pure Cu is mainly used to make various winding wires and those do not withstand high tensile load. For applications where a high load exists, pure Cu is not suitable. Alloyed Cu or composite wires are used for these purposes. The mechanical strengths can be improved by alloying with Zn, Ni, Al, Cd, Be, Pb, Sn, P, Fe etc, but the alloys have a lower conductivity than pure Cu.

Cu alloys are generally classified into two major groups – bronze and brass.

(a) Bronzes – Cu-Cd, Cu-Be alloys:

Bronzes have reasonably high strength, hardness, and wear resistance. Cu-Cd alloys are used for making contact wires. Cu-Be alloys contain 0.6-2%Be. They are precipitation-hardenable alloys and can be heat-treated and cold-worked to produce high strength. They are used for making current carrying springs, sliding contacts, brush-holders etc.

(b) Brass – Cu-Zn alloys:

Cu-Zn alloys consist of Cu with 5–40%Zn. Cu forms substitutional solid solutions with Zn up to ~35%Zn (α-phase). When the Zn content reaches ~40%, alloys with two phases, α-phase and β-phase, form. Brass has good mechanical properties, atmospheric corrosion resistance, formability, and weldability; and it is cheaper than pure Cu. The strength can be increased by cold deformation. Conductivity of brass is lower than pure Cu.

2.8.2.3 *Environmental stability and corrosion resistance*

Cables, wires for electricity transmission or thin metallic films in electronic devices are exposed to atmospheres that contain oxygen, water or even corrosive agents such as H_2S. Environmental stability is an important consideration.

Copper is different from iron based engineering metals that it combines corrosion resistance with high electrical and thermal conductivity. Cu is a relatively noble metal, hydrogen evolution is not usually a part of the corrosion process. For this reason, it is not corroded by acids unless oxygen or other oxidising agents (e.g. HNO_3) are present.

Copper and copper alloys are stable in urban, marine, and industrial atmospheres and waters. A black oxide forms on its surface that protects the metal from further corrosion. However, Cu and Cu alloys are easily attacked if the air (or water) contains ammonia, forming the green rust like those on copper tubes in old toilets. Also stress corrosion cracking (SCC) may take place in a warm, ammonia containing atmosphere.

Copper also oxidises at elevated temperatures to form two types of copper oxides:

$$4Cu + O_2 = 2Cu_2O, \qquad\qquad (2.46)$$

and
$$2Cu + O_2 = 2CuO. \qquad\qquad (2.47)$$

Copper oxides are semiconductors and harmful if build up in the electrical connection areas. Therefore, overheating should be avoided in electric circuits.

2.8.2.4 *Aluminium*

Aluminium (Al) is the second most common conducting material. Al and Al alloys have had increasing applications in electrical industries mainly because of economic considerations.

Al also has good conductivity ($\sigma_{Al} = 3.77 \times 10^7/\Omega\cdot m$). Compared with Cu ($\sigma_{Cu} = 5.95 \times 10^7$), the ratio of conductivity $\sigma_{Cu}/\sigma_{Al} = 1.6/1$. However, Al is much lighter than Cu. The density of Al, d_{Al}, is 2.70 g/cm^3 while d_{Cu} is 8.92 g/cm^3. Therefore, the conductivity per unit mass for Al and Cu is 2.1 to 1.

Al has some advantages and disadvantages to Cu:

1. Softer and weaker than Cu. Pure Al can be rolled into thin foils down to 6–7 μm thick, but cannot be drawn into very thin wires. The tensile strength of Al is lower than Cu.
2. Al oxidises very fast in air and forms a thin layer of dense, compact Al$_2$O$_3$, which can protect metal from further oxidation. Only in some special environments such as salt containing water or air (like seawater) can Al be seriously attacked.
3. Because Al is easy to be oxidised and the product Al$_2$O$_3$ has a high melting point (>2000°C), Al cannot be soldered by conventional melting methods, because Al$_2$O$_3$ will separate the melts. Some

special way such as ultrasonic vibration is used to break the oxide film and thus allow the melts to form a good bond.

Alloying elements are also used to improve the mechanical properties of Al. For instance, 0.3–0.5%Mg, 0.4–0.7%Si and/or 0.2–0.3%Fe can greatly improve the mechanical strength without losing much conductivity. Other alloying elements include Cu, Mn, Cr and Zn. Rapid solidification techniques have been used to produce Al alloys with improved microstructures and properties.

2.8.2.5 *Composite wires*

For conduction materials that need high strength, another method is to use composite cables. For example, steel-cored Al or Cu cables, made by winding up Al or Cu wires round a core of steel wire. Electroplating can be used to produce a copper layer on the surface of a steel rod.

Composite wires can also be made by so called "bimetallic" technique. For example to produce copper wire with a steel core, firstly copper is cast around a thick steel bar, then the Cu-steel bar is rolled into a rod. The rod can then be drawn down to the required size. The Cu-clad steel wires are particularly good for conducting high frequency current, where the current mainly flows through the surface Cu layer.

2.9 Materials for Electrical Resistors and Heating Elements

Electrical resistors and heating elements are another group of materials in applications involving electric conduction.

2.9.1 *Resistors*

Precision resistors in electronic devices require:

1) Suitable resistivity;
2) Low temperature coefficient of resistivity around room temperature or the temperature of use;
3) Wide service temperature range;

4) Long stability;
5) Low thermoelectric potential to copper;
6) Good mechanical property;
7) Low contact resistance;
8) High resistance to corrosion and oxidation, good weldability; and
9) High wear resistance.

Typical examples may include Manganin, 83Cu-13Mn-4Ni, Constantan, 55Cu-45Ni, and 72Fe-23Cr-5Al-0.5Co. These alloys have relatively stable microstructures and are chemically stable. They have TCR $< \pm 2 \times 10^{-7}$/K over the temperature range of 0–100°C. During their processing, a stress-relief anneal is usually applied to stabilise the microstructure and electrical resistance.

2.9.2 *Heating elements*

Heating elements are widely used in electric and electronic equipment to provide heat such as the resistance heating wires in furnaces. They require high resistivity and low TCR. To obtain the right resistance and to produce the required power, the correct material should be selected. The length and thickness of the conductors need to be properly designed. The resistivity at room temperature and service temperature may be quite different. Extra resistance may be needed during the low-temperature heating period.

High temperature oxidation resistance is more important than the electrical resistance and TCR for long-term applications. Almost all metals (except for a few noble metals like Au, Pt and Ag) react with oxygen to form oxides, which are thermodynamically more stable than the metals. Engineering alloys such as Fe, Cu, Ni, Co, Al and their alloys will oxidise at elevated temperatures. The realistic method to prevent oxidation failure is to form a protective layer of oxide. The oxide scale should be dense (not porous) and in this oxide the metallic ions or oxygen diffusion rates are slow. They should be strong and tough, have high melting points, and good adhesion to the metals. Cr_2O_3, Al_2O_3 and SiO_2 are oxides that possess these properties.

Cr containing alloys such as stainless steels and 80Ni-20Cr (nichrome) have been widely used at temperatures up to 1000 °C. Cr_2O_3 provides a good protective layer on this type of alloys. Above 1000 °C, substantial CrO_3 forms, which is volatile and non-protective, and the Cr-containing alloys no longer have good oxidation resistance.

Al containing alloys such as Ni-Cr-Al or Fe-Cr-Al alloys (e.g. 55Fe-37Cr-8Al) are used for higher temperatures heating up to 1200 °C. Al_2O_3 has a much better chemical stability than Cr_2O_3. However, Al_2O_3 is more brittle than Cr_2O_3. Scale spallation may take place during cooling stages. Small amounts of reactive elements such as Y or Ce are added into the alloys to improve oxide spallation resistance.

For temperatures higher than 1200 °C, ceramic materials have to be used, e.g. silicon carbide (SiC) or molybdenum disilicide ($MoSi_2$).

In electronic devices where vacuum or protective atmosphere can be applied, refractory metals such as W, Mo, Ta and graphite (C) are often used for heating elements. However, they cannot be used in oxidising atmosphere as they can be oxidised fast.

2.10 Case Study – Materials Selection for Electrical Contacts

The materials for electrical contacts are a special group of conduction materials. Firstly, they should have good electrical conductivity as the energy loss and heat generated can be kept to the minimum. There are also some other requirements for electrical contacts.

Electrical contacts may work under severe conditions when they need to break high current frequently, like those in switches working under a high voltage. They have to withstand sparks and arcs and usually suffer from (a) fast heating to high temperatures, (b) corrosion and oxidation caused by sparking, (c) erosion and evaporation from fusing and (d) mechanical wear of the working surface.

Because oxidation or corrosion, a layer of oxide, which has a much lower conductivity than metals, forms on the surface. These oxides result in a bad contact and generate large amounts of heat. Bad contacts may cause partial melting and welding of the contact materials.

2.10.1 *Heavily-loaded contacts*

This group of applications requires materials to have

1. High conductivity;
2. Good chemical stability; and
3. Good mechanical properties.

They are often used in pairs: for example, Ag with Ag + Pd, Ag with Ag + Cu, Ag with Cu + Cd, Cu with Cu + Cd, Ag with graphite etc. because of the contact friction and surface properties.

2.10.2 *Lightly-loaded contacts*

This group of applications mainly require constant and stable conductivity. Pure or alloyed noble metals such as Ag, Pt, Pd, Au and high melting-point metals, W and Mo are used for this purpose. Noble metals do not suffer from corrosion or oxidation, they are reliable and last longer, but are expensive. Ag and Au have relatively low melting points (962 °C and 1064 °C respectively).

W is cheaper and harder, and also has a very high melting temperature. But W oxidises at high temperatures in oxidising atmospheres. Sintered 50%W + 50%Ag or 60%Mo + 40%Ag have high strength and good conductivity, and have been widely used for both heavy- and light-loaded applications.

2.11 Case Study – Materials for Electrical Brushes

Electrical brushes are mainly used in conjunction with slip rings, commutators and/or other contact surfaces to maintain a stable and reliable electrical connection in rotary and linear sliding contact applications. Electrical brushes can have lots of shapes and configurations depending on the brush material used and the actual application cases. Typical shapes of electrical brushes include assembly, leaf spring, plunger, contact tip/button, bar stock, brush pad, tamped/shunted brush and solid rock stock. Figure 2.13 shows the typical carbon brushes. Electrical brushes designed for sliding contact are suitable for applications in which the contacting members provide the electrical connection or path for the transmission of power or signals across a rotary interface in motors or generators. Some electrical brushes are designed for retipping or replacement in various OEM units while others are designed for transmission or pick-up of electrical signals in testing, probing or instrumentation applications.

Materials for electrical brushes normally require very good frictional characteristics combined with high to moderately high conductivity. In general, electrical brush materials can be classified into four categories: graphite, electrographite, carbon-graphite and metal-graphite.

Fig. 2.13 Typical carbon brushes (http://www.engineersparadise.com/en/ipar/18318).

2.11.1 *Graphite*

Graphite brushes are mainly fabricated from natural, artificial or synthetic graphites. Source materials are ground to fine powders with well-controlled particle size and ash content. These green powders are mixed with or without binders and then moulded to produce hard/soft brushes. Graphite brushes have a low coefficient of friction (graphite is a good lubricant) and a cleaning action due to the presence of ash. They are widely used for high power equipment or sliding contact applications.

2.11.2 *Electrographite*

Electrographite brushes are produced from powders of graphite, coke, lampblack, pitch, tar and binder. The mixed powders are moulded in hydraulic presses into plates, baked at a typical temperature of 1000 °C, and finally sintered at a temperature typically ranging from 2500 to 3000 °C. The high temperature heat treatment process is known as graphitization. In this process, impurities are removed and amorphous carbon is converted into polycrystalline dense graphite with a relatively large crystallite size. These brushes having different hardness values can be used to provide good commutation in high speed machines. However they might not be suitable for applications where very high current densities or high mechanical strength are required.

2.11.3 *Carbon-graphite*

Carbon-graphite brushes are typically fabricated using a process that is similar with that for electrographite. Powders of graphite, carbon materials (coke and black) and binders (tars and pitches) are compacted and moulded into plates or pills and then heat treated to a temperature of around 1400 °C. These brushes have a high specific resistance thus their current-carrying capability is not very high compared with electrographite brushes. However, they have applications where mechanical strength and adverse atmospheric conditions need to be considered.

2.11.4 *Metal-graphite*

Metal-graphite brushes are fabricated from metal powders (copper or silver), natural/artificial graphite and resin through ordinary powder metallurgical process or plastic moulding techniques. In comparison with copper-graphite brushes, silver-graphite electrical brushes have better electrical performance and may be cost-effective. They provide low electrical noise levels, low and stable contact resistance, low coefficient of friction and high conductivity, and are finding applications where high current capacity and low contact voltage drop are required. Copper-graphite brush has excellent flexural strength and high conductivity. Metal-graphite brushes are suitable for low-voltage generators and slip ring assemblies.

Summary:

This chapter introduced the classical electron conduction theory: The electrons can be regarded as small negatively charged particles obeying Newton's laws. When an electric field is applied to a solid, free electrons are accelerated. They lose their kinetic energy by collisions with the atoms in the lattice and generate heat. The current depends upon the average electron velocity, which is determined by the applied electric field and the collision frequency. Electrons may move through an ideal crystal without resistance, but in real crystals electrons collide with phonons (vibration of the ions in the lattice) and various imperfection. The total resistivity is the sum of the residual and thermal contributions. Materials with good conductivity, adequate mechanical properties, corrosion resistance and manufacturing properties are used for electric current transmission, electrical contacts, electrical resistors and heating elements.

Important Concepts:

Current density, J: The electric current per unit area, in amperes per square meter.

Resistivity, ρ: The resistance of a specimen with unit length and unit cross sectional area.

Conductivity, σ: The reciprocal of resistivity, $1/\rho$.

Ohm's law: The current density is proportional to the conductivity and electric field, $J = \sigma \cdot E$.

Drift velocity of an electron, v_d: The average velocity of an electron.

Electron mobility, μ: The drift velocity of an electron in a unit field.

Relaxation time, τ: The average time between collisions of the conduction electrons with phonons, lattice defects, etc.

Mean free path, l_m: The average distance travelled by an electron between two collisions.

Residual resistivity, ρ_r: The temperature-independent part of the resistivity of a conductor. Defects and impurities are responsible for the residual resistivity.

Thermal resistivity ρ_T: The part of the resistivity caused by temperature. Thermal resistivity caused by the vibration of the positive ion-cores about their equilibrium positions.

TCR: Temperature coefficient of resistivity.

Matthiessen's rule: The total resistivity of a conductor is the sum of a temperature-dependent component ρ_T and a temperature-independent component ρ_r.

Hall effect: When a magnetic induction B is imposed perpendicular to the flow of current J, a voltage perpendicular to both B and J appears. This phenomenon is called Hall effect and the voltage is called Hall voltage.

Nordheim's rule: Expressed as $\rho_r(x) = A \cdot x(1-x)$ where x is the concentration of a given impurity, and $\rho_r(x)$ is the contribution due to the impurity. A is a constant that depends on the impurity and the conductor.

Resistivity ratio: The ratio of the resistivity at room temperature (298 K) to the resistivity at 4 K. Resistivity ratio is approximately ρ_{298}/ρ_r.

Wiedemann–Franz law: The ratio of the electrical conductivity to the thermal conductivity of a metal at a constant temperature is a constant.

Questions:

1. What are solid solution alloys? How can they be formed?
2. Briefly describe the factors that can affect the electrical resistance of a metallic conductor.
3. Write out expressions and briefly explain the physical meaning for the following items. Indicate what the symbols in the equations represent.
   (a) Classical electron conduction theory,
   (b) Drift velocity of an electron under an applied field E,
   (c) Mobility of an electron, as a function of its moving time between two collisions,
   (d) Mean free path of an electron,
   (e) Hall constant RH,
   (f) Resistivity ratio and its application,
   (g) Resistivity of a 2-phase metallic conductor, and
   (h) The wavelength of moving particles of mass m.   .
4. Discuss the various factors that affect the drift velocity of electrons in metals. Why does the conductivity of metals decrease at higher temperatures?
5. To what temperature must you raise pure Cu to obtain the same electrical resistivity as that you would get by alloying with 20 wt.%Zn? The Nordheim's constant $A = 1.8 \times 10^7$ $\Omega \cdot m$ for Zn in Cu, $TCR_{Cu} = 0.0068/K$ and $\rho_{298} = 1.67 \times 10^{-8}$ $\Omega \cdot m$.
6. Suppose the mobility of an electron in Ni is 0.06 $m^2/V \cdot s$. Estimate the fraction of the valance electrons that are carrying an electrical charge. Ni has a *fcc* structure, a valance of 2 and a lattice constant of 3.517 Å. $\sigma_{Ni} = 1.46 \times 10^7/\Omega \cdot m$.
7. A composite material consists of 50vol.% of parallel W fibres in an Al matrix. Calculate the resistivity of the composite at 20 °C, (a) parallel

to the fibres, and (b) perpendicular to the fibres. The resistivity of Al and W is 2.7×10^{-8} and 13.5×10^{-8} $\Omega \cdot m$ at 20 °C, respectively.

8. A current density of 10^8 A/m^2 is applied to a Ag wire. If all of the valence electrons serve as charge carriers, determine the carrier density and drift velocity of the electrons. Ag has a *fcc* structure with 4 atoms per cell, a valence of 1, and lattice constant of 4.0862 Å.

9. A Hall Probe made of a single crystal of InSb ($\mu = 8$ m^2/V·s) is 1 cm long, 0.5 cm wide and 0.1 cm thick. A voltage of 2 V is applied to the wide of the crystal and a Hall voltage of 5 μV is recorded from the long side of the crystal in a magnetic field. Calculate the strength of the field.

10. The residual conductivity of a piece of high purity Cu is decreased from 10^9 to $10^7/\Omega \cdot m$ by adding 1 ppm impurity of given kind. (a) Calculate the resistivity ratios for the Cu specimen with and without the impurity. (b) What will be the resistivity ratio if 20 ppm impurity is added?

11. To fabricate an electrical furnace with a heating power of 200 W working under 120 V voltage, how long the wire should be if a Nichrome wire of 1mm diameter is chosen? The furnace working temperature is around 1000 °C. $\rho_0 = 112 \times 10^{-8}$ $\Omega \cdot m$, TCR = 0.000085/K.

12. Calculate the resistivity of 95Cu-5Sn (wt%) at 20 and 50 °C. Useful data are listed below: Resistivity of pure Cu at 20 °C is 1.7×10^{-8} $\Omega \cdot m$, TCR of pure Cu is 0.0043/K, TCR of bronze (Cu-Sn) is ~0.001/K, and the solution resistivity coefficient of Sn in Cu is 2.5×10^{-6} $\Omega \cdot m$.

Chapter 3

Electron Energy in Solids

3.1 Introduction

In classical theory, the electrons are treated as small negatively charged particles with a certain mass and charge at the same time, and obey the classical Newton's laws and Maxwell–Boltzmann statistics. But people found serious problems trying to explain some phenomena, such as positive Hall coefficients and very small electronic specific heats. More difficulties will be seen in explaining semiconductor conduction behaviour. Therefore, Quantum Theory has been introduced to study the electrical properties of solids.

In quantum theory, an electron is no longer a particle with certain mass and charge, rather it behaves sometimes like a wave and sometimes like a particle. Its momentum and position can only be described in terms of probability.

3.1.1 Schroedinger's equation

Erwin Schroedinger (1887–1961) used a wave function ψ to describe a particle. The probability of finding the particle in a volume, dV, is given by $|\psi|^2 dV$. The equation related ψ with the energy of the particle:

$$\nabla^2\psi + (8\psi^2 \cdot m/h^2)(E - U)\psi = 0, \tag{3.1}$$

or

$$\hbar^2\nabla^2\psi + 2m(E - U)\psi = 0, \tag{3.2}$$

where $\hbar = h/2\pi$,

h = Planck's constant = 6.63×10^{-34} J·s,

E = total energy of the particle,

U = potential energy of the particle,

m = mass of the particle, and

∇ is the Laplasian operator and $\nabla^2\psi = \partial^2\psi/\partial x^2 + \partial^2\psi/\partial y^2 + \partial^2\psi/\partial z^2$

In one dimensional consideration, Schroedinger's equation can be written as:

$$\hbar^2\partial^2\psi/\partial x^2 + 2m(E - U)\psi = 0. \qquad (3.3)$$

This will enable us to find wave function ψ for any particles whose energy is known.

3.2 Quanta and Waves

In quantum theory, the energy of an electron is limited to discrete levels or bands. Thus the electron bound to the nucleus can exist only in a series of states with sharply defined energies. To change its energy, the electron must jump from one state to another. In this process, it can emit or absorb a photon of electro-magnetic radiation whose frequency is proportional to the energy difference between the two states. The photon is regarded as a packet of energy, or particle.

On the other hand, a moving particle can be treated as a packet of waves and can be assigned a wavelength which is proportional to the particle's moment.

3.2.1 *The Planck relation*

Max Planck (1900) discovered the relation of the energy and frequency of the photon (emitted or absorbed by a material):

$$E_p = h \cdot f. \qquad (3.4)$$

where f is the frequency of the radiation and h is Plank's constant. $h = 6.63 \times 10^{-34}$ J·s.

Electromagnetic radiation is being quantised into a particle like a photon of energy E_p.

The photons travel at the speed of light c, $c = 3 \times 10^8$ m/s,

And
$$c = \lambda \times f. \tag{3.5}$$

Therefore, Eq. (3.4) can be written as

$$E_p = h \cdot c / \lambda. \tag{3.6}$$

In Eq. (3.6), the unit of E_p is J, c is m/s, λ is m and h is J·s. If eV is used for E_p, Eq. (3.4) should be:

$$E_p = h \cdot c / (e \cdot \lambda). \tag{3.7}$$

3.2.2 *The De Broglie theory*

De Broglie (1924) developed a converse relationship that particles of mass m and velocity v act like waves having a wavelength λ:

$$\lambda = h/p, \tag{3.8}$$

or
$$\lambda = h/(m \cdot v), \tag{3.9}$$

where $p = mv$, is the momentum of the particle.

3.2.3 *The Bragg's law*

Davisson *et al* confirmed the wave nature of electrons by showing that a beam of electrons of the same energy could be diffracted by a crystal (see Figs. 3.1 to 3.3). The diffraction obeys Bragg's law:

$$n \cdot \lambda = 2d \times \sin\theta, \tag{3.10}$$

where λ is the wavelength of the electron beam, θ is the diffraction angle, d is the distance between the crystal planes (called d-spacing) and n = 1, 2, 3....

This is the principle of electron diffraction in a transmission electron microscope (TEM).

Fig. 3.1 Electron diffraction pattern from silicon single crystal (J.G. Du, W.H. Ko and D.J. Young, Sensors and Actuators A, **112**, 2004, pp116-121).

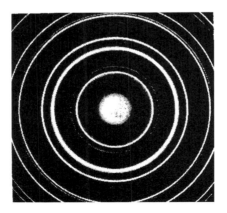

Fig. 3.2 A typical electron diffraction from polycrystalline Au (After J.F. Breedis).

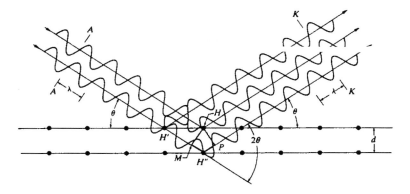

Fig. 3.3 Schematic of Bragg's law.

3.2.4 *The wave number, K*

A wave number, K, is used to describe the waves. K is a vector, has the same direction as the velocity v, and the magnitude is:

$$|K| = 2\pi/\lambda. \qquad (3.11)$$

Relating Eqs. (3.11) and (3.9), the kinetic energy of a free electron (which is also the total energy) will be:

$$E_K = 1/2 \ m{\cdot}v^2 = 1/2 \ m(h/m{\cdot}\lambda)^2 = (h^2{\cdot}K^2)/(8\pi^2{\cdot}m). \qquad (3.12)$$

Therefore, the relation between energy E_K and wave number K is parabolic (see Fig. 3.4).

3.3 Atomic Energy Levels

The electrons in atoms and molecules exist in well defined energy states. When an electron changes its state, the energy is emitted or absorbed in the form of a photon. The frequency of the photon f can be obtained from Eq. (3.4):

$$f = \Delta E/h = E_p/h.$$

Thus if the energy levels are known, the frequency may be calculated; or if the frequency of the photon is known, the energy levels may be calculated.

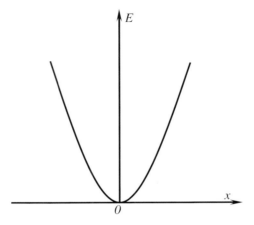

Fig. 3.4 The parabolic relation between the energy and wave number of free electrons.

3.3.1 *Bohr atomic model*

Hydrogen atom is the simplest atom. In 1913, Niels Bohr (1885–1962) pointed out that the electron in a hydrogen atom can only exist in one of a series of states. The energy is given by:

$$E_n = (-2\pi^2 \cdot m \cdot e^4)/(n^2 \cdot h^2) = -13.6/n^2 \text{ eV}. \tag{3.13}$$

E_n is the energy of the electron energy state, $n = 1, 2, 3, ...$, eV is electron volts. $n = 1$ corresponds to the state of the lowest energy, i.e., ground state.

For hydrogen, $E_1 = -13.6$ eV, $E_2 = -3.4$ eV and $E_3 = -1.51$ eV. For more complicated atoms, the energy levels are more complicated, but the principle is the same.

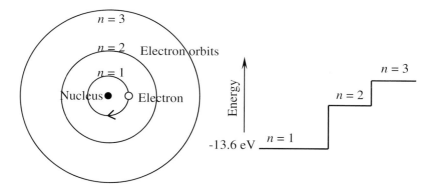

Fig. 3.5 Bohr atomic model and corresponding energy states.

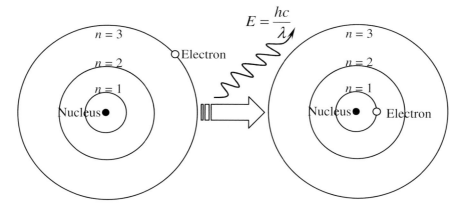

Fig. 3.6 Schematic showing an electron changing its state from higher energy level to lower energy level, leading to photo emission.

Electrons in atoms have defined energy states. When the states are changed, i.e., electrons undergo transitions from the orbits of different energy levels, emission or absorption of photon will take place and the frequency of radiation can be calculated using the Planck relation (see Fig. 3.6).

Example:

A hydrogen atom exists with its electron in $n = 3$ state. The electron undergoes a transition to $n = 2$ state:
1. Is energy emitted or absorbed in this transition?
2. Calculate the energy of the photon in J,
3. Calculate its frequency, and wavelength.

Answer:

(a) From $n = 3$ to $n = 2$, energy state goes from a higher to lower. So energy is emitted by this transition.

(b) The energy state is

$$E = -13.6/n^2 \text{ eV}$$
$$\Delta E = E_3 - E_2 = -13.6(1/9 - 1/4) = 1.89 \text{ eV} = 1.89 \times 1.6 \times 10^{-19}$$
$$= 3.02 \times 10^{-19} \text{ J}$$

(c) Frequency of the photon:

$$f = \Delta E/h = 3.02 \times 10^{-19} / 6.63 \times 10^{-34} = 4.55 \times 10^{14} \text{ /s}$$

and the wavelength of the photon is:

$$\lambda = c/f = 3 \times 10^8 / 4.55 \times 10^{14} = 6.59 \times 10^{-7} \text{ m} = 659 \text{ nm}$$

3.3.2 *Quantum numbers*

Quantum numbers are used to describe the energy state of an electron. There are four quantum numbers:

Principal quantum number, n. It describes the energy level of an electron. $n = 1, 2, 3, 4...$

Orbital quantum number, l. It represents the orbital angular momentum of the atomic electron. It defines the shape of the orbital. $l = 0, 1, ...$ to $n - 1$.

Magnetic quantum number, m_l. It is describing the orientation of the orbital. $m_l = -l, \ldots +l$.

Electron spin quantum number, m_s. It is describing the spin direction. $m_s = +1/2$ or $-1/2$

Each atomic state can be identified by a set of four quantum numbers: n, l, m_l and m_s.

3.3.3 *Pauli exclusion principle*

The Pauli Exclusion Principle (named after Wolfgang Ernst Pauli, 1900–1958) says that each energy state can be occupied by no more than one electron. Otherwise we may see all of the electrons in any atom crash together to the lowest energy state. Table 3.1 shows the electronic structure of several groups of metals. For example, sodium (Na) has the electronic structure as:

$$
\begin{array}{cccl}
\text{Na:} & 2 & 1s & (n = 1, l = 0, m_l = 0, m_s = \pm 1/2) \\
& 2 & 2s & (n = 2, l = 0, m_l = 0, m_s = \pm 1/2) \\
& 6 & 2p & (n = 2, l = 1, m_l = -1,0,+1, m_s = \pm 1/2) \\
& 1 & 3s & (n = 3, l = 0, m_l = 0, m_s = +1/2)
\end{array}
$$

A flame has relatively low energy, it can only excite the 3s electron, promoting it to a higher state. Energy changes are in the range of a few eV. The optical spectrum results from the downward transitions of this excited 3s electron, and has the wavelength in the range of 600 nm, showing an orange/yellow colour.

The electrons in the lower levels have low potential energies, they are tightly bound to the core of the atom. They can only be removed by bombardment with a high-energy beam of electrons or photons (e.g. X-ray). One of the higher-level electrons will descend into the empty state and the energy change of a much higher level. The energy will be released by photon emission, and the spectrum will be in the X-ray region.

Table 3.1 Electronic structure and electrical conductivity of various metals at 298 K.

Metal	Electronic structure	Electrical conductivity ($\Omega^{-1} \cdot m^{-1}$)
Alkali metals		
Li	$1s^22s^1$	1.07×10^7
Na	$1s^22s^22p^63s^1$	2.13×10^7
K	$1s^22s^22p^63s^23p^64s^1$	1.64×10^7
Rb	$......4s^24p^65s^1$	0.86×10^7
Cs	$......5s^25p^66s^1$	0.50×10^7
Alkali earths		
Be	$1s^22s^2$	2.50×10^7
Mg	$1s^22s^22p^63s^2$	2.25×10^7
Ca	$1s^22s^22p^63s^23p^64s^2$	3.16×10^7
Sr	$......4s^24p^65s^2$	0.43×10^7
Aluminium and Group IIIA		
B	$1s^22s^22p^1$	0.03×10^7
Al	$1s^22s^22p^63s^23p^1$	3.77×10^7
Ga	$...3s^23p^63d^{10}4s^24p^1$	0.66×10^7
In	$...4s^24p^64d^{10}5s^25p^1$	1.25×10^7
Tl	$...5s^25p^65d^{10}6s^26p^1$	0.56×10^7
Transition metals		
Sc	$1s^22s^22p^63s^23p^63d^14s^2$	0.77×10^7
Ti	$.................3d^24s^2$	0.24×10^7
V	$.................3d^34s^2$	0.40×10^7
Cr	$.................3d^54s^1$	0.77×10^7
Mn	$.................3d^54s^2$	0.11×10^7
Fe	$.................3d^64s^2$	1.00×10^7
Co	$.................3d^74s^2$	1.90×10^7
Ni	$.................3d^84s^2$	1.46×10^7
Copper and Group IB		
Cu	$1s^22s^22p^63s^23p^63d^{10}4s^1$	5.98×10^7
Ag	$...............4p^64d^{10}5s^1$	6.80×10^7
Au	$...............5p^65d^{10}6s^1$	4.26×10^7

3.4 The Lowest Energy Principle, Fermi Level and Fermi Distribution

The filling of electrons to the energy position follows the lowest energy principle: States of the lowest energy are filled first, then, the next lowest, and so on. Finally, all the electrons are accommodated.

The energy of the highest filled state at 0 K is called the Fermi level, or Fermi energy, E_f (named after Enrico Fermi, 1901–1954). The magnitude of E_f depends on the number of electrons per unit volume in the solid. At 0 K, all states up to E_f are full, and all states above E_f are empty. In other words, there are no transitions to states below E_f because they are full, and an electron can not change its state to above E_f unless enough energy is provided. At the Fermi level electrons moving through the crystal lattice, may be regarded as free electrons. The Fermi energy E_f can be expressed as:

$$E_f = \left(\frac{(hc)^2}{8mc^2}\right)\left(\frac{3}{\pi}\right)^{2/3} n^{2/3} , \tag{3.14}$$

where K_B = Boltzmann's constant = 8.62×10^{-5} eV/K or 1.38×10^{-23} J/K, E_f = Fermi level and T = absolute temperature.

At higher temperatures, however, the random thermal energy will empty a number of states below E_f by elevating some electrons to the higher energy states (Fig. 3.7). The energy distribution of electrons is called Fermi–Dirac distribution, $F(E)$, which gives the probability that a given available electron energy state will be occupied at a given temperature.

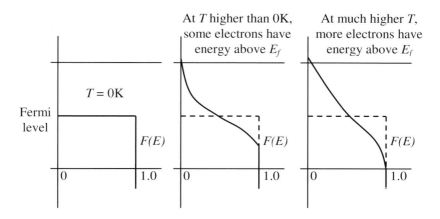

Fig. 3.7 Fermi distribution in a metal.

$$F(E) = \frac{1}{1 + \exp\left(\dfrac{E - E_f}{K_B T}\right)} \ , \tag{3.15}$$

This distribution is a function of temperature. A temperature increase increases $F(E)$ when $E > E_f$ and decreases $F(E)$ when $E < E_f$.

3.5 Energy Band in Solids and Forbidden Energy Gaps

The electrons of a single free-standing atom occupy atomic orbitals, which form a discrete set of energy levels. If several atoms are brought together into a molecule, their atomic orbitals split, producing a number of molecular orbitals proportional to the number of atoms. When a large number of atoms (of order 10^{20} or more) are brought together to form a solid, the number of orbitals becomes exceedingly large, and the difference in energy between them becomes very small.

The electronic band structure (or simply band structure) of a solid is the series of "forbidden" and "allowed" energy bands that it contains. Allowed bands may overlap, producing a single large band. The spaces between the energy bands are called Forbidden energy gaps because electrons may not have these energies. The band structure determines material's electronic and optical properties.

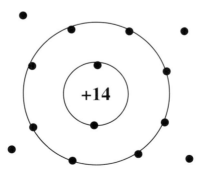

Fig. 3.7 Plane view of the atomic structure of silicon.

Fig. 3.8 Sharing of electrons in outer shell between atoms.

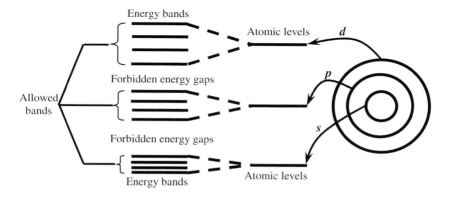

Fig. 3.9 Schematic showing the formation of energy bands.

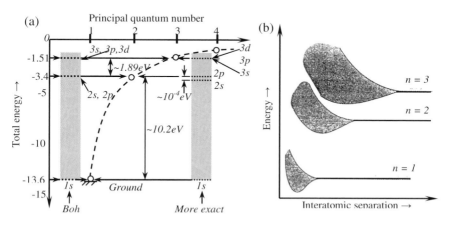

Fig. 3.10 (a) Energy level diagram for atomic H and (b) energy levels as a function of interatomic distance. The allowed energy levels have broadened into bands and the higher levels split first (after Rose, Shepard and Wulff, "Electronic Properties", John Wiley and Sons, 1966).

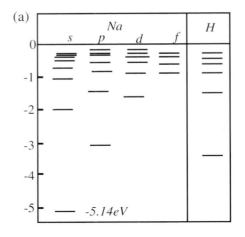

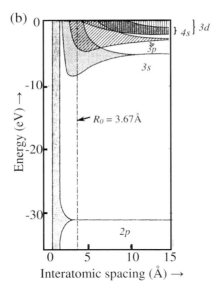

Fig. 3.11 (a) Energy level diagram for atomic Na and (b) Energy bands in metal Na, showing splitting of s, p, and d levels (after Rose, Shepard and Wulff, "Electronic Properties", John Wiley and Sons, 1966).

Figs. 3.7 to 3.9 illustrate how a band structure forms by using silicon as an example.

Figs. 3.10 and 3.11 are examples of solid hydrogen and sodium. The higher energy levels split first. There are several bands in Na that are overlap. The spaces between the bands are forbidden energy gaps because electrons may not have these energies in Na.

Because of the existence of the forbidden energy gaps between the energy bands, electrons in a solid will be subject to certain restrictions. Therefore, the continuity of the parabolic relation between E_K and K will be broken at the energy gaps. A large discontinuous energy is required for the electrons to pass from one state to the next.

3.6 Zone Model and Energy Well

3.6.1 *Zone model*

The energy gaps can also be explained from the view point of wave properties of the electron.

According to the Bragg diffraction law, electrons will be diffracted when the following equation is satisfied.

$$n \cdot \lambda = 2d \times \sin\theta. \tag{3.10}$$

Here λ is the wavelength of the electron beam, 2θ is the angle between the incident and diffraction beam, d is the distance between the crystal planes and n is an integer $= 1, 2, 3...$.

Because $|K| = 2\pi/\lambda$ (Eq. 3.11), we have

$$|K| = n \cdot \pi/(d \times \sin\theta). \tag{3.16}$$

This means that there is a series of values of $|K|$ for which electrons are diffracted and do not pass freely through the crystal.

When the energy is close to the forbidden region ($\pm K_1$, $\pm K_2$ ...), the rate of increase of energy decreases, then the slope actually goes to zero as K approaches the forbidden value. This behaviour is due to the increasingly strong diffraction effects as the critical value of K is approached. These K values correspond to the forbidden energies (forbidden zones) in the band structure, which means that no electrons can be present with such energies (see Fig. 3.12). This should be true whether the free electrons are from an impinging electron beam or from electrons in the crystal (e.g. valence electrons). This model is called the Zone Model.

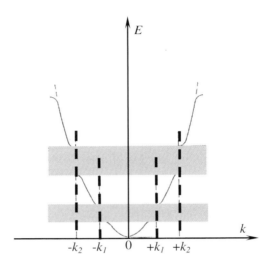

Fig. 3.12 The effect of Bragg diffraction on the $E(k)$ of the electron. The slope of the curve is zero at the forbidden energy values due to diffraction effects.

3.6.2 *Energy well*

The free electrons are the outer layer electrons of the individual atoms. The inner electrons are relatively undisturbed. When the free electrons travel through a crystal solid, the diffraction effects are due to the periodic positive charges in the lattice. A deep potential well is located at each ion position, due to the Coulomb force. This is called "Energy well".

The size of each well is a measure of the width of the energy gaps. The strength of the Bragg diffraction depends on the energy well size. Stronger diffraction leads to wider gaps.

3.7 Band Structures and Electrical Conductivity

The band structures vary for different groups of metals. Figure 3.13 uses examples to display how the electron band structures can influence the conduction behaviours of elements.

Mg: $1s^2 2s^2 2p^6 3s^2$

Magnesium is an alkali earth metal, it has two electrons in the outermost *s* band. We might expect them to have poor conductivity because there appear to be no unoccupied energy levels into which the electrons can be excited for conduction. However, because the *3s* band overlaps the *3p* band, there are a large number of unoccupied energy levels in the combined *3s* and *3p* band. Therefore, Mg has a good conductivity.

Al: $1s^2 2s^2 2p^6 3s^2 3p^1$

Aluminium follows Na and Mg in atomic number and has a partly filled *3p* band. The *3p* band serves as the conducting band and electrons easily enter the unoccupied levels in the *3p* band. Therefore, Al has a very good conductivity.

Fe: $1s^2 2s^2 2p^6 3s^2 3p^6 4s^2 3d^6$

The transition metals, like iron, have one or two electrons in their outermost *s* band and an unfilled *d* band. Electrons may enter the upper half of the overlapped bands. Therefore, the conductivity is not bad. However, the conductivity in this group is not high compared with some other groups, because there are complex interactions between the *s* and *d* bands.

Cu: $1s^2 2s^2 2p^6 3s^2 3p^6 3d^{10} 4s^1$

The metals in the IB group, i.e. copper, silver and gold, have one electron in their outermost *s* band. The inner *d* band electrons are tightly held by the atomic core and do not interact with the *s* band. Therefore, Cu, Ag and Au have very good conductivity.

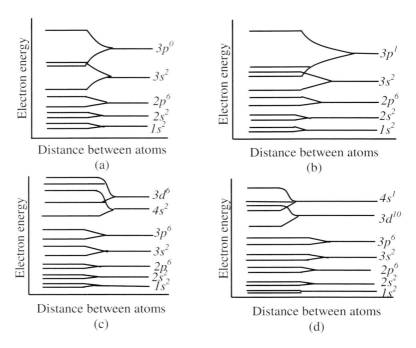

Fig. 3.13 Simplified band structures for selected metals. (a) Mg, (b) Al, (c) Fe and (d) Cu (after D Askeland, "The Science and Engineering of Materials", Chapman and Hall, 1990).

Energy band structure of metals and insulators

The band structure of metals (conductors) and insulators are very different. Conductors have partially filled valence bands in which electrons can move freely, while insulators have exactly filled valence bands in which electrons have no capacity to contribute to current unless they can get across the gap into the next unfilled band. In insulators, the energy gap is so wide that very few electrons can jump over it (Fig. 3.14).

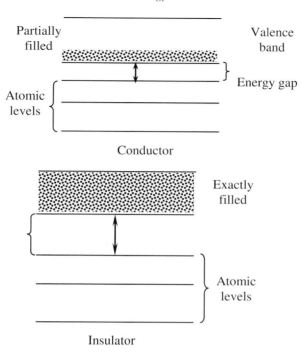

Fig. 3.14 Schematic drawing of electronic structure for a metal and an insulator.

Summary:

In Quantum theory, an electron is no longer a particle with certain weight and charge. It behaves sometimes like a wave and sometimes like a particle. Its momentum and position can only be described in terms of probability. The electrons bound to the nucleus of an atom can only exist in a series of defined energy states. Forbidden-energy gaps are located between these levels. To change the energy, an electron has to jump from one state to another. The electron energy level model can be extended to multi-atomic molecules and to form an energy band structure. Such a model can be used to explain the nature of conductors and insulators, and many of the electric, magnetic, and optical properties of materials.

Important Concepts:

Photon: A quantized particle of radiation which is absorbed or emitted instantaneously and wholly. The photon's energy $E_p = h \cdot v$.

Atomic Energy Levels: Discrete values of energy allowed to electrons which are bound to atomic nuclei.

Electron Energy State: A distinguishable quantum energy state, labelled by a characteristic set of numbers n, l, m_l and m_s.

Pauli Exclusion Principle: Only one electron may occupy any energy state.

De Broglie Relation: Particles may behave like waves with wavelength $\lambda = h/p$, where p is the particle momentum and h is Planck's constant.

Energy Band: Energy levels spaced so closed together that they may be treated as continuous band of allowed energy.

Forbidden Energy Gaps: The spaces between the energy bands, no electrons in solid may have these energies.

Wave Number: A vector quantity whose direction is that of propagation of the wave and whose magnitude is $2\pi/\lambda$.

The Lowest Energy Principle: The electrons in an atom will fill the state with the lowest energy first, then, the next lowest, and so on.

Fermi Level (E_f): The energy of the highest filled state in the highest energy band which contains electron in a metal at 0 K.

Energy Well: A potential well caused by the diffraction effects of the periodic positive charges in the crystal lattice.

Questions:

1. Name the four quantum numbers that are used to characterise every electron within an atom. Give the possible numbers of them and briefly explain what they define.

2. Using electron energy band structures to explain the different electrical conduction properties of metals and insulators.

3. Explain the electrical conductivities of Mg, Al, Fe and Cu with their band structures.

4. An electron has been accelerated by a potential difference of 10 kV. Calculate (a) its speed, and (b) its De Broglie wavelength.

5. A solid-state laser generator is used to characterise the photoemission properties of ZnO films. ZnO can be treated to have photoemissions of green radiation region ($\lambda \approx 600$ nm). What is the required energy level (in J & in eV) of the laser beam?

6. An electron beam is diffracted by the (111) plane of Ag which are 0.2359 nm apart. The strongest diffraction shows a Bragg's angle θ = 2.5°. Calculate (a) the De Broglie wavelength of the electron beam, (b) the speed of the electrons, and (c) the acceleration voltage.

7. Write out the Fermi distribution and explain (a) what is $F(E)$? (b) what is E? (c) what is E_f? (d) make a schematic energy-level drawing for the above distribution, and (e) show the effect of E and T on $F(E)$.

8. Potassium has its Fermi energy E_f = 2.1 eV. What energy corresponds to (a) a probability of occupancy of 0.98 and (b) a probability of 0.02 at 293 K?

9. Calculate the probability of an electron being thermally promoted to the conduction band in diamond at 25 °C. The energy difference between the Fermi level and conduction band is 2.8 eV.

10. From the electron structure of platinum (Pt), sketch the expected band structure and compare the resistivity, 10.5×10^{-8} $\Omega \cdot$m, with that of more common metals with a similar band structure.

Chapter 4

Electron Emission

4.1 Introduction and Work Function

In this chapter, we discuss the phenomena and mechanisms of electron emission. Electrons can be emitted from solids by four mechanisms: thermal, photoemission, field and secondary emission.

In metals, the valence electrons are free to move within the solid to conduct electricity. However, the electrons are bound to the solid by the field of the lattice of atoms or positive ions. They cannot escape, or emit, from the material unless enough energy is obtained. The energy with which they are held is called the work function, $e\phi$. In other words, the minimum amount of energy needed for an electron to leave a surface is called the work function.

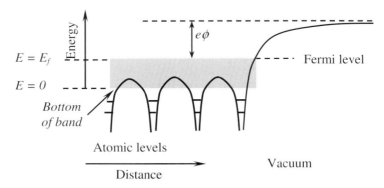

Fig. 4.1 The energy barrier at the surface to the emission of electrons (after Rose, Shepard and Wulff, "Electronic Properties", John Wiley and Sons, 1966).

Fig. 4.1 shows the potential energy of an electron plotted as a function of distance along a line of atom centres. The work required to remove an electron from Fermi level to the vacuum outside is $e\phi$. The magnitude of $e\phi$ depends on the Fermi level, the composition of the solid, and the structure of the emitting surface.

4.2 Thermionic Emission

4.2.1 *Theory*

Thermionic emission is the ejection of charge carriers (electrons or ions) from a heated conductor. The classical example of thermionic emission is the emission of electrons from a hot metal cathode into a vacuum (Edison effect, 1880). A basic thermionic device consists of a cathode (which is heated and negatively charged and serves as the electron emitter) and an anode (which is positively charged to draw off emitted electrons). These electrodes are situated in a vacuum system. When the cathode is heated to a high temperature, a certain number of electrons will gain additional thermal energy and have kinetic energies far above the average and well in excess of the work function. These electrons can escape from the solid. With increasing temperature, the fraction of high-energy electrons increases and a greater number escapes. Thermionic emission can be significant when temperature is over 1000 K. The thermionic emission of electrons is also known as thermal electron emission. However, thermionic emission is now used to describe any thermally excited charge emission process, including ion emission.

The correlation between the emission current and temperature can be described by the Richardson–Dushman equation:

$$J = A \cdot T^2 \cdot e^{(-e\phi/K_B T)}, \tag{4.1}$$

where J is the current density, T is absolute temperature, K_B is the Boltzman's constant, and A is the Richardson constant depending on the emitter material.

Constant A, work function $e\phi$ and the operating temperature T are the three factors that determine thermal emission. However, T is more important than the other two. This is why tungsten (W) is widely used as the thermal emitter material although its $e\phi$ is not low and A is not high. The Richardson constant A and work function $e\phi$ of some emitters are listed in Table 4.1.

Table 4.1 Richardson constant and work function of various materials.

Material	Work function (eV)	Richardson constant (A/cm^2·K^2)
Barium	2.1	60
Ba on W	1.6	1.5
Cesium	1.8	160
Cs oxide	0.8	~10^{-2}
Iriduim	5.4	170
Molybdenum	4.2	55
Nickel	4.6	30
Platinum	5.3	32
Rhenium	4.9	100
Tantalum	4.1	60
Thorium	3.4	70
Th on W	2.6	3.0
Thoria	2.5	3.0
Tungsten	4.5	60
LaB$_6$	2.7	40
TaC	3.1	0.3

4.2.2 *Thermal electron emitter materials*

4.2.2.1 *Metal*

Three metals give useful emission, *viz* tungsten, tantalum and rhenium.

Tungsten (W) is a traditional thermoemitter material. It has a work function $e\phi = 4.5$ eV, $A = 6 \times 10^5$ A/(m^2·K^2) = 60 A/(cm^2·K^2). The typical operating temperature T is around 2700 K. With Richardson equation, the electron current can be calculated as $J \approx 1.75$ A/cm^2. Typical tungsten cathode may have several configurations.

1. Thoriated tungsten: the emitting surface of this cathode is a monolayer of thorium atoms on tungsten carbide (W$_2$C). Thoriated tungsten, usually in the form of wire, contains about 0.5–1% thorium

oxide (ThO_2) uniformly dispersed throughout as fine particles. When the cathode is heated in vacuum, W_2C and ThO_2 react to form metallic W and Th ($2W_2C + ThO_2 \rightarrow 4W + Th + 2CO$).

2. Barium dispenser cathodes: the emitting surface of these cathodes is a monolayer of barium on a substrate of tungsten. These cathodes can be classified into two main groups, including "L" and impregnated cathodes.

3. Caesium on tungsten: tungsten must be operated in caesium vapour to maintain the activating layer.

The life time of a tungsten filament under the working condition of $J = 1.75$ A/cm^2, $T = 2700$ K, and a reasonably good vacuum is ~100 hours. Tungsten has a grain growth problem above 2000 K. ThO_2 is typically added to stabilise the grain boundaries of W.

Tantalum is ductile and can be made into many desired shapes, therefore it can be generally used for a configuration other than a wire. The cost and rarity of rhenium limit its use though this metal is relatively resistant to some harmful reactions.

4.2.2.2 *Oxide*

The best known oxide cathodes are thoria (ThO_2), gadolinia (Gd_2O_3), yttria (Y_2O_3) coated onto tungsten and barium strontium calcium oxide (BaO-SrO-CaO) deposited on pure nickel. Refractory oxides react considerably less than hot tungsten with the residual gases, and can be heated in air without permanent damage.

4.2.2.3 *Refractory compound*

Carbides, nitrides and borides have also been used for thermal electron emission. Lanthanum hexaboride (LaB_6) is an important electron source material. LaB_6 has a lower $e\phi$ (= 2.4 eV) and a similar A (= 4×10^5 A/m^2·K^2) compared with W. The emission current density J can reach 100 A/cm^2 at 2000 K, or 2 A/cm^2 at 1500 K. Therefore, LaB_6 can be used to provide much higher electron intensity and thus a higher brightness than W, or can last longer time at lower operating temperatures. Boron tends to migrate into almost any carrier metal at an excessive rate at higher temperatures. The best substrate metal is rhenium, or carburized tantalum. Surface contamination will reduce the life time. Therefore, it requires a higher operating vacuum.

Other compounds of practical interest are carbides, particularly those of uranium and zirconium. Their total emission density at ~2000 K is a few A/cm^2.

4.2.3 *Application*

Thermionic emission is widely used in tube circuit elements, X-ray tubes and thermionic energy conversion systems. Diodes and triodes are the basic building blocks of tube based electronic devices. However thermionic tubes need a relatively long time for warming up and require energy for heating. In addition, these tubes have a large size and a complex structure. Thus they had been replaced by transistors in most applications after 1960. In an X-ray tube, a beam of high energy electrons strikes a solid metallic target to produce rays.

Thermionic energy conversion is transforming thermal energy into useful electrical energy. A simple thermionic energy converter consists of a hot electron emitter and a cooler collector separated by a vacuum gap in which cesium (Cs) vapour is maintained. A voltage is generated due to the temperature difference between the emitter and collector. Typically, emitters operate at a temperature ranging from 1800–2100 K and collectors at 800–1100 K. The energy generator core could be based on flat or grooved metal surfaces as the electron emitter. Conventional thermionic energy converters use molybdenum (Mo) or tungsten (W) as the emitter. Thermionic energy conversion has been investigated as a method of space power generation in conjunction with nuclear reactors, solar power concentrators, fossil fuel or radioisotope heat source. However, before thermionic energy conversion gain significant commercial success, emitter materials with lower operation temperature (and longer operation lifetime) and collector materials with higher operation temperature should be developed. Recent research indicated that carbon-based materials (such as diamond film) may show the feasibility of energy conversion at temperatures considerably less than 1000 °C.

4.3 Photoemission

4.3.1 *Theory*

Electrons can be emitted from a solid (cathode) by interactions between an incident photon of electromagnetic radiation and an electron near the surface. This is called electron photoemission. Photoemission relies on the interaction between an incident photon of electromagnetic radiation and an electron near the surface of a metal or electromagnetic semiconductor. The energy of the incoming photon is entirely absorbed by the electron with which it interacts. Electron emission occurs when photon energy exceeds the difference between the Fermi level and the energy of a free electron in a vacuum. This difference is known as the photoelectric threshold energy and is identical to the work function in a metal.

4.3.2 *Photocathode materials*

Typical photocathodes are metals and semiconductors (as shown in Table 4.2). The most important properties for a photocathode are: quantum efficiency (QE), spectral response, operational lifetime, temporal response, saturation level, damage threshold, voltage hold-off, transverse energy spread of emitted electron beam, dark current and average current density. In general, semiconductors have high quantum efficiency but short life time and high sensitivity to contamination. Metallic photocathodes are strong and have long lifetime. Their main disadvantage is the low QE due to their high reflectivity and the shallow escape depth.

Photoelectric efficiency is defined as the number of electrons emitted per incident photon, or the current emitted per unit of light intensity. Photoelectric efficiency depends upon (1) the chemical and physical nature of the material, and (2) the wavelength of the incident radiation.

Table 4.2 Typical photocathode materials.

Materials		Threshold (nm)	Remarks
Metal	Ba	496	
	Ca	427	
	Cu	288	Used for RF gun
	Mg	339	Used for RF gun
	Nb	310	Robust & resistant to contamination
	Sm	459	
	Y	427	
Semiconductor	Ag-O-Cs		First compound photocathode; sensitive from 300 to 1200 nm
	Cs-Te	~320-354	Sensitive to UV
	Cs-I	~200	Sensitive to UV
	GaAs		Wide spectral response range from UV to 930 nm
	InGaAs		Extended sensitivity in infrared range compared to GaAs
	Na-K-Sb	620	Operation temperatures up to 175°C; ideal for photon counting applications
	Sb-Cs	620	Spectral response from UV to visible; mainly used in reflection-mode photocathodes
	Sb-Rb-Cs		High sensitivity and low dark current; used for ionizing radiation measurement in scintillation counters

4.3.3 *Application*

Photoemission is the basis of ionoscope and its successor, image orthicon (used for television camera) and also photomultiplier, a device used to detect low level electromagnetic radiation. Photomultiplier tubes can detect a candle at a distance of 10 km and have been used as scintillation counters and as detectors of faint light in the region of the visible spectrum in astronomy.

4.4 Field Emission

4.4.1 *Theory*

Field emission, also known as electron field emission, requires a high voltage difference between the cathode and the anode. Field emission in pure metals occurs in high electric fields: the gradients are typically higher than 1000 volts per micron (10^7–10^9 V/cm) and strongly dependent upon the work function.

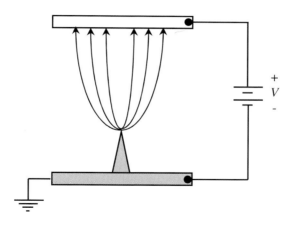

Fig. 4.2 Schematic of field electron emission.

Field emission is a unique quantum-mechanical effect of electrons tunnelling from a condensed matter (solid, liquid or individual atom) surface into vacuum or open air. The efficiency of this emission process is tens of millions of times higher than in other known emission processes. The theory of field emission from bulk metals was proposed by Fowler and Nordheim. The Fowler-Nordheim equations can be used as a rough approximation to describe emission from other materials.

4.4.2 *Device and material*

The actual emitter in a field emission device consists of an exceedingly fine wire. In order to produce high enough fields using reasonable potentials the wire is formed into a tip with the apex radius of curvature ranging from tens of angstroms to several micrometers. An ideal field emitter should be conductive, long and thin, robust, and cost-effective and have high mechanical strength. Mo and W tips produced by microfabrication techniques are commercially available and widely used. These two refractory metals have very high melting points, relatively low work function, high electrical conductivity and robustness.

Most recent studies indicated that quasi-one-dimensional nano-materials, including carbon nanotubes (CNTs), and aligned nano-rods/nanowires of copper sulphide (Cu_2S, CuS), silicon carbide (SiC), copper oxide (CuO), molybdenum oxide (MoO_2 & MoO_3) and zinc ox-ide (ZnO), are very promising candidates.

Particular interest is given to CNTs, a novel form of carbon discov-ered in 1991. Since 1995, CNTs have been recognised as the most prom-ising field electron emission material due to their excellent properties of:

1. Either metallic or semiconducting conductivity;
2. High geometrical field enhancement factor (1000);
3. A work function value lower than that of bulk graphite; and
4. Extremely low turn-on fields and high current densities.

Nowadays it is also possible to prepare very sharp emitters with a single atom at the end. In this case, electron emission comes from an area about twice the crystallographic size of a single atom.

4.4.3 *Application*

4.4.3.1 *Field emission microscopy*

Field emission microscopy (FEM) was invented by Erwin Müller in 1936. FEM employs field emission to study the fine surface structure and electronic property of materials. Images are obtained on the basis of the difference in work function of the various crystallographic planes on the surface.

A basic field emission microscope consists of a metallic sample in the form of a sharp tip and a fluorescent screen enclosed in ultrahigh vacuum (UHV). The sample is negatively charged (1–10 kV) relative to the screen. This gives the electric field near the tip apex to be the order of 10^{10} V/m. When the emitter surface is clean, this FEM image reflects: (a) the surface characteristics of the material from which the emitter is made, (b) the orientation of the material relative to the needle/wire axis, and (c) the shape of the emitter end form.

Refractory metals with high melting temperature, such as W, Mo, Pt and Ir, are conventional objects for FEM experiments since they can be fabricated in the shape of a sharp tip, used in UHV environments, and can tolerate the high electrostatic fields.

4.4.3.2 *Field emission display*

In a field emission display (FED), millions of electron emitters are arranged into regular arrays and placed in close proximity (0.2–2.0 mm) to a phosphor faceplate. A schematic diagram of a FED is shown in Fig. 4.3. The anode consists of a glass plate onto which a transparent indium tin oxide film and an anti-reflective coating are coated. Finally, red, green and blue phosphors are deposited using an electrophoretic deposition (EPD) method. Images are created by impinging electrons from the emitters onto phosphor pixels.

Microfabricated emitters such as silicon microtips and Spindt (named after Charles A. Spindt) field emission arrays (FEAs) have been applied. Spindt-type FEAs are relatively easy to manufacture and robust for FED applications. This type of cathodes was invented in the late 1960s and developed in the early 1970s by Spindt and coworkers. The original Spindt FEAs were composed of individual field emitters of small sharp molybdenum cones. Fabrication generally involves lithography (e.g., electron beam lithography and total internal reflection holographic lithography), deposition and etching. In addition, CNT-based emitters are being developed and vigorously pursued.

Other proposed applications of large-area field emission sources may include microwave generation, space-vehicle neutralization, X-ray generation, and (for array sources) multiple e-beam lithography. Recent attempts are to develop large-area emitters on flexible substrates, in line with wider trends towards plastic electronics.

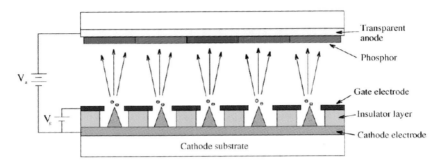

Fig. 4.3 Architecture of a Spindt cathode-based flat panel display (N.S. Xu and S. Ejaz Huq, Materials Science and Engineering R, **48**, 2005, pp47-189).

Fig. 4.4 Spindt type emission tips (A.A. Talin, K.A. Dean and J.E. Jaskie, Solid-State Electronics, **45**, 2001, pp963-979).

4.5 Secondary Electron Emission

Secondary electron emission is generated by the collisions of primary electrons with the conduction electrons of the solid. When high-energy electrons bombard a metal surface, secondary electrons can be ejected.

The kinetic energy of the secondary electrons does not depend on the kinetic energy of the incident beam and is usually about 10 eV.

The ratio of the number of emitted electrons I_s to the number of incident electrons I_p is called the secondary emission yield, δ

$$\delta = I_s/I_p. \qquad (4.2)$$

δ is a function of the kinetic energy of the incident beam (Fig. 4.5).

If the kinetic energy of the incident beam is low, only a few electrons will gain sufficient energy to escape from the emitter surface, so δ is low. If the high kinetic energies are very high, the beam penetrates the emitter before collisions occur. Many excited secondary electrons lose their energy by other collisions within the target before they reach the surface. Therefore, δ is also low. δ has a maximum value, which is a function of the energy of the incident electron beam.

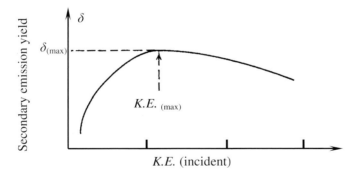

Fig. 4.5 Secondary electron emission yield δ as a function of the kinetic energy of the incident electrons (after Rose, Shepard and Wulff, Electronic Properties, John Wiley and Sons, 1966).

Summary:

Electrons may escape from the surface of a solid by absorption of electromagnetic radiation, by increasing the temperature of the solid, by high-energy electron bombardment or by applying a high electric field. The energy with which the electrons are held is called work function.

The Richardson–Dushman equation describes the emission of electrons due to thermal energy. Electron emission has applications in electronic technology and materials characterisation.

Important Concepts:

Work Function, $e\phi$. This is the energy difference between the free state and the highest full state in the valence band (Fermi level). $e\phi$ is commonly expressed in eV.

Thermal Electron Emission (also called thermionic emission): Electron emission by thermal excitation of electrons from the Fermi level to the free state.

Space Charge: The field of emitted electrons. At high current densities, this field must be added to the other field present. Space charge reduces electron emission.

Photoemission: Electron emission from a solid by excitation of electrons by radiation.

Photoelectric Efficiency (or Photoelectric Yield): The number of electrons emitted per incident photon.

Secondary electron emission: Electron emission due to collisions of primary electrons with the conduction electrons of the solid.

Electron Field Emission: Direct quantum mechanical tunnelling of the electrons through the energy barrier which is narrowed by the application of extremely high electric fields at the surface of the solid. No energy is required to excite the electrons across the work function barrier.

Questions:

1. Briefly explain the relationships of thermal electron emission with cathode temperature and applied voltage. In choosing a thermal electron emitter, what factors are important?

2. In choosing a photoemitter, what considerations are important? Having chosen the material, how could you improve the emission current?

3. The photoelectric threshold is the maximum wavelength of incident light that will produce photoemission. The threshold wavelength for a Na surface is 540 nm, calculate (a) the work function for Na, (b) the maximum kinetic energy when the impinging light has a wavelength of 200 nm.

4. For a Cs-coated oxide-Ag sandwich photoemitter, $e\phi = 0.9$ eV. (a) If light of 400 nm is incident on the emitter, what is the maximum velocity of the ejected electrons? (b) Calculate the longest wavelength that can eject photoelectrons.

5. What is the minimum frequency necessary to achieve photoemission from Ta? Work function $e\phi$ of Ta $= 4.1$ eV.

6. For the Richardson-Dushman equation, the value A for W is 6×10^5 A/(m·K)^2, and the work function is 4.5 eV.
   (a) For a cathode with an area of 2×10^{-5} m^2, what is the maximum emission current possible at 2000 K and 2500 K? What is the ratio of these two currants?
   (b) If the cathode is thoriated W ($e\phi = 2.7$ eV), to obtain the same emission as the ordinary W at 2500 K, at what temperature, approximately, the thoriated W should be operated?
   (c) For a Cs-coated W cathode, $e\phi = 1.36$ eV, $A = 0.032 \times 10^6$ A/(m·K)^2, what temperature is now necessary to get the same emission as in (b)?

7. Calculate the maximum energies in eV of emitted photoelectrons from Ag, Ca, Ni and Pt when the surface is illuminated with 250 nm radiation. $e\phi$ for Ag, Cs, Ni and Pt are 4.47, 1.90, 4.01 and 6.30 eV, respectively.

8. Briefly explain why one-dimensional nanostructures are ideal for field emission and application in flat display.

Chapter 5

Semiconductor, Properties and Materials

5.1 Introduction and Band Structure

5.1.1 *Introduction and brief history*

Semiconductors are the base of the electronic industry. They have had a major impact on the development of the modern technological society in which we live. Without semiconductor technology we would not have computers, modern electronic devices, instruments and equipment including stereos, TVs, VCRs, mobile phones and telecommunication systems.

Research in the semiconductor field can be tracked back to the middle of the 19th century. The first recorded observation was reported by M Faraday in 1833 that silver sulphide shows a negative TCR. But the research progress was very slow in the early years. During the period of two world wars and the time between them, military and commercial applications such as the radio, radar, power rectifier, infra-red detector and photoelectric devices etc. developed rapidly. In 1947 the transistor was developed at Bell Labs USA, which later led to an extensive investigation in solid-state electronics called "solid integrated circuit" (IC). Now, the largest ULSI (Ultra Large-Scale Integrated Circuits) contains $> 10^7$ components per chip. For instance, in 2007, Intel made the debut of 45 nm Si manufacturing technology and introduced the biggest change to transistors in 40 years. More than 2,000 45 nm transistors fit across the width of a human hair and the quad-core processor then has 820 million of transistors with a clock speed higher than 3 GHz. The small chip size makes it possible for systems to deliver the high level of performance with very low power consumption, and especially suitable for notebook type computers. The semiconductor industry has become one of the largest industries in the world since the 1980s.

5.1.2 *Energy band structure in metals, insulators and semiconductors*

As we discussed in chapter 3, the electrons in a solid are constrained to have energies that lie in a number of energy bands. The band structure determines the electrical properties of the materials (Fig. 5.1).

A metal has a partially filled band - the conduction band. Because the band is not fully filled, the electrons have sufficient energy to move when an electric filed is applied. The potential fields of ions on the lattice sites create resistance to the move of electrons, but do not stop the electron from moving.

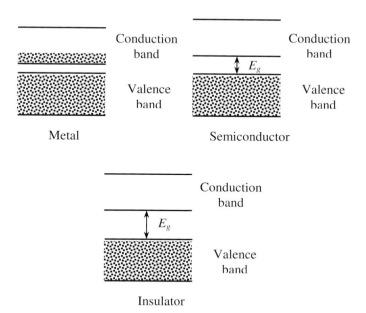

Fig. 5.1 Schematics of the energy band structures of a metal, a semiconductor and an insulator.

An insulator has the highest occupied band completely filled with electrons. This is the valence band. In the valence band, electrons must move through the potential field of the lattice, therefore it is very difficult. There is a large forbidden energy gap (2–10 eV) between the valence band and the next band. An electron in an insulator can not be freed from the lattice potential unless it receives energy high enough to promote it across the energy gap.

Semiconductors are between metals and insulators. The valence band is completely filled, but the gap is rather narrow ($\leq$ 2 eV). The narrow gap allows some of the electrons to be thermally excited from the filled valance band to the vacant conduction band, where they are free to move under an electric field. The Fermi level, E_f, is located in the centre of the band gap. In other words,

$$E_f = E_g/2. \tag{5.1}$$

5.2 Band Model

5.2.1 *Band model and hole conduction*

A material in which conduction occurs because of electron excitation is called an intrinsic semiconductor. A two dimensional cubic crystal model is shown in the Fig. 5.2. Conduction in an intrinsic semiconductor is caused by electron excitation. Each electron that moves to the conduction band leaves behind a vacant state called the "hole" in the valence band. The hole behaves like a particle of positive charge. Therefore, each thermal excitation of an electron liberates not one, but two charge carriers.

The concept of a hole in the valence band is similar to a vacancy in a crystal. The motion of a hole to the right is equivalent to an electron moving to the left, since negative charge has been transferred to the left. This is called "hole conduction".

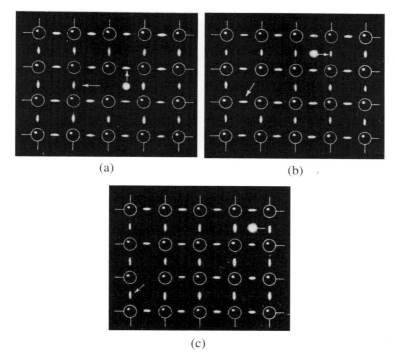

(a)　　　　　　　　　(b)

(c)

Fig. 5.2 Creation and motion of a conduction electron and hole in a semiconductor. (a) an electron breaks away from the covalent bond, leaving a vacant bonding state, or a hole, (b) the conduction electron moves to the right; the hole to the down and left, (c) the electron and hole continue to move away (after Rose, Shepard and Wulff, "Electronic Properties", John Wiley and Sons, 1966).

5.2.2 *Type of bandgap*

Bandgap in semiconducting materials could have two types: direct and indirect. The direct bandgap is characterised as: the minimum of the conduction band lies directly above the maximum of the valence band in momentum space. In a direct bandgap semiconductor, electrons at the conduction-band minimum can combine directly with holes at the valence band maximum, while conserving momentum. The energy of the recombination across the bandgap will be emitted in the form of a photon of light. This is radiative recombination, also called spontaneous emission.

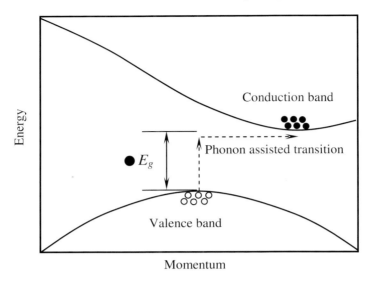

Fig. 5.3 Schematic of an indirect bandgap.

Indirect bandgap is a bandgap in which the minimum energy in the conduction band is shifted by a k-vector relative to the valence band, i.e., the minimum of the conduction band does not occur at the same wave vector as the maximum of the valence band. The k-vector difference represents a difference in momentum. Semiconductors that have an indirect bandgap are inefficient at emitting light.

5.2.3 *Elementary semiconductors*

Silicon (Si) and germanium (Ge) are two of the most important elementary semiconducting elements. They are in the group IV-A in the periodic table and have a valence of four. Both of them have a diamond cubic crystal structure with highly directional covalent bonds (Fig. 5.4). Each Si or Ge atom contributes four valence electrons. These electrons form pairs to bond the atoms together in the crystal lattice.

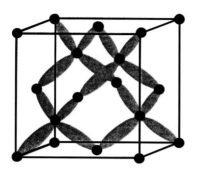 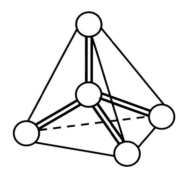

Fig. 5.4 Diamond cubic crystal structure. Diamond carbon, Si and Ge all have this crystal structure.

5.2.4 *Compound semiconductors*

5.2.4.1 *Binary compounds*

Elements that do not have a valence of four may also form semiconductors providing the average number of valence electrons per atom is four. Many binary compounds are made up of elements whose combined valence is 8, like 4 + 4 in Si and Ge. They have a similar diamond cubic (DC) lattice and similar semiconductor properties, if their electrochemical properties are similar.

A trivalent (3) and a pentavalent (5) element can form a so called III-V semiconductor compound, which has a composition of MX with M being a 3+ valence element and X being a 5+ valence element. A typical example is gallium arsenide (GaAs). A divalent (2) and a hexavalent (6) form a II-VI semiconductor such as ZnS, ZnSe, ZnO, CdTe and HgSe.

5.2.4.2 *Ternary compounds*

Typical examples of ternary semiconductors may include GaAlAs, HgCdTe and GaAsP. The formation principle is keeping the number of valence electrons per atom equal to four. For example in GaAlAs, the molar fraction of Ga + Al = 1/2 and As = 1/2, so the III and V elements in the compound remain equal.

Table 5.1 lists the common elemental and compound semiconductors.

Table 5.1 Some elemental and compound semiconductors.

Semiconductor	Bandgap (eV)	Applications
Silicon (Si)	1.1	Conduction controlled devices Integrated circuits Detectors and solar cells
Germanium (Ge)	0.67	Photo-detectors
Gallium arsenide (GaAs)	1.4	Conduction controlled devices Optoelectric devices like lasers and LEDs Solar cells
Gallium phosphide (GaP)	2.2	Light emitting diodes
Indium phosphide (InP)	1.3	Solar cells Substrates for thin films of other semiconductor materials
Cadmium telluride (CdTe)	1.5	Solar cells X-ray photo-detectors
Cadmium sulphide (CdS)	2.4	Solar cells
Gallium aluminium arsenide (GaAlAs)*	-	Laser crystals LEDs
Mercury cadmium telluride (HgCdTe)*	-	Infra-red photo-detectors
Gallium arsenide phosphide (GaAsP)*	-	Light emitting diodes

* In these ternary alloys the bandgap depends on the ratio of gallium to aluminium etc, and so cannot be given as a single value (source: C.R.M. Grovenor, "Materials for Semiconductor Devices", IOM).

5.3 Intrinsic Semiconductor

Intrinsic semiconductor is such a material in which conduction occurs because of electron excitation. Intrinsic semiconductors do not contain impurities. They do contain electrons as well as holes. The electron density equals the hole density since the thermal activation of an electron from the valence band to the conduction band yields a free electron in the conduction band as well as a free hole in the valence band. When an electric field is applied, the electrons move in one direction and the holes in the other direction. The fraction of the electrons in the valence band that will be promoted across a band gap of E_g at temperature T is proportional to $\exp(-E_g/2K_BT)$. The number of charge carriers increase with the increasing temperature and decreasing E_g.

5.3.1 *Conductivity*

Electrical conduction in intrinsic semiconductors is performed by both the electrons and holes. The hole acts like a positively charged particle. When an electric field is applied, the electrons move in one direction and the holes in the other direction, the total current density J_{total} can be expressed as:

$$J_{total} = J_n + J_p. \qquad (5.2)$$

Because $$J = n \cdot e \cdot v_d, \qquad (2.13)$$

$$J = n \cdot e \cdot v_n + p \cdot e \cdot v_p, \qquad (5.3)$$

where n = number of conduction electrons per unit volume, p = number of conduction holes per unit volume, e = absolute value of electron or hole charge, and v_n and v_p = drift velocities of electrons and holes, respectively.

Use Ohm's law $$J = \sigma \cdot E, \qquad (2.15)$$

$$\sigma = J/E = n \cdot e \cdot v_n/E + p \cdot e \cdot v_p/E, \qquad (5.4)$$

Electron mobility μ is the electron drift velocity in a unit field and

$$\mu = v_d/E, \qquad (2.12)$$

$$v_n/E = \mu_n \text{ (electron mobility)}, \qquad (5.5)$$

$$v_p/E = \mu_p \text{ (hole mobility)}, \qquad (5.6)$$

$$\sigma = n \cdot e \cdot \mu_n + p \cdot e \cdot \mu_p. \qquad (5.7)$$

In intrinsic semiconductors, electrons and holes are created in pairs.

$$n = p = n_i. \tag{5.8}$$

where n_i is the intrinsic carrier concentration in unit volume.

$$\sigma = n_i \times e \times (\mu_n + \mu_p). \tag{5.9}$$

In intrinsic semiconductors, the mobility of electrons is greater than that of holes:

$$\mu_n > \mu_p.$$

Table 5.2 lists the carrier density and mobility of Si and Ge at 300 K.

Table 5.2 Some physical properties of Si and Ge at 300 K.

Intrinsic Semiconductors	Si	Ge
Energy gap (eV)	1.12	0.67
Electron mobility, μ_n (m^2/V·s)	0.135	0.39
Hole mobility, μ_p (m^2/V·s)	0.048	0.19
Intrinsic carrier density, n (/m^3)	1.5×10^{16}	2.4×10^{19}
Intrinsic resistivity, ρ (Ω·m)	2300	0.46
Density (g/cm^3)	2.33	5.32
Melting point (°C)	1410	937
Lattice parameter (nm)	0.543	0.566

5.3.2 *Temperature effect*

In intrinsic semiconductors, the number (density) of carriers is determined by temperature.

$$n = n_n = n_p = n_0 \times exp(-E_g/2kT), \tag{5.10}$$

where E_g is the band gap energy, n_0 is a constant.

Combine Eqs. (5.8) and (5.10)

$$\sigma = n_0 \times e \times (\mu_n + \mu_p) \times exp(-E_g/2K_B \cdot T), \tag{5.11}$$

or $$\sigma = \sigma_0 \times exp(-E_g/2K_BT),$$ (5.12)

and $$\sigma_0 = n_0 \times e \times (\mu_n + \mu_p).$$ (5.13)

σ_0 is a constant that depends on the carrier density and mobility.

According to Eq. (5.12), a plot of $ln\sigma$ versus T^{-1} will show a linear relation with a slope of $-1/2E_g/K_B$. Therefore, E_g can be experimentally measured. Fig. 5.5 shows such plots for Si and Ge. Fig. 5.6 shows the correlation between bandgap and electron excitation (conductivity).

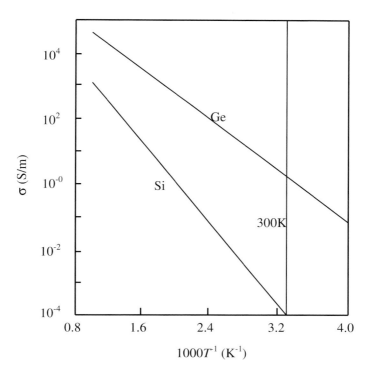

Fig. 5.5 Conductivity σ versus *1/T* for Si and Ge (source: L.A.A. Warnes, "Electronic Materials", Macmillan Education Ltd. 1990).

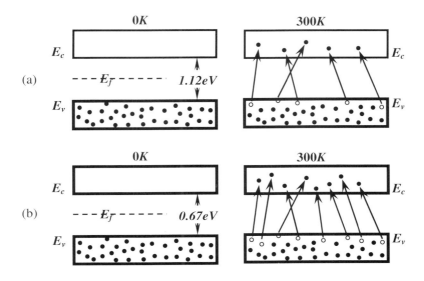

Fig. 5.6 Schematics showing electron excitation in Si (a) and Ge (b).

5.4 Extrinsic Semiconductor

One of the main reasons that semiconductors are useful in electronics is that their properties can be altered in a controllable way by adding small amounts of impurities. These impurities are called dopants. Heavily doping a semiconductor can increase its conductivity by a factor greater than a billion. In modern integrated circuits, for instance, heavily-doped polycrystalline silicon is often used as a replacement for metals. An extrinsic semiconductor is a semiconductor that has been doped with impurities to modify the number and type of free charge carriers present.

5.4.1 *n-type extrinsic semiconductors*

The purpose of *n*-type doping is to produce an abundance of mobile or "carrier" electrons in the material. If an atom with five valence electrons, such as phosphorus (P), arsenic (As), or antimony (Sb), is incorporated into the crystal lattice in place of a Si atom, then that atom will have four covalent bonds and one unbonded electron. This extra electron is only weakly bound to the atom; and this new state of bonding energy is rather low, ~0.01 eV. It can then be easily excited into the conduction band. At room temperatures, virtually all such electrons are excited into the conduction band. Since excitation of these weakly bound electrons does not result in the formation of a hole, the number of electrons in such a material far exceeds the number of thermally generated holes. In this case the electrons are the majority carriers and the holes are the minority carriers.

At room temperatures the thermal energy of an electron, E, equals 0.025 eV, the free electron can be excited by the thermal energy to the conduction band, resulting in conductivity. An impurity of this kind is called a donor. It donates conducting electrons without producing holes in the valence band. This type of semiconductor is called *n*-type extrinsic semiconductor.

Energy level near the conduction band which binds the extra electrons to the donor atoms is called the Donor level. The energy difference between the donor level and conduction band, ΔE, is equal to the binding energy of the extra electron to the donor atom.

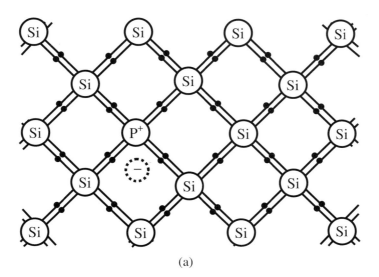

(a)

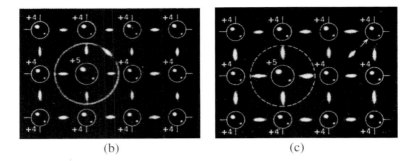

(b)   (c)

Fig. 5.7 A donor impurity atom in a semiconductor (a) schematic of doping of P into Si, (b) A 5+ valent impurity atom with donor state and (c) the electron has left the donor state and entered the conduction band (after Rose, Shepard and Wulff, "Electronic Properties", John Wiley and Sons, 1966).

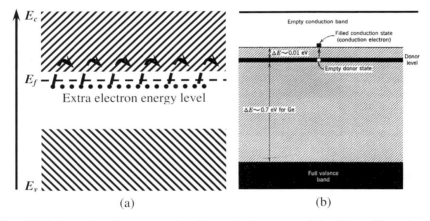

(a)   (b)

Fig. 5.8 Schematics of *n*-type semiconductor band structure (after Rose, Shepard and Wulff, "Electronic Properties", John Wiley and Sons, 1966).

   In an *n*-type extrinsic semiconductor, conducting electrons will be outnumbering the holes in the valence band. So the electrons are the majority carriers and the holes are minority carriers.

5.4.2 *p-type extrinsic semiconductors*

The purpose of *p*-type doping is to create an abundance of holes as the charge carriers. When a trivalent atom (such as boron) is substituted into the silicon crystal lattice, one electron is missing from one of the four covalent bonds normal for the silicon lattice. The dopant atom will accept an electron from a neighbouring atom's covalent bond to complete the fourth bond, causing the loss of one bond from the neighbouring atom and resulting in the formation of a "hole". Each hole is associated with a nearby negative-charged dopant ion, and the semiconductor remains electrically neutral as a whole.

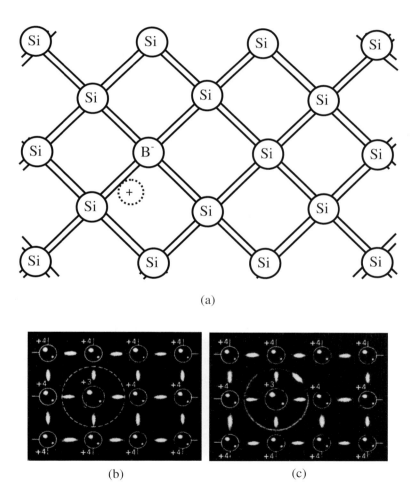

(a)

(b)   (c)

Fig. 5.9 An acceptor impurity atom in semiconductor.

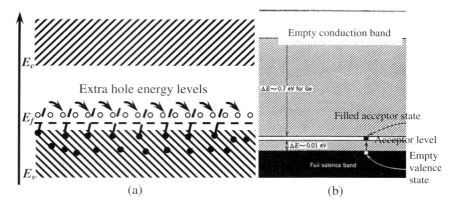

Fig. 5.10 Schematics of p-type semiconductor band structure (after Rose, Shepard and Wulff, "Electronic Properties", John Wiley and Sons, 1966).

A hole behaves as a quantity of positive charge. When a sufficiently large number of acceptor atoms are added, the holes greatly outnumber the thermally-excited electrons. Thus, the holes are the majority carriers, while electrons are the minority carriers in *p*-type materials.

The hole is attracted to the charge. A set of quantum states is established (Figs. 5.9 and 5.10). The energy state of the hole is about 0.01 eV above the valence band. Thus a hole can be created in the valence band merely by thermal excitation. This type of semiconductor is called the *p*-type extrinsic semiconductor.

The energy level near the valence band that attracts electrons from the valence band is called the Acceptor level. The ΔE between the acceptor level and the valence band is equal to the binding energy between the hole and the acceptor atom. The liberation of the hole from the acceptor atoms is equivalent to the excitation of an electron from the valence band to the acceptor level.

5.4.3 *Doping*

In many intrinsic semiconductors, the impurity concentration is smaller than one part per million (1 ppm = 10^{-6} = 0.0001%). For extrinsic semiconductor materials, the impurity level (doping level) is usually from 100 to 1000 ppm (i.e, 0.01–0.1%), still very low compared to structural materials. This is why electronic materials need a very high purity.

For Si or Ge based semiconductors, examples of donor impurities (elements) are P, Sb and As; while acceptor elements are B, Al, Ga and In. The energy level established by these elements is close enough to the conduction or valence bands to give significant carrier concentration at normal temperature. Table 5.3 shows the excitation (ionisation) energy of different dopants in Si and Ge. The resistivity of Si decreases with the dopant concentration (Fig. 5.11).

Doping in compound semiconductors can be carried out in the same way as for Si or Ge. For instance, gallium arsenide (GaAs) can be doped into *n*-type by replacing As with tellurium (Te, valence 6). Each Te donates one free electron by creating a donor state near the bottom of the conduction band. Adding bivalent Zn to substitute for some of the trivalent Ga creates holes and so produces *p*-type semiconductors.

There is an additional complexity in doping III-V compounds like GaP. Group IV like Si can substitute either Ga or P. These are called "amphoteric" (two-way) dopants, and can produce both donor and acceptor levels.

Table 5.3 Dopant excitation energies.

Semiconductor	Dopant and type	Excitation energy (eV)
Ge	B (p)	0.010
	Al (p)	0.010
	Ga (p)	0.011
	P (n)	0.012
	As (n)	0.013
	Sb (n)	0.0096
Si	B (p)	0.045
	Al (p)	0.057
	Ga (p)	0.065
	P (n)	0.044
	As (n)	0.055
	Sb (n)	0.039

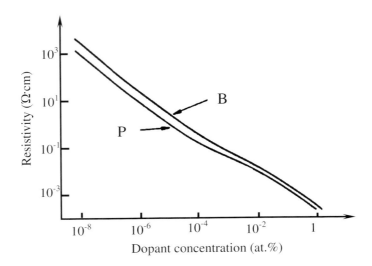

Fig. 5.11 The resistivity of a Si crystal as a function of doping concentrations (C.R.M. Grovenor, "Materials for Semiconductor Devices", IOM).

Effective doping can also be achieved with atoms of different size. Nitrogen in GaP as a dopant is not because of the different valence (N and P have the same valence of 5) but the size of the N is very different from P (0.07 to 0.11 nm). Substitution of N for P distorts the local potential in GaP and creates a new shallow state. This type of effect is useful for controlling the photoelectronic properties of semiconductors.

5.4.4 *Conductivity in extrinsic semiconductors*

In *n*-type semiconductors, electrons are the majority carriers. Each donor atom gives one electron, thus the carrier density obeys:

$$n_n = N_d, \tag{5.14}$$

where N_d is the donor concentration in the semiconductor. The conductivity:

$$\sigma_n = N_d \cdot e \cdot \mu_n. \tag{5.15}$$

In *p*-type semiconductors, each acceptor atom creates one hole. Holes are the majority carriers and

$$p_p = N_a, \tag{5.16}$$

where N_a is the acceptor concentration. The conductivity

$$\sigma_p = N_a \cdot e \cdot \mu_p. \tag{5.17}$$

5.4.5 *Mass action law*

In semiconductors, conducting electrons and holes are constantly being generated and recombined. At a constant temperature, the product of the free electron and hole concentrations is a constant:

$$n_n \times n_p = n_i^2, \tag{5.18}$$

where n_i is the intrinsic concentration of carriers in a semiconductor, and a constant at a given temperature.

This relation is valid for both intrinsic and extrinsic semiconductors. In extrinsic semiconductors, one type (*n* or *p*) of carrier is increased and the other type reduced through recombination, the product of the two (*n* and *p*) is still a constant. This is called the Mass Action Law.

5.4.6 *Temperature effect*

The carrier density in an extrinsic semiconductor is also sensitive to temperature. There are three main effects:

1. At low temperatures (<100 K), the carrier concentration increases with temperature following $\exp(-\Delta E/K_B T)$. The dopant atoms are not fully ionised in this region as $K_B T \ll \Delta E$. This is called "extrinsic regime".
2. As the temperature rises, the donor levels may become exhausted, or the acceptor levels saturated. The carrier concentration becomes

Introduction to Electronic Materials for Engineers

relatively insensitive to the temperature. This is called "exhaustion regime".

3. As temperature increases further, electrons are directly excited from the valence band to the conduction band in large number, since sufficient thermal energy is now available. Conduction therefore becomes intrinsic as the number of both holes and electrons increases (See Fig. 5.10). The carrier concentration increases with temperature following $\exp(-\Delta E/2K_B T)$. This is called "intrinsic regime". In this regime, the quantities of excited electrons and holes exceed by far the limited number of the extrinsic carriers.

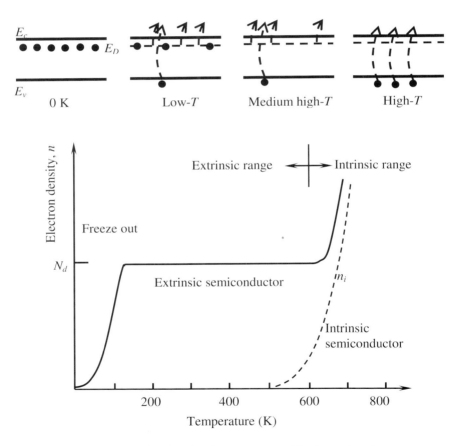

Fig. 5.12 Schematics showing the temperature effect in semiconductors.

The temperature at which intrinsic conduction becomes important sets an upper limit to the working temperature of a semiconductor device. For Ge based extrinsic semiconductors, the energy gap for intrinsic conduction is ~0.67 eV and the upper limit is about 100°C. For Si based extrinsic semiconductors, the gap is ~1.12 eV and the temperature limit is ~200°C. This is an important advantage of Si to Ge.

5.5 Impurities and Defects in Semiconductors

5.5.1 *Impurities*

Impurities, in the forms of single atom or complex between atom and other impurity, can greatly influence the electronic property of semiconductors. According to their effects on the semiconductor's electronic property, these impurities can be categorised as:

1. Acceptor impurities
2. Donor impurities
3. Amphoteric impurities
4. Neutral impurities
5. Deep state impurities

Typical impurities in Si and Ge can be discussed as below.

1. Atoms from group-III & V: They are mainly substitution impurities, have one energy level and small ionization energy (completely ionized at room temperature), and act as shallow acceptors and/or donors.
2. Group-IV: C, Sn, Si in Ge and Ge in Si are neutral impurities, thus don't have obvious effect on electronic properties.
3. Atoms from group-IB and transition metals: These impurities in semiconductors normally have multiple energy levels. Their ionization energy is relatively large (incomplete ionization at room temperature).
4. Au in Si acts as deep donor and deep acceptor, therefore are amphoteric impurities.

5.5.2 *Lattice defects and grain boundaries*

In an ideal lattice each atom is at its designated position. Deviations from the ideal structure are called defects. Defects like dislocations and grain boundaries in crystal lattices can also be electrically active. Dislocations are line defects along which the crystal lattice is shifted. Three different types of dislocations can be formed: edge, screw and 60° dislocations.

The boundaries of crystal grains are called grain boundaries. In the grain boundary area in a covalently bonded semiconductor, some bonds will be strained and some will be broken (Fig. 5.13). The unpaired electrons and strained bonds can both create electronic states in the semiconductor band gap, like those created from the addition of dopant atoms. Such defects can have a large impact on the electric properties. They can act as barriers for transport or as carrier sinks. The effect of dislocations is rather similar to that of grain boundaries, but is usually not so strong.

Crystal-2

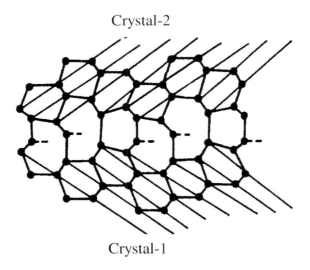

Crystal-1

Fig. 5.13 A sketch of the structure of a grain boundary in a covalently bonded semiconductor (C.R.M. Grovenor, "Materials for Semiconductor Devices", IOM).

Grain boundary effects on the electrical properties are demonstrated by comparison of single crystal materials with polycrystalline materials. There are three major effects:

1. Polycrystalline semiconductors usually have much higher resistivity than single crystal semiconductors (with the same dopant concentration) because the grain boundaries often provide a potential barrier for passing of the free carriers.
2. The minority free carrier lifetime (e.g. the time a hole will survive in a *n*-type semiconductor, which is important in photoelectronic devices) is lower in polycrystalline materials. This is because that the process of recombination of an electron with a hole is promoted by the presence of deep states in the semiconductor band-gap. These states exist in some grain boundary areas.
3. Grain boundaries and dislocations can act as short circuit paths running through semiconductors, particularly if heavy metal impurities are segregated to the defects. Short circuit paths can be very damaging to the devices.

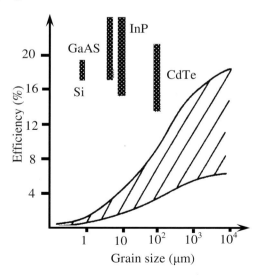

Fig. 5.14 The efficiency of solar cells made of single crystals and polycrystalline semi-

conductors with different grain size (source: C.R.M. Grovenor, "Materials for Semiconductor Devices", IOM).

All the three effects are harmful, and likely to decrease the speed and efficiency of the carrier migration. Fig. 5.14 shows the efficiency versus grain size plot for solar cells. The single crystals have a much higher efficiency than the polycrystalline semiconductors. Therefore, single crystals are widely used in device manufacturing.

5.6 *p-n* Junctions and Their Typical Characteristics

5.6.1 *p-n junction*

Most of the semiconductor devices are based on the principle of modifying the conducting properties of a semiconductor block in a carefully controlled method. The *p-n* junction is a simple device (Fig. 5.15).

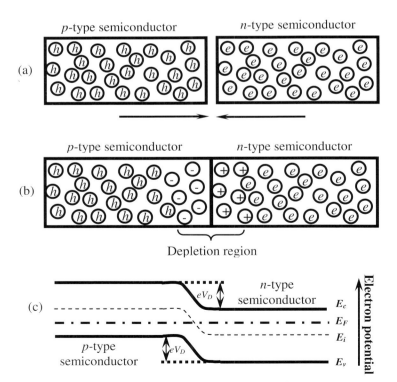

$$eV_D = K_B T \ln \frac{\left(N_D - N_A\right)_n \left(N_A - N_D\right)_p}{n_i^2}$$

(h) hole (e) electron

$(-)$ Negative ion from filled hole $(+)$ Positive ion from removed electron

Fig. 5.15 The formation of a p-n junction in a semiconductor crystal. (a) the excess of electrons and holes in the two halves, (b) the flow of these carriers to equalise the electrochemical potential and the formation of the depletion regions, and (c) shows the energy level and junction barrier.

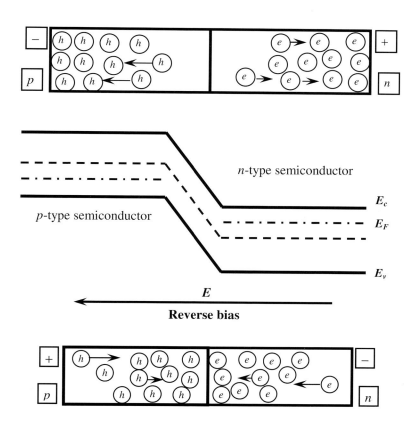

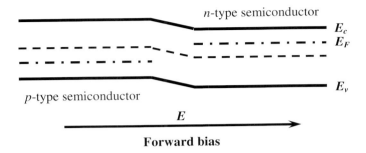

Fig. 5.16 A *p-n* junction under forward and reverse bias.

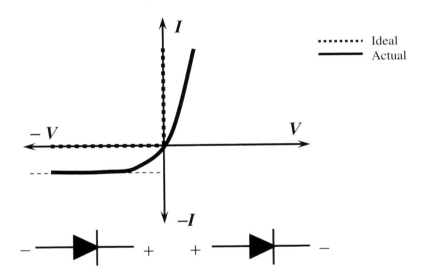

Fig. 5.17 Electrical characteristics of an ideal rectifier.

A semiconductor crystal is half heavily *p*-doped and half heavily *n*-doped. At the junction, a sudden transition occurs. The *p*-part has excess holes and *n*-part has excess electrons. Therefore, the interface is not in equilibrium as the electrochemical potential is not the same. Charge will flow from one side to another. The flow of charge will produce regions on both sides called "depletion regions", because they have lost some of their free carriers. The electrochemical potential is now equal on both sides, but the depletion regions are not electrically neutral. The *p*-type half has a net negative charge and *n*-type side has positive charge.

These charge regions create an electric field across the junction. The potential difference on either side of the junction is called the junction barrier height.

When a voltage (bias) s applied across the *p-n* junction (left-side *p*-type and right-side *n*-type), as shown in Fig. 5.16:

1. With a positive voltage from the *n*-side, "reverse" biased, the majority carriers are quickly exhausted. The junction becomes polarised, and very little current goes through.
2. With a positive voltage from the *p*-side, "forward" biased, excess majority carriers enter the *p-n* region. They can recombine continuously near the junction. Current flows continuously.

This is a rectifier. The *I-V* plot with the symbol is shown in Fig. 5.17.

5.6.2 *Rectifier equation*

The current and voltage applied to a rectifier can be expressed as:

$$I(V) = I_s \times [exp(e \cdot V / K_B \cdot T) - 1], \tag{5.19}$$

where I_s = reverse saturation current (due to the flow of minority carriers), and V = applied voltage.

When V is positive (forward biased):

$$I(V) = I_s \times [exp(e \cdot V / K_B \cdot T) - 1] \sim I_s \times exp(e \cdot V / K_B \cdot T). \tag{5.20}$$

When V is negative (reverse biased):

$$I(V) \sim -I_s. \tag{5.21}$$

At 300 K, I_s are ~10^{-6} and ~10^{-10} A/mm^2 for Ge and Si, respectively. Practically, $I(V)$ might be much greater than I_s because of surface defects and leakage.

5.6.3 *Break-down characteristics of p-n junctions*

Break-down behaviour of *p-n* junction is an important parameter for determination of device performance. Two mechanisms are mainly used to explain this process, i.e., avalanche break-down and quantum tunnelling break-down (Zener effect). The working mechanism is governed by the doping level in *p-* and *n*-regions.

The break-down characteristics can be used for reference voltage diodes (Zener diodes), avalanche photodiode (APD), IMPact ionization avalanche transit time diode and tunnel diode (or Esaki diode).

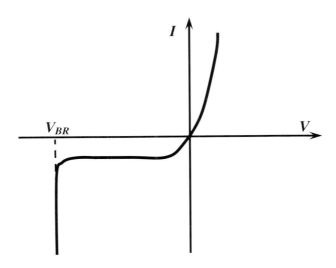

Fig. 5.18 Break-down characteristics of *p-n* junctions.

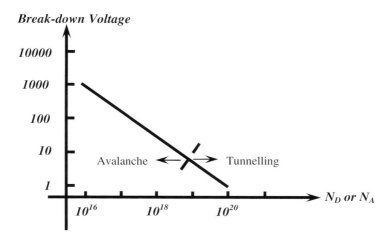

Fig. 5.19 Doping level in *p*- & *n*-regions on type of break-down.

5.7 Doping Processes

Bulk extrinsic semiconductors can be produced by melting the intrinsic semiconductors and adding a small amount of doping elements to them. Doping layers can also be commonly achieved by diffusion and ion implantation processes.

5.7.1 *Diffusion processes and depth control*

Atomic diffusion is used for introducing dopants into semiconductors in a controlled way. Typically, a crystal of *p*-doped silicon has a small region covered by an *n*-type element. The *n*-type dopant must first compensate the existing *p*-type dopant before an *n*-type region is created. This process is called "type conversion".

The diffusion process is a thermally activated process driven by concentration gradients. It usually follows a vacancy mechanism, and the process is controlled by the diffusion coefficient (D) of the dopant in the host semiconductor crystal.

$$J = -D \times (dC/dx),\qquad\qquad (5.22)$$

where J is the net flow of atoms, D is the diffusion coefficient, dC/dx is the concentration gradient of the dopant. For a thermally activated process,

$$D = D_0 \times e^{(-Q/K_B \cdot T)}.\qquad\qquad (5.23)$$

where Q = diffusion activation energy, K_B = Boltzmann's constant, T = absolute temperature, and D_0 is a constant called frequency factor, which is related to the frequency of the atomic vibration.

From Eqs. (5.22) and (5.23), we can see that the diffusion rate increases with the increasing dopant concentration gradient dC/dx, increasing temperature and decreasing activation energy.

When Si wafers are put in a furnace with a controlled concentration of gaseous dopant sources, the surface concentration of dopant, C_s, is assumed to be a constant. According to the Fick's 2nd law:

$$C(x,t) = C_S erfc(\frac{x}{2\sqrt{D \times t}}),\qquad\qquad (5.24)$$

where $C(x,t)$ is the dopant concentration at a distance x below the Si wafer surface, t is the diffusion time. "*erfc*" is a special function called "complex error function" and can be found in tables, and D is the diffusion coefficient. Therefore, $C(x,t)$ can be calculated.

If the dopant surface concentration is not a constant, the diffusion profiles will be different. Fig. 5.20 shows the diffusion profiles by using a constant source and a limited source. Constant-source diffusion is also known as unlimited-source diffusion. Concentration of dopant on the surface of the wafer remains constant during the diffusion process, i.e. while some dopant atoms diffuse into the substrate additional dopant atoms are continuously supplied to the surface of the wafer. Limited-source diffusion is also known as drive-in process. The concentration of dopant on the surface decreases during the diffusion process, i.e. while some dopant atoms diffuse into the substrate no new dopant atoms are supplied to the surface of the wafer. The diffusion processes follow thermodynamic law. It can only be controlled under the diffusion laws to produce suitable dopant concentration profiles.

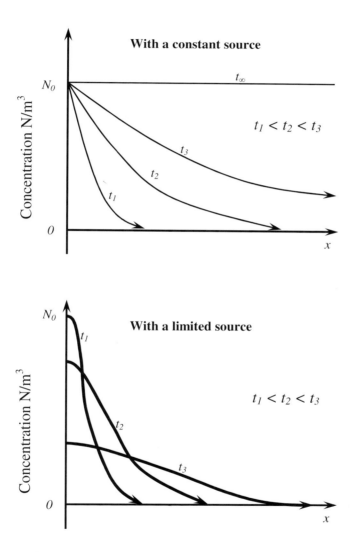

Fig. 5.20 Diffusion concentration profiles (x is the diffusion depth) (after Braithwaite and Weaver, "Electronic Materials", Butterworths, 1990).

Main objectives of ion implantation in semiconductor industry are:

1. The desired quantity of dopant should be implanted;
2. The dopant should be implanted at the correct depth;
3. The dopant implanted should be restricted to the desired areas of the wafer;
4. The implanted dopants should be electrically active; and
5. There should be no damage to the implanted areas.

5.7.2 *Ion implantation*

Ion implantation is a type of surface doping technique. The dopants are introduced by bombardment with energetic dopant ions such as As^+, P^+ and B^+. William Shockley first recognized the potential of ion implantation for doping semiconductor materials, and his 1954 patent application demonstrates a remarkable understanding of the relevant process issues long before implantation entered mass production.

Potential energy is used for accelerating the ions, usually 10 to 500 keV. Fig. 5.21 is a schematic showing the ion implantation equipment. The depth of ion penetration is determined by the energy of the incident ion beam. The distribution (or profile) of the implanted species roughly obeys a Gaussian function (Fig. 5.22). Typical implanted depths are from a few nm to 1 μm (Table 5.4).

5.7.3 *Comparison of diffusion and ion implantation*

The thermal diffusion process has a major drawback. It has difficulties to produce dopings with complicated profiles. If more than one diffusion anneal is needed, each anneal will cause re-distributions of other elements, and will also increase the contamination level. With ion implantation, the contamination level can be kept very low. Ion implantation also has the following advantages:

1. It is not an equilibrium process, therefore, it is not limited by the phase diagrams.

2. A very wide range of dopant profiles can be obtained by using several implants of the same dopant with different energies.
3. It is a well controlled process. The depth x_0 and deviation s can be accurately controlled.
4. It is a low temperature process, well defined masks can be used to produce accurate implantation.

However, ion implantation also has some disadvantages:

1. Implanted species may come to rest in interstitial positions in the lattice where they will not create carriers; and
2. Considerable lattice damage is caused by the radiation bombardment. Sometimes the damage is so severe that a region of the semiconductor becomes amorphous.

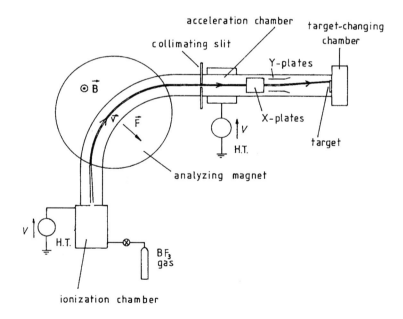

Fig. 5.21 A schematic drawing of ion implantation equipment (L.A.A. Warnes, "Electronic Materials", Macmillan Education Ltd. 1990).

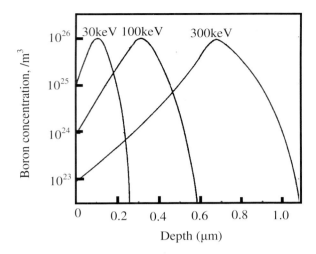

Fig. 5.22 Implantation profiles for B in Si (L.A.A. Warnes, "Electronic Materials", Macmillan Education Ltd. 1990).

Table 5.4 Ion implantation parameters for silicon.

Ion parameter		B^+		P^+		As^+	
		x_0	σ	x_0	σ	x_0	σ
Energy (keV)	10	0.04	0.02	0.015	0.008	0.011	0.004
	30	0.11	0.04	0.04	0.02	0.023	0.009
	100	0.31	0.07	0.14	0.05	0.07	0.03
	300	0.66	0.11	0.41	0.11	0.19	0.07

Note: All distances in µm (L.A.A. Warnes, "Electronic Materials", Macmillan Education Ltd. 1990).

Annealing is needed to remove the defects and to allow the implanted dopant atoms to diffuse to the substitutional sites, where they become electrically active. Annealing also promotes crystallisation of the amorphous materials.

5.8 Metal-Semiconductor Contacts

Semiconductor circuits in a Si chip need electrical connections between them and to the outside world. There are two ways to make a contact.

The traditional way to connect semiconductor devices is to attach a metal wire onto the surface of the semiconductor material. Modern devices are interconnected by thin metal films which is called semiconductor metallisation. The films run over the surface of the semiconductor, making contacts between the devices and the outside. In both connection methods, we have the similar considerations such as the contact nature, adhesion, compatibility, and patternability.

5.8.1 *Contact nature*

When a metal contacts a semiconductor, the two materials usually have different electrochemical potentials, just like the *p-n* junction. A simplified way to characterise the electrochemical potentials is:

For a metal, we use work function ϕ_m. This is the ability of the metal to hold electrons in it.

For a semiconductor, we use electron affinity χ_s. This is the ability of a semiconductor to hold charges in it, which depends on the dopant concentration and materials.

To equalise the electrochemical potentials, charge will flow between the metal and semiconductor, as we described in *p-n* junction.

Table 5.5 Work functions of typical metals.

Metals	Work function (volt)
Ag	4.26
Al	4.28
Au	5.1
Cr	4.5
Mo	4.6
Ni	5.15
Pd	5.12
Pt	5.65
Ti	4.33
W	4.55

Table 5.6 Electron affinities of typical semiconductors.

Semiconductor	Electron affinity (volt)
Si	4.01
Ge	4.13
GaAs	4.07
AlAs	3.5

5.8.1.1 *Schottky contacts*

When $\phi_m > \chi_s$, there is a large barrier at the metal/n-type semiconductor interface. Electrons cannot flow from metal to semiconductor unless the barrier is overcome. Electrons can flow from n-type semiconductor to metal if a positive potential is applied. This type of contacts is called "Schottky Barriers", as shown in Fig. 5.23(a).

5.8.1.2 *Ohm contacts*

When $\phi_m < \chi_s$, there is almost no potential barrier. No resistance to current flow in either direction, and current flow will obey the Ohm's law. This type of contact is called the "Ohmic Contact" (Fig. 5.23(b)). It is a passive component of electronic circuit.

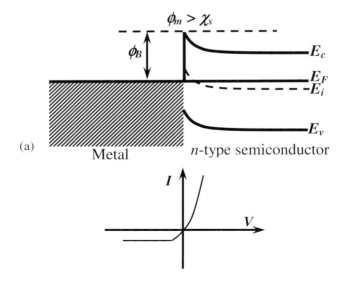

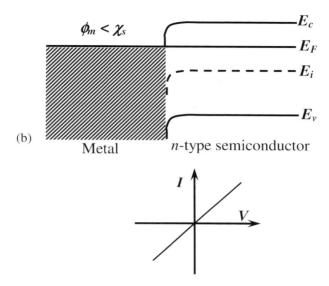

Fig. 5.23 Metal-semiconductor contact. (a) Schottky contact, carriers can be excited from the n-type semiconductor into the metal, but not in the other direction. (b) Ohmic contact, where there is no barrier.

Most metal/semiconductor junctions show Schottky rectifying behaviour. One exception is indium (In) which makes good ohmic contact to several compound semiconductors. For instance, In-CdS gives a true ohmic contact and a linear *I-V* relation.

Many ohmic contacts are prepared by annealing alloy contact layers which contain an element that is a dopant in the semiconductor, like Au alloyed with Sb, Ge, Zn and Sn. During the annealing, this element diffuses into the semiconductor, producing a heavily doped region under the contact. This heavily doped layer makes the depletion region thinner and easier for free carriers to pass through the barrier.

5.8.2 *Other properties and contact materials*

Aluminium is widely used as the contact and metallisation material. Copper is also used for this purpose to obtain lower resistivity, but with more complicated processes.

In microelectronic devices, metal-Si electrical connections are required in certain areas but not everywhere. For the areas on which electrical connection is not required, a SiO_2 layer is built up. Fortunately, the metallisation layer (Al) has good adhesion to both the Si surface and SiO_2 layer.

Resistance of the metal contacts is also a consideration. Certain thickness is required to reduce the resistance, and therefore, to reduce the heat generated by the current flow. As for Al contacts, ~1 μm thickness is usually enough. Internal stress may cause the layers to peel off if they are too thick.

5.9 Simple Devices Using Semiconductors

5.9.1 *MOSFET*

Most integrated circuits are assemblies of MOS (Metal-Oxide-Silicon) devices such as MOSFET (Metal-Oxide-Silicon Field Effect Transistor). They are easier to fabricate than bipolar devices. The simplified structure of a typical MOSFET is illustrated in Fig. 5.24.

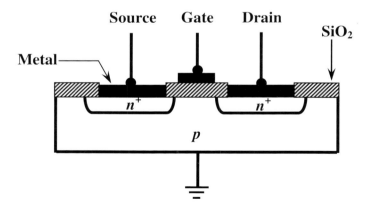

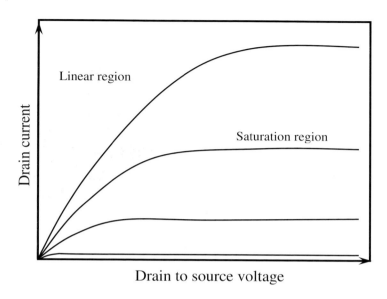

Fig. 5.24 A schematic of metal-oxide-silicon (MOS) field effect transistor and its current-voltage (*I-V*) characteristics.

A MOSFET consists of source and drain embedded in a *p*-type material. The gate terminal is connected to a metal layer that is separated from the *p*-type material by an insulator. The flow of electrons from the source to the drain is controlled by the gate voltage. If the input voltage applied to the gate is positive, free electrons will be attracted from the *n*-regions and the *p*-region to the underside of the SiO_2 layer, at the gate region. The abundance of electrons under the gate forms an *n*-channel between the two *n*-regions, thus providing a conductive path for the current to flow from the source to the drain. In this case, the MOSFET is said to be "on". If the input voltage at the gate is negative, the electrons in the *p*-region under the gate are repelled, and no *n*-channel is formed and no current will flow, thus turning the MOSFET "off".

5.9.2 *Solar cell*

A typical solar cell consists of a *p-n* junction connected with two ohmic contacts. One of the contacts has to be transparent to let the light in.

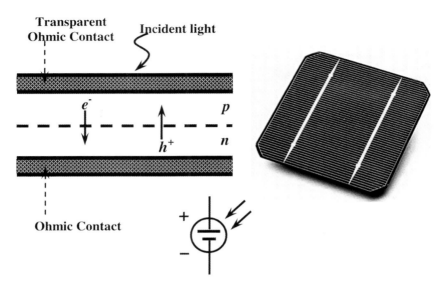

Fig. 5.25 The structure and schematic symbol of a solar cell. A very similar structure can be used as a photodiode to measure the intensity of the incident light. A solar cell made from a monocrystalline silicon wafer was also shown at the right side (http://www.eere.energy.gov/solar/pv_systems.html).

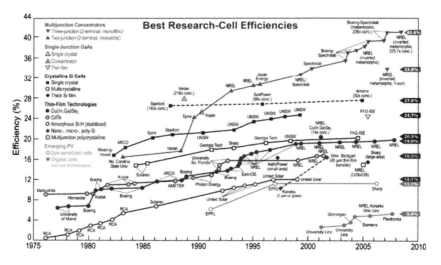

Fig. 5.26 Evolution of the conversion efficiency of solar cells tested in lab (National Renewable Energy Laboratory/NREL, 2008).

When the incident light's energy is greater than the band-gap E_g, a large number of electron-hole pairs are created. This is because an incident photon excited an electron from the filled valence band to the conduction band, leaving a free hole behind in the valence band. The electrons will pass into the *n*-type material and the hole into the *p*-type material, they are then separated by the potential gradient at the junction. By collecting this current, power is generated from the incident light (Fig. 5.25).

Current solar cells are dominated by Si (either single-crystal or poly-crystalline) with nearly 90% of the market. In particular, polycrystalline-Si has a market share of ~53%, followed by single-crystal Si of 33%. However, silicon has an indirect bandgap, this means that its absorption coefficient is much lower than that of a direct bandgap semiconductor such as CdTe and GaAs. Currently, materials, including amorphous silicon (a-Si:H), GaAs, CdTe and CuInGaSe$_2$ (CIGS), are being actively developed for the fabrication of thin film solar cells with advantages of cheap production and high efficiency.

Fig. 5.26 depicts the evolution of conversion efficiency for various solar cells over the past thirty years. The efficiency of single-crystal Si cells has reached 24.7%, while that of poly-Si cells is 20.3%. At this moment, the highest record efficiency for all solar cell technologies is at 40.8% by a so-called tandem cell. This cell stacks three *p-n* junctions made of Ga$_x$In$_{1-x}$As or Ga$_x$In$_{1-x}$P with different compositions

5.9.3 *Light emitting diode (LED)*

An LED works in an opposite way to use the *p-n* junction as that of solar cell. Light is generated with electricity. By applying a forward bias to a *p-n* junction, a high density of electrons can be introduced into the *p*-type semiconductor, where they recombine with holes. This annihilation reaction releases light with a photon energy roughly equal to the band-gap of the semiconductor.

By choosing a semiconductor with the right bandgap, the diode can emit light in the visible range of the electromagnetic spectrum (Fig. 5.27).

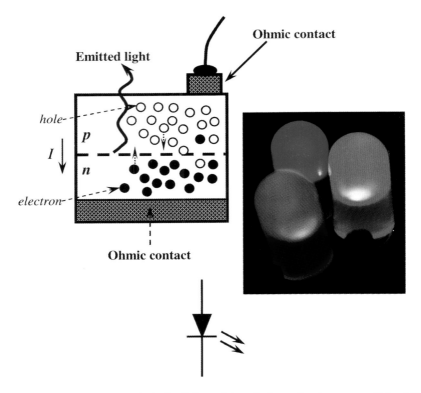

Fig. 5.27 The structure, symbol of a light emitting diode; and red, green and blue LEDs of the 5mm type (http://en.wikipedia.org/wiki/File:RBG-LED.jpg).

5.10 Integrating Microelectronic Circuits

5.10.1 *The concept of integrated circuitry*

A modern microelectronic system contains a large number of simple devices. The power of a computer comes from connecting more and more devices together. But the speed of a computer depends on the communication paths between the devices. The shorter of the paths, the faster the computer. This leads to the development of integrated circuitry (IC).

An IC chip is a collection of components connected to form a complete electronic circuit that is manufactured on a single piece of semiconductor material (where the devices and their interconnections become "writing" in a tiny piece of silicon chip). Jack Kilby of Texas Instruments is credited with conceiving and constructing the first IC in 1958. In the Kilby IC, the various semiconductor components (transistors, diodes, resistors or capacitors) were interconnected with so-called "flying wires". In 1959, Robert Noyce of Fairchild applied the idea of an IC in which the semiconductor components are interconnected within the chip using a planar fabrication process, thus eliminating the flying wires.

IC complexity has advanced from small-scale integration (SSI) in the 1960s, to medium-scale (MSI), to large-scale integration (LSI), to very large-scale integration (VLSI), which characterizes devices containing 10^5 or more components per chip, and to ultra large scale integrated (ULSI) circuit contains $> 10^7$ components in a few mm size Si chip. The individual feature is smaller than 0.1 μm.

Compared with the old electronic technologies such as vacuum tubes or discrete transistors, the integrated circuitry clearly has the following advantages:

1. much smaller in size,
2. much higher speed,
3. big reduction in energy consumption,
4. consume much less material,
5. much lower cost, and
6. great improvement in reliability.

All these features cannot be over-emphasised. Without integrated circuitry technology, our personal computers, audio-video equipment, modern electronic devices, appliances and telecommunication systems are all impossible.

Modern integrated circuits are mainly made from Si semiconductor. Silicon dioxide (SiO_2) is the key feature of this technology. There are two types of integrated circuits, one with rather simple repeated structure and the other with complex array of devices (Fig. 5.28).

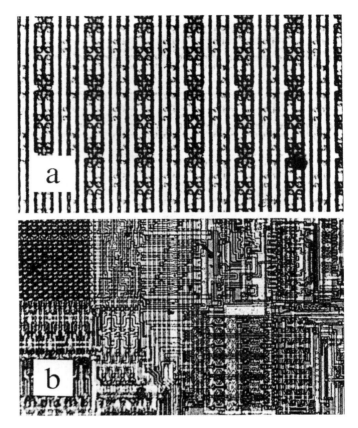

Fig. 5.28 The structure of the top layer Al conducting track on Si chips. (a) A simple repeated structure, (b) a complex array of devices (C.R.M. Grovenor, "Materials for Semiconductor Devices", IOM).

The processing technologies for both are similar, and will be discussed in the next few sections. The main processing steps include:

- Oxidation
- Photolithography
- Diffusion
- Epitaxial deposition
- Metallization
- Passivation

➢ Backside grinding
➢ Backside metallization
➢ Electrical probing
➢ Die separation

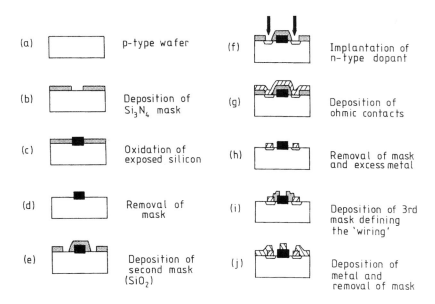

(a) p-type wafer

(b) Deposition of Si_3N_4 mask

(c) Oxidation of exposed silicon

(d) Removal of mask

(e) Deposition of second mask (SiO_2)

(f) Implantation of n-type dopant

(g) Deposition of ohmic contacts

(h) Removal of mask and excess metal

(i) Deposition of 3rd mask defining the 'wiring'

(j) Deposition of metal and removal of mask

Fig. 5.29 Some of the stages in the fabrication of an array of MOSFET (C.R.M. Grovenor, "Materials for Semiconductor Devices", IOM).

Fig. 5.29 is showing the typical stages required for the fabrication of a simple MOSFET.

5.10.2 *Reliability of semiconductor devices*

Reliability is always an important consideration in electrical and electronic industries. Modern computers or electronic devices consist of millions and millions of circuits. The failure of one unit may cause the device and the whole equipment to stop working or become unstable.

Integrated circuitry has two advantages related to the improved reliability. Firstly, they operate at near room temperatures, therefore have much longer life than that of heated filament type of devices. Secondly, the modern devices with integrated circuits do not use solder-interconnections, which are prone to various soldering defects. These two factors result in a greatly improved reliability in modern electronic devices. Therefore, the failure possibilities per function were decreased with the increasing scale of the integrated circuits.

However, this increase cannot continue indefinitely. When the integration scale reaches a certain limit, new reliability problems are created due to the complexity of the device. Because the increase of the integration scale is mainly caused by reducing the component size, there will always be limits to various processing technologies such as the resolution of the lithography and etching. Attempting to push these to their limits will result in a decrease of the reliability.

5.11 Growth Techniques for Typical Semiconducting Materials

Si is the most important semiconductor material. It is currently used to produce most of the semiconductor devices, especially the integrated solid circuits.

5.11.1 *Single crystal Si manufacturing process*

The raw material for Si is quartz sand. It is a type of fairly pure SiO_2. The typical production routine for single crystal silicon is:

(1) Reduction of quartz sand:

$$SiO_2 \text{ (s)} + 2C \text{ (s)} \rightarrow Si \text{ (s)} + 2CO \text{ (g)} \uparrow \qquad (5.25)$$

(2) Purifying Si by distillation:

$$Si \text{ (s)} + 3HCl \text{ (g)} \rightarrow SiHCl_3 \text{ (l)} + H_2 \text{ (g)} \uparrow \qquad (5.26)$$

(3) $SiHCl_3$ is reduced to produce polycrystalline, electronic grade silicon (EGS):

$$2SiHCl_3 \text{ (l)} + 2H_2 \text{ (g)} \rightarrow 2Si \text{ (s)} + 6HCl \text{ (g)} \uparrow \qquad (5.27)$$

The EGS Si has a very high purity, but it is polycrystalline, and is not suitable for semiconductor devices because rapid diffusion of dopants will take place through the grain boundary areas.

(4) Produce single crystal Si by melting and crystal growth with Czochralski Grower.

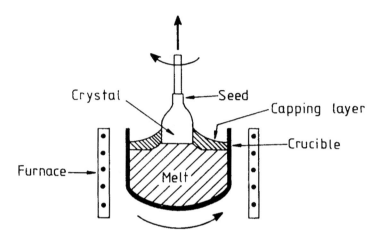

Fig. 5.30 A schematic show of the single crystal growing process in a Czochralski crystal puller (C.R.M. Grovenor, "Materials for Semiconductor Devices", IOM).

The Czochralski process is a method of crystal growth used to obtain single crystals of semiconductors (e.g. silicon, germanium and gallium arsenide), metals (e.g. palladium, platinum, silver, gold), salts and some man made gemstones. The process is named after Jan Czochralski, who discovered the method in 1916 while investigating the crystallization rates of metals. Czochralski Grower is illustrated in Fig. 5.30. The Si charge is placed in a silica crucible of very high purity, surrounded by a graphite heating unit (also has a high purity). A pure argon atmosphere is used to protect Si from oxidation. The temperature of the Si melt is controlled at ~1420°C, only 10°C above its melting point. A small, defect-free, Si crystal seed is lowered into contact with the Si melt and carefully withdrawn to form a cylindrical single-crystal rod up to 400 mm in diameter and 2 m long. The drawing speed is ~2 mm/min.

Zone melting is a method of separation by melting in which a molten zone traverses a long ingot of impure metal or chemical. In its common use for purification, the molten region melts impure solid at its forward edge and leaves a wake of purer material solidified behind it as it moves through the ingot. The impurities concentrate in the melt, and are moved to one end of the ingot.

The float zone (FZ) process is another method for growing single-crystal silicon. It involves the passing of a molten zone through a polysilicon rod that approximately has the same dimensions as the final ingot. The purity of an ingot produced by the FZ process is higher than that of an ingot produced by the Czochralski process. As such, devices that require ultra pure starting silicon substrates should use wafers produced using the FZ method.

5.11.2 *Si wafer production*

A single crystal Si ingot is usually ground and polished to make it perfectly cylindrical with a smooth surface. These cylinders are then carefully aligned to the required orientation and cut into slices (wafers) of ~0.5 mm thick with a circular saw. One side of the slices need to be polished to a high finishing grade. Now the Si wafers are ready for producing integrated circuits.

5.12 Metallisation

5.12.1 *PVD and CVD processes*

Single Si crystals can also be grown from the vapour phase. Crystal growth from a melt is a nearly equilibrium process. Supersaturation of the melts is small. The phase diagram is a reasonable guide to the distribution of elements in the crystals. Growth from vapour phase is not an equilibrium process. The kinetics of transport of the deposit species is more important in determining the growth rate and composition.

There are two methods to form a vapour: (1) Evaporation or sublimation (physical vapour deposition, PVD), and (2) Chemical vapour deposition (CVD). The PVD processes use high temperature to form vapour from a liquid or solid surface, then deposit the vapour on a cooler substrate nearby. In the CVD processes, volatile compound that contains the element(s) to be deposited is broken down over a heated substrate by chemical reduction or thermal decomposition. The element is released from a supersaturated vapour and rapidly condenses to form a crystalline deposit.

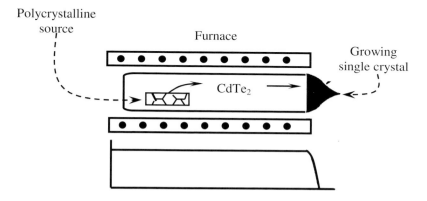

Fig. 5.31 A sketch of the vapour phase growth of cadmium telluride crystals (C.R.M. Grovenor, "Materials for Semiconductor Devices", IOM).

Vapour deposition is a non-equilibrium process. It is affected strongly by the reaction kinetics. The growth rate of the crystals depends on the speed of the arrival of the deposit species on the substrate. The composition generally equals the concentration of this phase in the gas vapour. Fig. 5.31 is a schematic drawing of vapour phase growth of cadmium telluride crystals.

5.12.2 *Metallisation*

The deposition of a conductive material, to form the interconnection leads between the circuit component parts and the bonding pads on the surface of the chip, is referred to as the metallization process. Materials such as aluminium, aluminium alloys, platinum, titanium, tungsten, molybdenum, and gold are used for the various metallization processes. Of these, aluminium is the most commonly used metallization material.

Metallisation can be performed by vacuum evaporation, sputtering or chemical vapour deposition. Vacuum evaporation is used to deposit thin layers of metals on semiconductor devices. Fig. 5.32 (a) is a schematic diagram of a vacuum evaporator. The metal to be evaporated is placed in a boat on a small heater. The temperature of the heater is then raised until the metal vaporises. A substrate with a covering mask is placed on the top of the heater. The atoms or molecules from the thermal vaporization source reach the substrate without collisions to the residual gas molecules in the deposition chamber. This type of PVD process requires relatively good vacuum, normally better than 10^{-4} Torr. The source metal can also be heated by electron beam bombardment heating. Common heating techniques for evaporation are: resistive heating; high energy electron beams; low energy electron beams; and inductive heating.

Sputtering is another type of vacuum deposition method. The physical sputtering process, involves the physical (not thermal) vaporization of atoms from a surface by momentum transfer from bombarding energetic atomic sized particles. The energetic particles are usually ions of a gaseous material accelerated in an electric field. Sputtering was first observed by Grove in 1852 and Pulker in 1858 using von Guericke-type oil-sealed piston vacuum pumps.

Sputter deposition, or simply sputtering, is the deposition of particles whose origin is from a target surface being sputtered. Sputter deposition of films was first reported by Wright in 1877. The deposition chamber is evacuated and then an inert gas (pure argon) is introduced to have a suitable working pressure (10^{-3}–10^{-2} torr). The cathode is made of the metal that is going to deposit on the substrate. When a high direct-current voltage is applied between the electrodes, the inert gas is ionised and the positive ions accelerated to the cathode. On striking the cathode, these ions will collide with the cathode atoms, giving them sufficient energy to be ejected. The ejected atoms will travel through space and deposit on the substrate (anode).

(a)

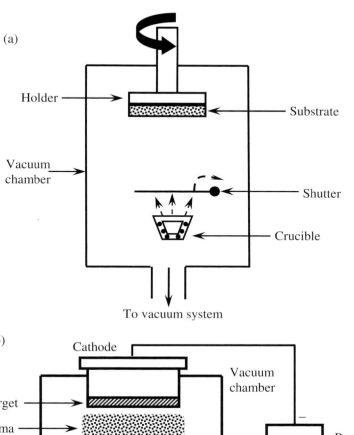

Holder

Substrate

Vacuum
chamber

Shutter

Crucible

To vacuum system

(b)

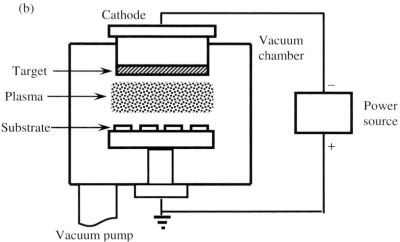

Cathode

Vacuum
chamber

Target

Plasma

Substrate

Power
source

Vacuum pump

Fig. 5.32 Schematic diagrams of a vacuum evaporator (a) and sputtering device (b).

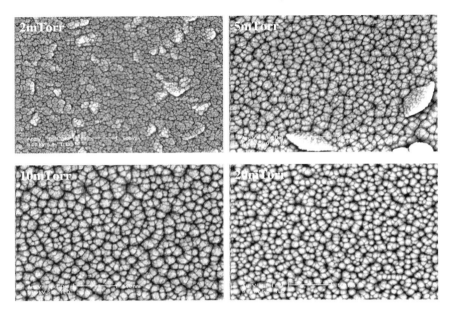

Fig. 5.33 Influence of argon working gas pressure on the surface morphology of ZnO films deposited on glass substrates by unbalance magnetron sputtering from a ZnO ceramic target.

In radio frequency (RF) sputtering, an RF instead of direct current (DC) is applied to the cathode. Non-conductive materials can be deposited. This advantage is very important for coatings and thin films of electronic materials including carbides, nitrides and oxides. Figure 5.33 shows the ZnO thin films deposited by magnetron sputter with different working gas pressure.

Chemical vapour deposition (CVD) can be defined as the deposition of a solid on a heated surface from a chemical reaction in the vapour phase. It belongs to the class of vapour-transfer processes which is atomistic in nature. In other words, the deposition species are atoms or molecules or a combination of these. CVD can also be used to deposit thin metallic films. They are not single crystals, but a polycrystalline metallic conducting layer. For example, tungsten (W), molybdenum (Mo) and aluminium (Al) can be deposited with the following reactions:

$$WF_6 \text{ (g)} + 3H_2 \text{ (g)} \rightarrow W \text{ (s)} + 6HF \text{ (g)} \text{ at } 300°C, \quad (5.28)$$

$$MoCl_5 \text{ (g)} + 5/2 \, H_2 \text{ (g)} \rightarrow Mo \text{ (s)} + 5HCL \text{ (g)} \text{ at } 800°C, \quad (5.29)$$

$$(CH_3)_2CH\text{-}CH_2Al \rightarrow Al \text{ (s)} + 3(CH_3)_2C{=}CH_2 \text{ (g)} + 3/2 \, H_2 \text{ (g)}. \quad (5.30)$$

5.13 Oxidation of Silicon and Gallium Arsenic

5.13.1 *Conventional oxidation processing of Si*

One of the most important advantages of silicon as a semiconducting material comes from its native oxide (SiO_2). SiO_2 is a good insulator, dense, mechanically strong and tough. A more important merit of SiO_2 is that it can grow into a silicon surface, bonding extremely well to the underlying silicon. SiO_2 also has very good chemical and thermal stabilities. It has a high melting point, does not dissolve in water, ordinary acids and basic solutions. Furthermore, it dissolves in hydrofluoric acid (HF) solution. This property has been widely used for selective removal in order to expose regions of the Si substrate. These are the reasons that Si is widely used as the substrate materials for integrated circuits.

Table 5.7 Properties of SiO_2.

Density	2.0-2.3 g/cm^3
Electrical conductivity	Varies widely
Breakdown field	10 MV/cm
Thermal conductivity	0.01 W/cm·K
Thermal diffusivity	0.009 cm^2/sec
CTE	0.5 ppm/K
Refractive index	1.46
Dielectric constant	3.9
Energy gap	~9 eV

SiO_2 films with different thickness are produced by oxidation (dry or wet) for different applications. Thick oxide films (0.1–1 μm), normally grown by wet oxidation, are used for device isolation and as masking layers, thin oxide layer (< 100 nm) are finding their applications as gate dielectrics, flash memory tunnel oxides and dynamic random access memory (DRAM) capacitor oxides.

Oxidation of Si for the formation of SiO_2 can be performed in dry oxygen or in water-vapour containing atmospheres. The process then is normally described as "dry oxidation" or "wet oxidation".

$$Dry\ oxidation:\ Si + O_2 \rightarrow SiO_2$$
$$Wet\ oxidation:\ Si + 2H_2O \rightarrow SiO_2 + 2H_2$$

Dry oxidation normally leads to perfect Si-SiO_2 interfacial characteristics and can be used for the fabrication of critical insulating region (MOSFET). In comparison with dry oxidation, higher oxide growth rate can be expected with wet oxidation under identical conditions.

The oxidation processing is a complicated chemical-physical process. The Deal–Grove model describes thermal oxidation of silicon for oxide thicknesses ranging from 30 to 2000 nm, oxidant partial pressures between 0.1 and 1.0 atm, temperatures from 700–1300 °C, under both pure oxygen and water vapour. In the initial stage, the oxidation process involves adsorption of oxygen atoms on the surface, chemical reactions between absorbed oxygen and silicon, and oxide crystals nucleation and growth. The thickness of the oxide is in a few Å to a few hundred Å range. Chemical reactions are the limiting step for the process. The reaction follows a linear rate law:

$$x = K_L \times (t + t'), \tag{5.31}$$

where x = the thickness of the oxide as a function of the oxidation time t, K_L = a linear rate constant, and t' = a constant corresponding to the initial thickness of the oxide.

Both K_L and t' depend on the materials, temperature, oxygen pressure etc. For a single crystal, they also relate to the crystal orientation. The order for the oxidation rate is: (110) > (111) > (311) > (511) > (100) and this order parallels the areal density of Si atoms. In single crystal Si, K_L in <111> orientation is 2/3 larger than in <100> orientation.

After the initial stage, (thickness > a few hundred Å), diffusion takes over the control and the oxide growth obeys a parabolic rate law:

$$x^2 = K_P t + C, \qquad (5.32)$$

or
$$x = K_P'\sqrt{t} + C,' \qquad (5.33)$$

where x = the thickness of the oxide, K_P and K_P' are parabolic rate constants, depending on the temperature and oxygenpartial pressure; C and C' are constants.

The transition from linear to parabolic is often not clear. The relations between K_P and oxidation temperature T obeys an Arrhenius relation:

$$K_P = K_0 \times exp(-Q/R \cdot T), \qquad (5.34)$$

where Q is the activation energy for the diffusion process, R is gas constant and K_0 is a constant.

For the growth of oxides with a thickness less than 30 nm, the Deal – Grove model for oxidation is not working well. The oxidation rate was observed to be faster than that predicted by the Deal–Grove model. Many effects have been proposed to explain the growth rate enhancement, such as field-enhanced oxidation, structural effects (microchannels, stress effects modifying the oxidant diffusivity) and changes in the oxygen solubility in the oxide.

Thermal oxidation of silicon at pressures greater than 1 atm has also been used to accelerate the oxide growth at lower temperatures, thereby minimizing dopant redistribution in the silicon. Furthermore, oxidation induced stacking faults can be reduced when processing at higher pressures. High pressure oxidation has found applications mainly in the areas where thick oxides are needed.

5.13.2 *Ultra thin SiO_2 films*

In the past a few decades, SiO_2 insulating films have decreased their thickness from hundreds of nanometers to less than 2 nm today, to maintain the high drive current and gate capacitance required of scaled MOSFETs. A further thickness reduction appears to be required. For example, a SiO_2 insulating film with a thickness of 1.2 nm is needed for

0.5 mm devices and 0.5 nm for 0.12 mm ones. On the other hand, the continuous thickness reduction requires a higher quality of the insulating layers, such as integrity, density and insulating property.

SiO$_2$ films produced by conventional thermal oxidation may have difficulties to meet these critical requirements since their sub-oxide transition layer is already ~1–1.5 nm and their average density is much lower than that of the thick SiO$_2$ film. Consequently the dielectric breakdown of thin thermally-grown SiO$_2$ film is significantly contemplated even at a typical operating voltage, hence cannot be used for construction of mini-semiconductor devices. However, it is generally believed that high quality SiO$_2$ films would still attract great attention due to their high compatibility with Si and their potentials as a unique buffer material filling the gap between the conventional oxide films and high-k materials. Therefore, fabrication of ultra-thin SiO$_2$ layers/films with superior performance using modified oxidation techniques, such as low-pressure and/or low-temperature oxidation, rapid thermal oxidation (RTO), in-situ steam generation (ISSG) oxidation, laser and plasma oxidation and VUV photo-oxidation, is still an active field in semiconductor industry.

Rapid thermal oxidation currently is one of the primary thermal techniques for growth of thin SiO$_2$ films. In this process, the Si wafers are heated electrically by a current passing through them or heated optically by lamp radiation. This technique requires low thermal budget and the Si wafer can be heated and cooled quickly. RTO processes are generally conducted at a wide temperature range from 500 to 1100 °C in various atmospheres, including O$_2$, O$_3$, N$_2$/O$_2$, N$_2$O or NO.

Oxidation at low temperatures is another way to synthesize thin or ultra-thin SiO$_2$ films since it can reduce the interface thickness and achieve more precise control of the oxidation process. However, the oxidation process becomes too slow. Recently, it has been shown that oxidation of Si at low temperatures (< 500 °C) could be enhanced by UV or vacuum UV (VUV) radiation. It was believed that if the photons released by the lamp have a higher energy than the bond energy of O$_2$, O$_2$ will be readily dissociated to generate excited state oxygen atoms which can react with Si to form SiO$_2$ at low temperatures. Plasma oxidation has also been employed to produce ultra-thin SiO$_2$ films in O$_2$ and/or N$_2$O atmospheres at room temperature. In plasma, various active oxygen ion species will be formed, leading to ionic transport through the growing film. The electric field, generated due to the different mobility of electron and ion, will accelerate the ions moving toward Si surface and travel across the growing film through a combined drift-diffusion mechanism, resulting in a higher diffusion flux for reaction at the interface.

Ultra-thin SiO_2 films, 1.2 to 5 nm, had been successfully prepared under well-controlled processing conditions using RTO, photo-assisted or plasma oxidation techniques. Extensive structural and electrical characterizations confirmed that these films possess high density, good wafer uniformity, excellent surface smoothness, high dielectric breakdown field strength and low leakage current density.

5.13.3 *Oxidation of GaAs*

Since the 1960s, oxidation of GaAs was quite active with an attempt to develop oxide-masked III-V semiconductors. These experiments were conducted in a wide temperature range, and characterizations with extensive techniques, such as XPS, AES, SIMS and TEM, were focused on the analysis of the phase compositions of the thermally grown oxide scales. The main thermal oxidation product was found to be Ga_2O_3 with a small amount of $GaAsO_4$, As_2O_5, As_2O_3 or As at the oxide/GaAs interface, depending on the temperature. These studies also turned out that the structure and orientation of the surfaces have a considerable influence on the quality and structure of the oxide layers.

The process of GaAs oxidation is so complex that even after several years of work there are important issues that are still a matter of controversy. Consequently, the GaAs metal insulator semiconductor field effect transistor (MISFET) technology did not develop very well. The challenges in the fabrication of high quality oxide layers on GaAs stimulate researchers to find out the most suitable techniques and conditions to solve the interface related problems.

Compositional studies using XPS showed that electron cyclotron resonance (ECR) plasma oxidation grown oxides are nearly stoichiometric and close to $GaAsO_4$ while thermally grown oxides are closer to Ga_2O_3

having significant amounts of As^{+3} oxidation states. Capacitance-voltage measurements showed that the interfaces produced by thermal and ECR plasma oxidations of GaAs have similar electrical properties. Both interfaces were equally unpassivated in terms of high levels of interface electronic states and Fermi level pinning. Wet oxidation at 500–520 $^{\circ}$C was also tried to grow native Ga oxide on GaAs surface. The results were not very promising since these films still suffered from a high GaAs-oxide interface recombination rate.

A new method using ultraviolet and ozone oxidation – oxinitridation (plasma nitridation after oxidation) has been developed to produce nanometer-scale gate insulating layers on GaAs surfaces and then to fabricate depletion-type recessed gate GaAs-MISFETs. The devices exhibited smaller leakage current than the simple oxide gate device and showed good pinch-off, no hysteresis, higher breakdown voltage and higher transconductance with no dip at the flatband voltage, suggesting the existence of very little interface charge. In addition, in-situ or ex-situ deposition of Ga_2O_3 on GaAs using MBE might be one way for the implementation of thermodynamically stable oxide-GaAs interfaces.

5.14 Lithography and Etching

5.14.1 *Lithography*

The integrated circuitry (IC) processing involves making patterned layers of oxide, nitride or metals by putting masks on the Si wafer. This patterning operation is called "lithography".

Lithography is crucial to the production of VLSI. Photolithography or optical lithography is a process used in semiconductor device fabrication to transfer a pattern from a photomask (also called reticle) to the surface of a substrate, such as single crystal silicon wafer, glass, sapphire and metal. The size of the devices depends on how accurately the masks can be made and how accurately they can be aligned to each other on the wafer surface. Even a simple device requires several masks, each one must be placed in the right position to the previous ones. Photolithography typically involves the steps listed below:

1. Substrate preparation;
2. Photoresist application (photosensitive polymeric material);
3. Soft-baking (to drive off excess solvent);
4. Exposure (photomask, projection with lamp or excimer laser);
5. Developing (removes either the exposed or the unexposed regions);
6. Hard-baking (in order to solidify the remaining photoresist to better serve as a protecting layer); and
7. Etching

and various other chemical treatments such as thinning agents, edge-bead removal etc. in repeated steps on an initially flat substrate.

The photoresist is a type of material that changes its structure and properties under the exposure of either UV light or laser. If it is a negative-acting photoresist, the areas that are exposed to UV light polymerize (i.e., harden) and thus are insoluble during development, whereas the unexposed areas are washed away. This results in a negative image of the photomask being formed in the photoresist. An alternative to the negative image forming photoresist is a positive-reacting photoresist, where the material behaves in the opposite way. Areas exposed to UV light become unpolymerized, or soluble when immersed in chemical solvents.

Until the advent of VLSI circuits in the mid 1980s, the negative photoresist, because of its superior developing characteristics, was the resist most commonly used. However, due to its poor resolution capability, it could no longer provide the requirements demanded by the high-density features of VLSI circuits. As a result, the semiconductor industry has transitioned to the positive photoresist because of its superior resolution capability.

In a complex integrated circuit, (for example, CMOS) a wafer will go through the photolithographic area up to 50 times. For Thin-Film-Transistor (TFT) processing, much fewer photolithographical processes are usually required.

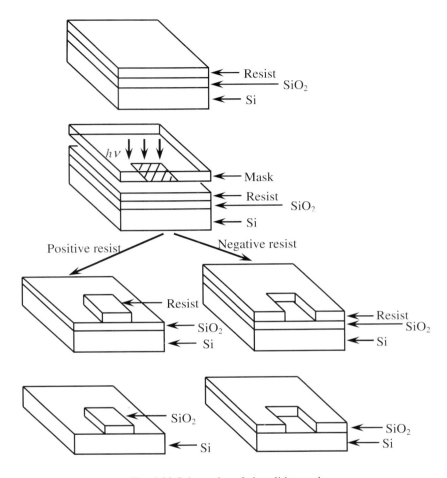

Fig. 5.33 Schematics of photolithography.

In an optical lithographical system, light passes through the transparent areas of the mask to achieve a perfect pattern transfer. But, with growing distance between mask and wafer (proximity printing), interference patterns occur, ending in an aerial image with a smooth distribution of the light intensity with its peak in the centre of the slit and tails beyond the area defined by the mask. The resolution (critical dimension) is a function of the radiation wavelength. ICs are usually patterned with near UV radiation sources. Current state-of-the-art photolithography tools use deep ultraviolet (DUV) light with wavelengths of 248 and 193 nm, which allow minimum resist feature sizes down to 50 nm. High-index immersion lithography is the newest extension of 193 nm lithography to be considered. In 2006, features less than 30 nm have been demonstrated by IBM using this technique. Extreme ultraviolet lithography (EUV) is also under development.

The requirements of optics and materials are becoming increasingly crucial for even smaller patterns, such the optical lithography seems to be approaching its technological limit. Therefore, non-optical lithography techniques such as electron beam, X-ray, and ion lithography have been attracting attentions in an effort to replace or mix and match with conventional optical lithography.

5.14.2 *Etching*

Etching is the selective removal of material to form patterns. There are two types of approaches: (1) chemical methods use reactions to change the solid into a gas or solution (wet etching) and (2) physical methods remove the atoms on the solid with the force of energetic ion bombardment (dry etching). The etching technology is becoming more and more sophisticated in order to assure that etching occurs only where it is supposed to, and not elsewhere. Extensive use is made of selective etches which preferentially remove one material: ideally, the etch should etch the pattern into the layer and then stop, with no attack on the layer below. Compensation sometimes is necessary for the inevitable variations in etch rate across a large wafer. Additionally, the shape of the side walls of the etched region becomes increasingly important as the feature size on ICs decreases.

During wet etching, material is dissolved when immersed in a chemical solution. Buffered hydrofluoric acid (HF) was commonly used to etch silicon dioxide over a silicon substrate. Common problems associated with wet etching are:

- The etching process is isotropic thus leads to large bias when etching thick films. Single crystal materials, such as Si, exhibit anisotropic etching in certain chemicals.
- It normally requires the disposal of large amounts of toxic chemical waste thus is not environmentally friendly.

146   *Introduction to Electronic Materials for Engineers*

- Wet etching also suffers from the downtime created by bath changes.

Because of the above-mentioned problems, wet etching has been replaced by dry etching for pattern transfer gradually.

During dry etching, material is sputtered or dissolved using reactive ions or a vapour phase etchant.

Reactive ion etching (RIE) uses a plasma that is generated by an RF power source to break the gas molecules into ions. The ions are accelerated towards and react at the surface of the material being etched, forming another gaseous material. The low pressure operation provides good mass transfer which reduces the micro-and macro-loading effects.

Sputter etching is essentially RIE without reactive ions. The systems used are very similar in principle to sputtering deposition systems. The main difference is that the substrate is now subjected to the ion bombardment instead of the material target used in sputter deposition.

Vapour phase etching: In this process the wafer to be etched is placed inside a chamber, in which one or more gases are introduced. The material to be etched is dissolved at the surface in a chemical reaction with the gas molecules. The two most common vapour phase etching technologies are silicon dioxide etching using hydrogen fluoride (HF) and silicon etching using xenon diflouride (XeF_2), both of which are isotropic in nature.

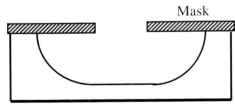

Fig. 5.34 A schematic of isotropic etching.

5.15 Packaging and Packing Materials

The semiconductor devices are produced on the surface of thin wafers. Each wafer contains many complete units. They are separated by cutting with a diamond saw or laser beam into small chips. The size of an individual chip is usually from one to a few millimetres. With all the metal interconnections and insulator layers, they are delicate and difficult to handle. They must be supported by a substrate, then connected by fine wires or solders to contact pins, then sealed into a ceramic, metal or plastic package. The term of "packaging" mainly stands for encapsulation. However, interconnections are sometimes involved when discussing the problems of packaging.

5.15.1 *Requirements of device packaging*

The reliability and efficiency of a semiconductor device is greatly influenced by the packaging technology, which has had rapid development in the recent years. Device packaging generally serve four functions:

1. A substantial lead system providing contacts to and from the chip and transferring electronic functionality from the chip to the outside world;
2. A physical protection providing good mechanical strength for handling and transportation;
3. An environmental protection to semiconductor and its device; and
4. An avenue for dissipation of heat generated in the device.

To achieve these functions, a package commonly has the following parts of die-attachment area, inner and outer leads, chip/package connection, and enclosures.

5.15.2 *Package types*

Three types of packaging technologies are used: single chip can, dual-in-line package (DIL), and hybrid circuit package.

A schematic drawing of the single chip can packaging is shown in Fig. 5.35. This is the lowest level of electronic packaging. A single chip is bonded to a substrate with wires contacting metal pins. The can is evacuated or filled with an inert gas before being finally sealed. This type of packaging is relatively simple, and does not usually have heat releasing problem.

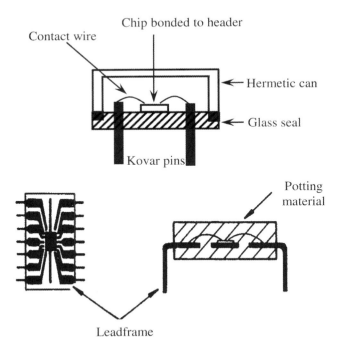

Fig. 5.35 The structure of a single chip can package.

Fig. 5.36 The structure of a dual-in-line (DIL) package (C.R.M. Grovenor, "Materials for Semiconductor Devices", IOM; http://en.wikipedia.org/wiki/Dual_in-line_package).

A dual in-line package (DIL or DIP) is an electronic device package with a rectangular housing and two parallel rows of electrical connecting pins, usually protruding from the longer sides of the package and bent downward. DIPs may be used for integrated circuits (ICs, "chips"), like microprocessors, or for arrays of discrete components such as resistors or toggle switches. They can be mounted on a printed circuit board (PCB) either directly using through-hole technology, or using inexpensive sockets to allow for easy replacement of the device. Dual-in-line package (DIL) is schematically shown in Fig. 5.36. The whole package is finally potted in a block of polymer material.

5.15.3 *Packaging materials*

Metals/alloys, ceramics, polymers and glasses are commonly used for electronic device packaging. They have their own advantages and shortcomings.

Metals/alloys are primarily used as electrical and thermal conductors in the package. They are also used to act as heat sinks for power devices and as shields for RF applications. Commonly used metallic materials for electronic packaging applications may include Al, Au, Pd, Fe-Ni alloy (Kovar, similar CTE as Si), Pb-Sn, Au-Sn, Au-Ge, Ag-Sn, Sn-Sb, Sn-Ag-Cu.

Ceramics are used as insulating materials for substrates that provide a structural base that electrically isolates lines and pads. Ceramics such as alumina or beryllia are often used as the substrate materials, especially when heat dissipation from the chip is likely to be a problem. The CTE of alumina closely matches that of silicon and several metal packaging alloys (Kovar and Alloy 42). This close match prevents differential stresses that can lead to mechanical failure.

Polymers have the advantages of relatively low cost and easy to process in large volumes. Plastic epoxy is the most common packaging material, in which the whole chip and wires are embedded. Fillers such as silica or alumina are added to the resin before it is cured so that the thermal expansion coefficient of the plastic is closer to the chip. Polymers are also used as insulating adhesives to glue components to a substrate or board to provide mechanical strength. However, many improvements are needed for polymer packaging materials. The properties of polymers that need to be improved include a low dielectric constant, high electrical resistance, high thermal stability and thermal conductivity and low moisture absorption.

5.16 Future Development of Semiconductors

The rapid development in the past forty years of the semiconductor industry is likely to continue in the following areas:

1. Larger scale of integrated circuitry, higher packing density and more complicated devices,
2. Higher operating speeds, and
3. Improved and new semiconductor materials.

The first area has already been studied actively for decades. ULSI is the result of that development. The size of an individual feature is limited by the conventional ultraviolet light lithography and is currently around 1 μm. The two approaches which can overcome this limit are electron beam lithography and X-ray lithography. The resolution limit due to the wavelength of UV light can be overcome. The sub-micron technology will also increase the operating speed. This is very important for microwave field effect transistors. The small gate width decreases the electron transit time and enables the transistor to operate at a higher frequency.

One direction of development is to improve the material quality of silicon based semiconductors. The refinement of fabrication techniques results in reduction of impurity elements and defects in the material. Amorphous silicon is being used to replace single-crystal silicon for certain applications such as high-efficiency photovoltaic cells. Through the use of amorphous silicon, the cost of materials will also be reduced.

With regards to new semiconductor materials, the following properties are necessary:

1. It must have a fairly large bandgap energy, at least 1 eV. Ideally, the bandgap should be tuned so that the desired electronic and optical properties can be achieved;
2. The carrier mobilities should be as high as possible;
3. It must be easy to make single crystals and can be doped both *n*-type and *p*-type; and
4. Nanostructured electronic materials, especially, low-dimensional materials, such as quantum dots, nanorods, nanowires, or nanobelts and superlattices, can be prepared.

Table 5.8 compares the properties of elemental and some compound semiconductors. III-V and II-VI compounds are the possible materials that may replace silicon. However, there are very few candidates that have higher mobility than silicon – only GaAs and InP. At the present time, certain III-V semiconductors are being developed and used for special applications where they have better performance than silicon.

Apart from a higher electron mobility, GaAs has a higher bandgap energy than Si. This means that GaAs can operate at a higher temperature ~350°C, while Si has a temperature limit of ~250°C. GaAs circuits consume less power, have lower noise, can radiate and detect light more efficiently in the visible range and have an easily engineered band gap. When GaAs is alloyed with GaP the band gap can be altered from 1.42–2.26 eV, corresponding to wavelengths from red to green. GaAs is used to fabricate two terminal microwave devices (transferred electron devices, or TED). It provides the basis of microwave power sources in excess of 100 GHz. GaAs is also used to make metal-semiconductor FET (MESFET) with high operating speed. III-V semiconductors have been used to produce optoelectronic devices, lasers, light-emitting diodes (LED), photodetectors and integrated optical systems.

Why have the III-V semiconductors not replaced silicon? The main reasons are that they are more difficult to grow as large single crystals and do not form adherent oxides for masking purposes. These make it more costly to process III-V semiconductors into integrated circuits. There is still a long way to go before silicon can be largely replaced.

Table 5.8 Bandgap energy and mobility of several semiconductors at 300 K

Semiconductor	Crystal form	Melting point (K)	Bandgap (eV)	Mobility (m^2/V·s)	
				Electron	Hole
Diamond	A4	4300	5.4	0.18	0.14
Si	A4	1683	1.12	0.14	0.05
Ge	A4	1210	0.67	0.38	0.18
B-SiC	B3	3070	2.3	0.01	0.002
AlAs	B3	1870	2.16	0.12	0.04
AlSb	B3	1330	1.60	0.09	0.055
GaP	B3	1750	2.24	0.03	0.015
GaAs	B3	1510	1.43	0.85	0.04
InP	B3	1330	1.35	0.50	0.015
α-SiC	B4	3070	2.86	0.04	0.002
ZnS	B4	2100	3.67	0.02	0.01
ZnSe	B4	1793	2.58	0.054	0.003
CdS	B4	1748	2.42	0.04	0.002
CdSe	B4	1512	1.74	0.065	0.004
CdTe	B4	1200	1.44	0.12	0.006

Notes:
A4 – Diamond
B3 – Sphalerite (Zincblende)
B4 – Wurtzite (Zincite)
Mobilities given are highest values from Hall effect data
Melting points are sublimation temperature in some cases

Summary:

The energy band theory can be used to explain the conductivity of semi-conductors. Intrinsic semiconductors have an energy band gap in the order of 1 eV, much smaller than those of insulators. By doping the intrinsic semiconductors with impurity atoms to make them extrinsic, the level of energy required to cause semiconductors to be conductive is greatly reduced. Extrinsic semiconductors can be *n*-type or *p*-type. The *n*-type and *p*-type semiconductors have electrons and holes as their majority carriers, respectively. By fabricating *p-n* junctions in a single crystal semiconductor, such as Si, various types of electronic devices such as diodes and transistors can be made. Modern microelectronic technology has developed to such an extent that millions of circuits can be placed on a single silicon "chip" with which the highly sophisticated and powerful computers and electronic equipment are constructed.

Important Concepts:

Hole: A positive charge carrier with a charge of 1.6×10^{-19} C.

Intrinsic semiconductor: A material which has a conductivity between that of insulators and conductors. Its valence band is full and is separated from the conduction band by an energy gap small enough to be surmounted by thermal excitation. Current carriers are electrons in the conduction band and holes in the valence band in equal amounts.

Doping: Addition of controlled amounts of impurities to increase the number of charge carriers in a semiconductor.

n-type extrinsic semiconductor: A semiconductor containing donor impurities which donate electrons to the conduction band by thermal excitation. The majority carriers are electrons.

p-type extrinsic semiconductor: A semiconductor containing acceptor impurities which accommodate electrons that have been thermally excited from the full valence band. The resulting holes in the valence band are the majority carriers.

Donor level: Energy levels near the conduction band which bind the electrons to the donor atoms.

Acceptor level: Energy levels near the valence band that attract electrons from the valence band.

Donor exhaustion: When all of the extrinsic donor levels in a *n*-type semiconductor are filled.

Acceptor saturation: When all of the extrinsic acceptor levels in a *p*-type semiconductor are filled.

p-n junction: An abrupt boundary between *p*-type and *n*-type regions within a single crystal or a semiconducting material.

Rectifier/diode: An electron device which conducts electric current in one direction only; in the other direction, the device is an insulator.

Zener diode: A *p-n* junction device which, with a high reverse bias, causes a current to flow.

Bias: The voltage applied to a rectifying junction.

Transistor: A two-junction device. One junction is forward biased, injecting minority carriers into the base. The second junction is reverse biased to a relatively high voltage. The current through the second junction is controlled by the current through the first. The middle section of a *n-p-n* or *p-n-p* transistor is called base.

Contact potential: The potential difference between two solids due to the difference in the Fermi levels. Equal to the difference in work function of the two solids.

Schottky contact: A type of metal-semiconductor contact where an energy barrier exists at the interface. It has a rectifying nature like *p-n* junction.

Ohmic contact: A type of metal-semiconductor contact where no potential barrier exists. Current can flow in either direction when a voltage is applied.

MOSFET: Metal-oxide-silicon field effect transistor.

LED: Light emitting diode.

Ion implantation: A process to introduce dopants by bombardment with energetic ions. The depth of the implantation is mainly determined by energy of the incident beam. The distribution of the implanted species roughly obeys Gaussian function.

Czochralski technique: The process to withdraw a single crystal from the melt as fast as it grows.

PVD: Physical vapour deposition, to form a vapour by vaporisation or sublimation from a liquid or solid, then deposit the vapour on a substrate surface.

CVD: Chemical vapour deposition, a volatile compound, that contains the element to be deposited, is broken down over a heated substrate by chemical reaction or thermal decomposition and then condenses to form a deposit on the substrate surface.

Lithography: The technique of making patterned masks on the surface of a semiconductor wafer to produce integrated circuits.

Etching: The selective removal of material to form patterns.

Semiconductor device packaging: The technique to pack the integrated circuits and devices to produce electronic components that are both durable and easy to connect to the other components in a large electronic system.

Microelectronics: Miniaturisation of electronic devices into an extremely small size.

Questions:

1. Compare diffusion and ion implantation semiconductor doping processes, briefly explain the differences from (1) the process nature, (2) the control of doping depth, and (3) the doping distributions. (4) Also summary the advantages and disadvantages of ion implantation over diffusion processes.
2. Explain the differences between Schottky and Ohmic contacts in terms of (i) energy mechanisms and (ii) conduction behaviour.
3. You are asked to characterise a new semiconductor. If its conductivity at 20°C is 250 $(\Omega \cdot m)^{-1}$ and at 100°C is 1100 $(\Omega \cdot m)^{-1}$, what is its band gap energy, E_g, in eV and in J?
4. For semiconductor tin (Sn), determine
   (a) the number of charge carriers per cubic meter,

(b) the fraction of the total electrons in the valance band that are excited into the conduction band, and

(c) the constant n_0 at room temperature (25°C).

Sn has diamond cubic structure and a lattice constant of 0.6491 nm, $\sigma = 0.9 \times 10^7 \ /\Omega\cdot m$, $E_g = 0.08$ eV, $\mu_n = 0.25$ and $\mu_p = 0.24 \ m^2/V\cdot s$.

5. Phosphorus-doped silicon has an electrical resistivity of 8.33×10^{-5} $\Omega\cdot m$ at room temperature. Assume the mobilities of charge carriers to be the constants 0.135 $m^2/V\cdot s$ for electrons and 0.048 $m^2/V\cdot s$ for holes.

(a) What are the minority carriers?

(b) Calculate the majority carrier concentration in the exhaustion zone.

(c) Calculate the atomic percentage of P in Si.

Si has a diamond cubic structure with 8 atoms in a unit cell and a lattice parameter of 0.5431 nm.

6. 0.1 g of B has been added to 10 kg pure Si.

(a) What type of this semiconductor is?

(b) Calculate the carrier density, and

(c) The electrical resistivity in the saturation regime.

Si has a diamond cubic structure with a lattice constant of 0.5431 nm. The atomic mass of Si and B are 28.1 and 10.8 g/mol, respectively. $\mu_n = 0.135$ and $\mu_p = 0.048 \ m^2/V\cdot s$. ($1.4 \times 10^{24}/m^3$, 9.44×10^{-5} $\Omega\cdot m$)

7. Suppose 2.42 kg of Ga are combined with 2.58 kg As to produce 5 kg GaAs.

(a) Will this produce a p-type or an n-type semiconductor?

(b) Calculate the number of extrinsic charge carriers per cubic cm using a lattice constant $a0 = 0.563$ nm.

(c) Calculate the conductivity of this product in the saturation regime.

8. In a sample of extrinsic silicon, the doping is such that there are 10^{22} electrons/m^3 and 2.3×10^{10} holes/m^3.

(a) What are the minority carriers?

(b) Calculate the fraction of the conductivity contributed by the minority carriers.

Electron mobility μ_n = 0.135 m^2/V·s, and the hole mobility μ_p = 0.048 m^2/V·s at 300 K.

9. Oxidation of Si at 1000°C obeys a parabolic rate law of $y^2 = K_P t + C$ where y is the mass gain of unit area and K_P is the parabolic rate constant. A mass gain of 2 mg was measured after 2 hours oxidation on a newly polished Si wafer with 16 cm^2 surface area. Calculate
   (a) the parabolic rate constant KP,
   (b) the amount of oxide formed on the specimen, and
   (c) the time needed to produce an oxide layer of 1 μm thick.
   The atomic mass of Si is 28 and the density of SiO$_2$ is 2.2 g/cm^3.

10. When Eu is added to ZnS, donor levels are created. When electrons drop back from the conduction band to the donor levels, the emitted photons are in the red spectrum with λ = 680 nm. Calculate the energy difference between (1) the donor level and the conduction band, and (2) the donor level and the valence band. (3) Plot a diagram to show the energy state of the doped ZnS. The energy gap E_g in ZnS is 3.54 eV.

11. The diffusion coefficients for Al ion in Al$_2$O$_3$ are 7.48×10^{-23} m^2/s at 1000 °C and 2.48×10^{-14} m²/s at 1500 °C. Calculate the diffusion activation energy, Q, and the diffusion coefficient D_0. R = 8.314 J/(K·mol).

Chapter 6

Magnetic Properties and Materials

6.1 Introduction

Magnetism is a phenomenon by which materials exert an attractive or repulsive force on other materials. It has a very long history because of the existence of a natural mineral called "magnetite", an iron oxide, Fe_3O_4. Magnetite is sometimes found in large quantities in beach sand. Such mineral sands or iron sands or black sands are found in various places such as California and the west coast of New Zealand. It is believed that the Chinese used magnets to make compasses around 400 BC. It was known as one of "the four great inventions" in the ancient China. The word "magnet" is derived from a Greek word of a place "Magnesia", where magnetite was widely distributed. In old English, it was called "lodestone", meaning "waystone". People used it to find directions.

Iron, steels and the mineral lodestone are well known materials exhibiting easily detectable magnetic properties. However, all materials are influenced to greater or lesser degree by the presence of a magnetic field.

The first truly scientific study of magnetism was made by an Englishman William Gilbert (1540–1603). He experimented with "lodestones" and iron magnets, formed a clear picture of the earth's magnetic field and cleared away many superstitions about magnets. For about 200 years after his study, there was no fundamental discovery in this field. Applications were also developed slowly. Around 1800, people made compound magnets of many magnetised steel strips, which could lift 28 times of their own weight of iron! Although the only way to make magnets was to rub steel with a lodestone or another magnet, until the electromagnetic field was discovered by Hans Oersted in ~1820.

It was not generally recognised that the development of new magnetic materials was also responsible for the revolutionary developments in modern electric and electronic industries. Magnetic materials play very important roles in almost all the equipment that use modern technology, for example, ferrite magnets in TV, memory cores in computers, permanent magnets in motors, superconducting magnets in particle accelerators, to name a few. We would have no audio/video equipment without suitable magnetic materials.

Magnetic materials are functional materials, but sometimes they are also used in large quantities (many tons) such as for the core materials in power transformers. Like semiconductor materials, the quality of magnetic materials strongly influences the performance, efficiency, energy consumption, size and reliability of electric and electronic equipment. Great progress has been made in quality improvement of the magnetic materials in the recent years. Fig. 6.1 shows the progress in permanent magnetic materials during the last 100 years.

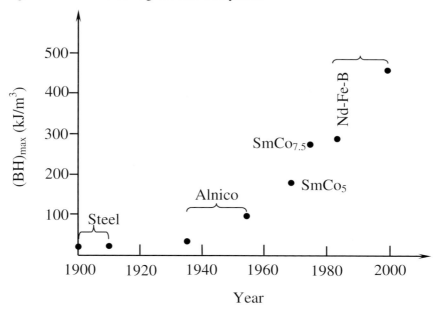

Fig. 6.1 Progress in magnetic materials measured by the maximum energy products $(BH)_{max}$.

6.2 Fundamentals

6.2.1 *Magnetic flux and permeability*

Compare an electric circuit with a magnetic circuit (Fig. 6.2): if a voltage, V, is applied to a conductor, the current I that flows in it is related to the conductivity σ of the material:

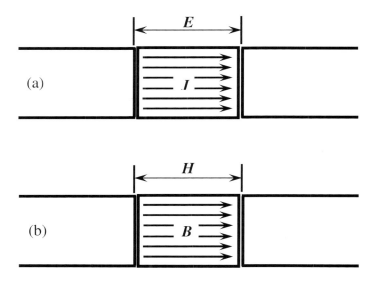

Fig. 6.2 Analogy between (a) electric circuit and (b) magnetic circuit.

$$V = I \cdot R, R = \rho \cdot l/A = l/(\sigma \cdot A),$$

$$V = (I/A)(l/\sigma),$$

$$\sigma = (I/A) \times (1/E), \tag{6.1}$$

$E = V/l$, $\sigma = J/E =$ (current density)/(electrical field).

In an analogous way, if a magnetic field H is applied across a gap, the magnetic flux (or induction) B in the gap is proportional to the permeability μ of the material in the gap.

$$\mu = B/H = \text{(flux density)/(magnetic field)}, \tag{6.2}$$

where μ is called magnetic permeability, which is analogous to the conductivity σ in electric conduction.

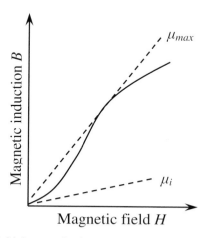

Fig. 6.3 *B-H* initial magnetisation curve for a ferromagnetic material.

However, magnetic permeability is different from electric conductivity in the following ways:

- σ in an electric circuit is a constant (independent of E and I), while μ in a magnetic field changes with H. J/E is a constant for linear materials, metals in particular, while B/H is not a constant. Initial μ, μ_i, and maximum μ, μ_{max}, are used to characterise μ.
- The magnetic flux can persist after the field is removed while the electric current does not. This flux is called "remanent induction".

6.2.2 *Basic units*

In the SI (meter-kilogram-second) system, H is in A/m, B is in weber/m^2 = tesla (T) = volt-second/m^2, and μ is in weber/A·m = henry/m.

In the SI system, μ in vacuum = μ_0 = $4\pi \times 10^{-7}$ henry/m. However, Gaussian system is still popular in physics. In the Gaussian units, H is in oersted, B is in gauss and μ is in gauss/oersted. The following table compares the units used in the two systems:

Table 6.1 The units used in SI and Gaussian systems.

	SI	Gaussian
H	1 A/m	$= 4\pi \times 10^{-3}$ oersted
B	1 T	$= 10^4$ gauss
μ_0	$4\pi \times 10^{-7}$ henry/m	$= 1$ gauss/oersted

6.2.3 *Magnetisation and relative permeability*

From Eq. (6.2), in vacuum we have $B = \mu_0 \cdot H$.

If a material is inserted into the gap, Eq. (6.2) should be written as:

$$B = \mu \cdot H = \mu_0 (H + M), \qquad (6.3)$$

where M is the magnetisation of the solid. The magnetic induction (flux), B, is the sum of the applied field H and the external field M that arises from the magnetisation of the materials in the gap.

For ferromagnetic materials, $\mu_0 \cdot M$ is much larger than $\mu_0 \cdot H$.

$$B \approx \mu_0 \cdot M, \qquad (6.4)$$

$\mu_0 \cdot M$ can be used to replace B.

For non-ferro/ferrimagnetic materials: $\mu_0 \cdot M << \mu_0 \cdot H$ thus $B \approx \mu_0 \cdot H$.

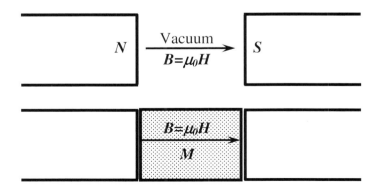

Fig. 6.4 Magnetisation of materials.

Relative permeability, μ_r, is defined as the ratio of μ/μ_0:

$$\mu_r = \mu/\mu_0, \tag{6.5}$$

$$B = \mu_0 \cdot \mu_r \cdot H, \tag{6.6}$$

Remember μ is not a constant (while σ is a constant), it changes as the material is magnetised.

6.2.4 *Magnetic susceptibility*

The magnetic properties of a material are characterised not only by the magnitude and sign of M (magnetisation of the material), but also by the way in which M varies with H. The ratio of M to H is called magnetic susceptibility, χ.

$$\chi = M/H. \tag{6.7}$$

Typical *M-H* plots (magnetisation curves) are shown in Fig. 6.4.

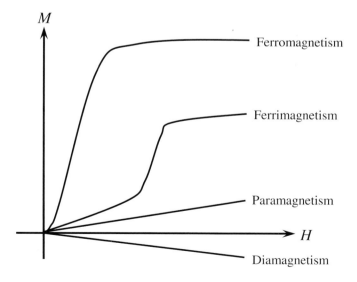

Fig. 6.4 Typical magnetisation curves for different magnetic material.

Table 6.2 Example of magnetisms.

Class	Sample	χ
Diamagnetic	H_2O	-9.0×10^{-6}
Paramagnetic	Al	2.2×10^{-5}
Ferromagnetic	Fe	3,000
Antiferromagtic	Tb	9.5×10^{-2}
Ferrimagnetic	$MnZn(Fe_2O_4)_2$	2,500

6.2.5 *The origin of magnetism, Bohr magneton*

Magnetic fields and forces are originated from the movement of the basic electric charge, the electron. When electrons move in a conducting wire, a magnetic field, H, is produced around the wire (Fig. 6.5):

$$H = 0.4 \times \pi \cdot n \cdot I/l, \qquad (6.8)$$

where n = the number of turns, I = current, and l = length of the wire.

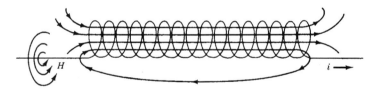

Fig. 6.5 A magnetic field created around a coil.

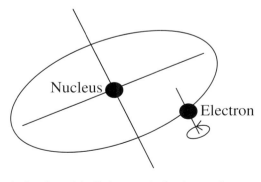

Fig. 6.6 A schematic drawing of the Bohr atom indicating an electron spinning on its own axis and revolving about its nucleus. These are the origins of the magnetism in materials.

Magnetism in materials is due to the motion of electrons, by the intrinsic spin and orbital motion about the nuclei (Fig. 6.6). Bohr suggested a fundamental quantity later called the Bohr magneton, which is the strength of the magnetic field associated with an isolated electron. This magnetic field is a constant and can be calculated as follows:

$$\text{Bohr magneton} = e \cdot h/(4\pi \cdot m) = 9.27 \times 10^{-24} \text{ A} \cdot \text{m}^2, \qquad (6.9)$$

where e = the charge of an electron, h is Planck's constant, and m is the mass of an electron.

For most elements, the magnetic moments from the paired electron spin and rotation are cancelled in atoms. Only unpaired spins of electrons can be the source of ferromagnetism. Fe, Co and Ni all have unpaired $3d$ electrons. Elements with unpaired $4d$ electrons, the lanthanide series, sometimes also show ferromagnetism. A simplified calculation for the maximum magnetisation of these elements can be made by using the total unpaired $3d$ electrons as Bohr electrons.

6.2.6 *Types of magnetism*

There are five different types of magnetic behaviours of materials: diamagnetism, paramagnetism, ferromagnetism, ferrimagnetism and antiferromagnetism. Fig. 6.4 displays the typical magnetisation curves of the different magnetic behaviours.

6.2.6.1 *Diamagnetism*

Magnetization is negative, i.e., $\chi < 0$. External magnetic field acting on atoms slightly unbalances their orbiting electrons and creates small magnetic dipoles within the atoms, which oppose the applied field, and this action produces a negative magnetic effect to the applied field. Diamagnetic effect is very small. The magnetic susceptibility χ is around 10^{-6} to 10^{-5}.

Inert gases, many organic compounds, some metals (Bi, Zn, Ag) and nonmetals (S, P, Si) are typical examples of diamagnetic materials. Diamagnetism has no significant engineering importance.

6.2.6.2 *Paramagnetism*

Many solids that have small but positive magnetic susceptibilities are called paramagnetic. An applied field aligns the individual magnetic dipoles of the atoms or molecules and slightly increases B. This effect is greater than diamagnetic effect but still very small. The magnetic susceptibility χ ranges from $10^{-6}–10^{-2}$. Temperature reduces the paramagnetic effects. The correlation between magnetic susceptibility and temperature can be described by Curie–Weiss law. Alkaline metals, such as Li, Na, K and Rh, can show such a weak, positive, attractive reaction to an applied field.

Curie–Weiss law: $\chi_p = C/(T - T_p)$,

where: C: Curie constant

T: Temperature

T_p: Paramagnetic Curie temperature.

Diamagnetic and paramagnetic effects are all induced by an applied field. When field is removed, the effect disappears. Diamagnetic and paramagnetic properties have limited engineering applications.

Table 6.3 Magnetic susceptibilities of some diamagnetic and paramagnetic materials.

Diamagnetic	Magnetic susceptibility $(\chi \times 10^{-6})$	Paramagnetic	Magnetic susceptibility $(\chi \times 10^{-6})$
Cadmium	-0.18	Aluminium	+0.65
Copper	-0.086	Calcium	+1.10
Silver	-0.20	Oxygen	+106.2
Tin	-0.25	Platinum	+1.10
Zinc	-0.157	Titanium	+1.25

6.2.6.3 *Ferromagnetism*

A group of materials have very different magnetisation properties from the first two groups. Firstly, the susceptibility χ is positive and very large, about 10^7 times greater than χ in paramagnetic materials ($10^1 < \chi < 10^6$). This means that a very large magnetisation will be created by the material under an applied field. Secondly, large magnetic fields can be retained after the applied field is removed. This magnetic behaviour is called ferromagnetism, and the materials are called ferromagnetic materials. These properties are of great engineering importance.

The most important ferromagnetic elements are Fe, Co and Ni. A rare-earth element gadolinium (Gd) is also ferromagnetic below 16°C, but has little engineering application. Fe, Co, and Ni are transition metals and have unpaired inner $3d$ electrons (see Fig. 6.7). Fe atom has four unpaired $3d$ electrons, Co has three unpaired $3d$ electrons, and Ni has two unpaired $3d$ electrons. The spins of the $3d$ electrons of adjacent atoms align in a parallel direction by a phenomenon called "spontaneous magnetisation". This parallel alignment of atomic magnetic dipoles occurs in microscopic regions called "magnetic domains".

If the domains are randomly oriented, there will be no net magnetisation in a bulk sample. If the domains are aligned in a magnetic field, the magnetic induction of the specimen will be very strong.

Table 6.4 Magnetic moments in the atoms of $3d$ transition elements.

Atom	Number of electrons	Unpaired 3d electrons	4s electrons
V	23	3	2
Cr	24	5	1
Mn	25	5	2
Fe	26	4	2
Co	27	3	2
Ni	28	2	2
Cu	29	0	1

Atom	Electronic configuration 3d orbitals
V	[↑][ ] [↑][ ] [↑][ ] [ ][ ] [ ][ ]
Cr	[↑][ ] [↑][ ] [↑][ ] [↑][ ] [↑][ ]
Mn	[↑][ ] [↑][ ] [↑][ ] [↑][ ] [↑][ ]
Fe	[↑][↓] [↑][ ] [↑][ ] [↑][ ] [↑][ ]
Co	[↑][↓] [↑][↓] [↑][ ] [↑][ ] [↑][ ]
Ni	[↑][↓] [↑][↓] [↑][↓] [↑][ ] [↑][ ]
Cu	[↑][↓] [↑][↓] [↑][↓] [↑][↓] [↑][↓]

6.2.6.4 *Electron structure and magnetic behaviour*

Table 6.4 shows the electronic configurations of some $3d$ transition elements. Why are Fe, Co and Ni ferromagnetic materials while Cr and Mn are not, while they all have unpaired $3d$ electrons? This is due to the "magnetic exchange interaction energy". This energy associates with the coupling of individual magnetic dipoles into a single magnetic domain. Only when the exchange energy is positive, can the material be ferromagnetic.

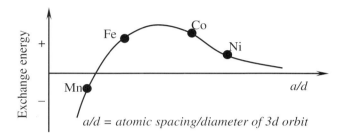

Fig. 6.7 Magnetic exchange interaction energy as a function of the ratio of atomic spacing to the diameter of the $3d$ orbit for some transition elements (after W.F. Smith, "Foundations of Materials Science and Engineering", McGraw-Hill, 1993).

This exchange energy is related to the ratio of the atomic spacing to the $3d$ orbit. The ratio must be in the range of 1.4 to 2.7 (Fig. 6.7). Fe, Co and Ni satisfy the above conditions, but Cr and Mn do not.

An example to show the magnetisation of ferromagnetic materials:

Use Bohr Magneton to

(a) Calculate the total effect of the Bohr Magnetons in Fe, Co and Ni - assume all unpaired electrons are Bohr Magnetons;
(b) Compare with the saturated M obtained from experimental measurements (see how far the Bohr theory is from reality)

Fe has a *bcc* structure and a lattice constant a_0 of 0.2866 nm, Ni has a *fcc* structure and $a_0 = 0.3517$ nm, Co has a *hcp* (hexagonal close-packed) structure, $a_0 = 0.2507$ nm and $c_0 = 0.4069$ nm.

Solution:

(a) Assuming the four unpaired $3d$ electrons in one Fe atom each contributes magnetisation of a Bohr Magneton, the total magnetisation of Fe in a unit volume (m^3) can be calculated as:

M_{Fe} = (2 atoms/cell) × (4 magnetons/atom) × 9.27 × 10^{-24} A·m²/(0.2866
  × 10^{-9} m)³ = 3.15 × 10^6 A/m.

The magnetisation of Ni and Co is calculated in the same way.

M_{Ni} = (4 atoms/cell) × (2 magnetons/atom) × 9.27 × 10^{-24} A·m²/
  (0.35167 × 10^{-9} m)³ = 1.71 × 10^6 A/m.

M_{Co} = (6 atoms/cell) × (3 magnetons/atom) × 9.27 × 10^{-24} A·m²/[3 ×
  (0.25071 × 10^{-9} m)² × 0.40686 × 10^{-9} × cos30°] = 2.51×10^6 A/m.

The unit of M is A×m²/m³ = A/m.

(b) Compare the M with the experimental results:

Elements	Fe	Co	Ni
Calculated M_s (A/m × 10^6)	3.15	2.51	1.71
Experimental M_s (A/m × 10^6)	1.714	1.422	0.484
Ratio M_{Exp}/M_{Cal}	0.54	0.56	0.28

The results indicated that each unpaired $3d$ electron in Fe and Co provides magnetisation of more than ½ Bohr magneton, while slightly less than 1/3 of magneton for Ni.

6.2.6.5 *Antiferromagnetism*

In the presence of a magnetic field, in some materials the magnetic dipoles of atoms align themselves in opposite directions. Therefore, the atoms do not show net magnetic moment. Manganese (Mn), chromium (Cr), MnO, CrO and CoO exhibit this behaviour, because they have a negative exchange energy.

6.2.6.6 *Ferrimagnetism*

In some ceramic compounds, different ions have different magnitude of magnetic moments. When these magnetic moments are aligned in an antiparallel manner, there is a net magnetic moment in one direction. Magnetic susceptibility, χ, normally is ranging from 10^0 to 10^4.

Ferromagnetism   Antiferromagnetism   Ferrimagnetism

Fig. 6.8 Alignment of magnetic dipoles for different types of magnetism: (a) ferromagnetism, (b) antiferromagnetism, and (c) ferrimagnetism.

Fig. 6.8 is a schematic drawing illustrating the arrangements of dipoles in ferromagnetic, antiferromagnetic and ferrimagnetic materials. Typical examples of ferrites can be Fe_3O_4 group, $BaO\text{-}6Fe_2O_3$, $SrFe_{12}O_{17}$ and $NiO\text{-}Fe_2O_3$.

These ceramic magnetic materials are called ferrites. Their magnetic properties are very much like those of ferromagnets. But they have much lower electrical conductivity than metallic magnets because of their ceramic nature. These properties make them very useful in many electronic applications.

6.2.7 *Effect of temperature on ferromagnetism - Curie Point*

At any temperature above 0 K, thermal energy causes the magnetic dipoles in a ferromagnetic material to deviate from the perfect parallel alignment. In other words, the exchange energy that aligns the dipoles is counter-balanced by the randomising effects of the thermal energy. Therefore, the maximum magnetisation decreases with temperature.

When the temperature increases to a certain level, the ferromagnetism in this material completely disappears and the material becomes paramagnetic. This temperature is called "Curie temperature" or "Curie point" (T_c). Fig. 6.9 shows the plots of saturation magnetisation versus temperature for Fe, Co and Ni.

The Curie temperature depends on the type of material and can be changed by alloying elements. For Gd, Ni, Fe and Co, T_{curie} = 16, 358, 770 and 1131°C, respectively.

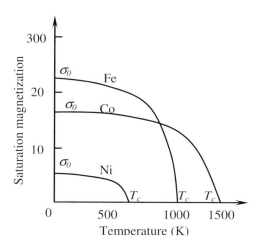

Fig. 6.9 Saturation magnetisation of Fe, Co, and Ni as a function of temperature.

6.2.8 *Ferromagnetic anisotropy and magnetostriction*

6.2.8.1 *Ferromagnetic anisotropy*

The *B-H* curves of a single crystal of a ferromagnetic material are different for different crystal orientations to the magnetic field.

Fig. 6.10 shows that one direction is more easily magnetised than the others. Saturation along this direction occurs at a lower applied field. For instance, in *bcc* Fe, <100> is the easiest direction for magnetisation and <111> is the most difficult direction. In *fcc* Ni, <111> is the easiest direction and <100> is the most difficult direction.

In a polycrystalline material, different grains approach saturation differently. The grains with easy orientations saturate at lower applied fields. The grains with hard orientations rotate their moment into the field direction only at higher fields. The work required to overcome this anisotropy is called "magnetocrystalline energy", or "magnetocrystalline anisotropic energy".

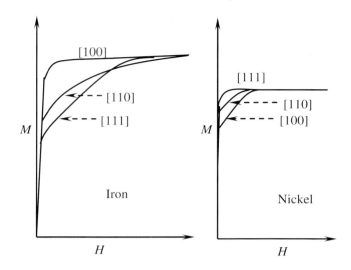

Fig. 6.10 Magnetisation *vs.* applied field for field direction along the [100], [110] and [111] directions of Fe and Ni single crystals (after Rose, Shepard and Wulff, "Electronic Properties", John Wiley and Sons, 1966).

6.2.8.2 *Magnetostriction*

When the electron spin dipole moments of atoms are rotated into alignment during magnetisation, the spacing between the atoms changes since they may attract or repel each other. Therefore, the shape and dimension of a ferromagnetic material change. This phenomenon is called "magnetostriction" (Fig. 6.11). The effect was first identified in 1842 by James Joule when observing a nickel sample.

Magnetostriction is a reversible strain along the magnetisation axis. It is anisotropic, not only because that the magnetisation curve is anisotropic for different orientations, but also because the elastic deformation of the crystal is anisotropic. It saturates at a high field (Fig. 6.12). Magnetostriction and magnetisation usually saturate at the same time.

Magnetostriction can be measured with a simple relation:

$$\lambda_s = \Delta l / l, \tag{6.10}$$

where Δl = extension (+) or contraction (-) of a specimen of length l in the direction of an applied field when the field is raised from zero to a value of saturation.

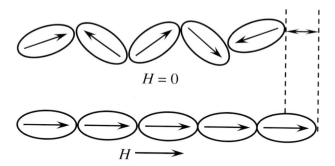

$H = 0$

$H \longrightarrow$

Fig. 6.11 A schematic of magnetostriction.

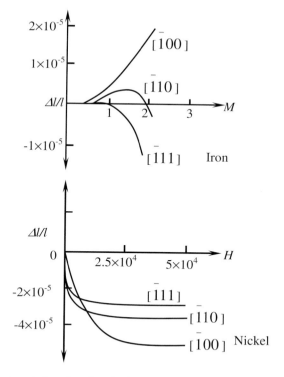

Fig. 6.12 Magnetostriction data for single crystal Fe and Ni showing anisotropy (after Rose, Shepard and Wulff, "Electronic Properties", John Wiley and Sons, 1966).

6.2.8.3 *Ferromagnetic domains and domain movement*

6.2.8.3.1 Ferromagnetic domains

In ferromagnetic materials, there is a structure called "magnetic domains". Below the Curie temperature, the magnetic dipole moments of atoms tend to align themselves in a parallel direction in this small volume regions, as schematically shown in Fig. 6.13. However, the materisl does not show magnetisation if the domains have not been aligned.

6.2.8.3.2 Domain movement during magnetisation

When a piece of Fe or Ni is cooled down slowly from above its Curie temperature, the magnetic domains are randomly orientated. When an external magnetic field is applied, the magnetic domains show the following two-stage reactions:

First stage: when an external magnetic field is applied, the domains with moments parallel to the applied field grow, at the expense of those with less favourable orientations. In this stage, the domain growth takes place by domain wall movement. The magnetisation (or induction) increases rapidly as the applied field increases.

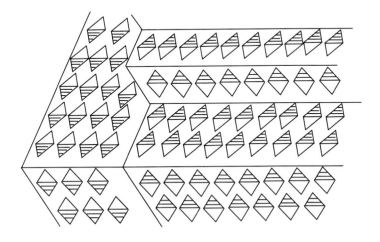

Fig. 6.13 Schematic drawing of magnetic domains in a ferromagnetic material (after Rose, Shepard and Wulff, "Electronic Properties", John Wiley and Sons, 1966).

Second stage: when domain wall growth has finished, if the applied field is continuously increased, domain rotation occurs.

The wall movement requires less energy than domain rotation, so it takes place first. The domain rotation requires considerably higher energy, so it takes place later. During the domain rotation stage, B increases slowly with H. Figs. 6.14 and 6.15 show the domain growth and rotation processes during magnetisation.

When the applied field is removed, some of the magnetisation is lost because of the tendency for the domains to rotate back to their original alignment, but the rest of the domains retain their magnetisation direction and the specimen remains magnetised.

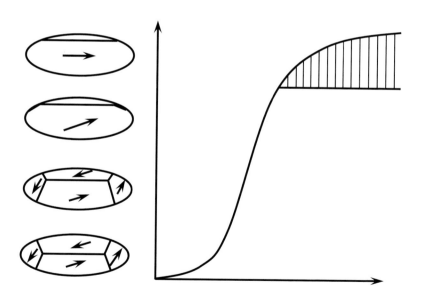

Fig. 6.14 Magnetic domain growth and rotation as a ferromagnetic material is magnetised by an applied magnetic field (after Rose, Shepard and Wulff, "Electronic Properties", John Wiley and Sons, 1966).

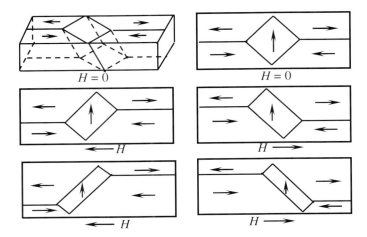

Fig. 6.15 Movement of domain boundaries in a Fe crystal during magnetisation (courtesy General Electric Research Laboratory).

6.2.9 *Magnetic energy and domain structure - Types of energy that determines the structure of domains*

The domain structure of a ferromagnetic material is determined by the related energies. The most stable structure is obtained when the overall potential energy is a minimum. The total magnetic energy is the sum of the following five energies:

1. Magnetic exchange energy,
2. Magnetostatic energy,
3. Magnetocrystalline energy,
4. Domain wall energy, and
5. Magnetostriction energy.

6.2.9.1 *Magnetic exchange energy*

This is the energy associated with the coupling of individual dipoles into a single domain. It is a potential energy within a domain and is minimised when all the dipoles are aligned in one direction. This energy can be positive or negative, depending on the atomic structure. When the ratio of atomic space to the $3d$ orbit is between 1.4 to 2.7, the exchange energy is positive. Only positive exchange energy results in the dipole alignment in one direction (and therefore, showing ferromagnetism).

6.2.9.2 *Magnetostatic energy*

This is the magnetic potential energy due to the external magnetic field surrounding a magnet. It is produced by its external field. This potential energy can be minimised by small domain formation and splitting. For a unit volume of a material, a single domain structure has the highest magnetostatic energy. By dividing the large domain, the intensity and extent of the external magnetic field are reduced (Fig. 6.16). Therefore, the formation of multiple domains reduces the magnetostatic energy.

6.2.9.3 *Magnetocrystalline anisotropy energy*

Magnetocrystalline (anisotropy) energy is the energy required during the magnetisation of a ferromagnetic material to rotate the magnetic domains because of crystalline anisotropy.

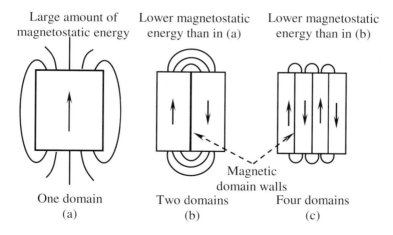

Fig. 6.16 Schematic illustration showing how reducing the domain size decreases the magnetostatic energy by reducing the external magnetic field (after W.F. Smith, "Foundations of Materials Science and Engineering", McGraw-Hill, 1993).

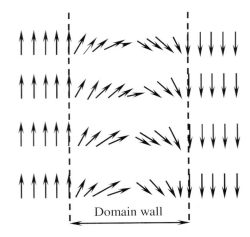

Fig. 6.17 Magnetic dipole arrangements at domain wall.

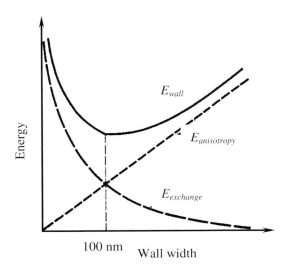

Fig. 6.18 Relationship among magnetic exchange energy, magnetocrystalline aniso-tropic energy and wall width.

For instance, the difference in magnetisation of the hard [111] direction and easy [100] direction in *bcc* Fe is about 1.4×10^4 J/m^3.

6.2.9.4 *Domain wall energy*

Domain wall is defined as the boundary between two domains whose overall magnetic moments are in different orientations. The potential energy associated with the disorder of dipole moments in the wall area is called domain wall energy.

Fig. 6.17 shows the dipole arrangements at a domain wall. Compared with grain boundaries, where the crystal orientation changes from one orientation to another, domain walls are much thicker. In grain boundary areas, the crystal orientation changes in approximately three atoms' distance, while a domain changes its orientation gradually, with a domain wall of ~300 atoms wide.

The reason for the large width of a domain wall is due to a balance between two factors: exchange and magnetocrystalline energy. When there is a small difference in orientation between the two dipoles, the exchange energy is minimised. The exchange energy tends to widen the domain wall, thus the orientation difference can be small. However, the wider the wall, the greater will be the number of dipoles lying in directions different from those of easy magnetisation and the magnetocrystalline anisotropy energy will be increased. Therefore, the equilibrium wall width will be reached when the sum of the two energies is minimum (Fig. 6.18).

6.2.9.5 *Magnetostriction energy*

The energy due to the mechanical stress caused by magnetostriction in a ferromagnetic material is called magnetostriction energy.

In cubic magnetic materials (Fe, Ni, Co), the formation of triangle shape domains, called domain closures, at the ends of the domains, reduces the magnetostriction energy associated with the external field, and hence lower the energy of the system (Fig. 6.19). It appears that large domains with less domain wall length would have a lower energy. However, magnetostriction will be greater for larger domains. Small domains have smaller magnetostriction energy. Therefore, the equilibrium domain is reached again when the sum of the two energies is a minimum.

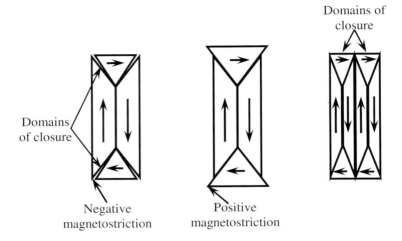

Fig. 6.19 Magnetostriction in cubic magnetic materials. (a) negative and (b) positive magnetostriction pulling apart the domain boundaries. (c) Lowering the magnetostrictive stresses by creating smaller domain size structure (after W.F. Smith, "Foundations of Materials Science and Engineering", McGraw-Hill, 1993).

To conclude, the domain structure in ferromagnetic materials is determined by the total contributions of the five energies, exchange, magnetostatic, magnetocrystalline, domain wall and magnetostriction energies. The equilibrium domain structure is reached when the sum of these energies is a minimum.

6.2.10 *Magnetisation and demagnetisation of magnetic materials, hysteresis loop*

To study the effect of an applied magnetic field H on the magnetic induction B of a ferro- or ferri-magnetic material, one can magnetise and demagnetise the material and plot the B *vs.* H (Fig. 6.20). This is called hysteresis loop. They are widely used to describe the magnetic properties of a ferromagnetic or ferrimagnetic material. The important magnetic properties that can be found in the plot are:

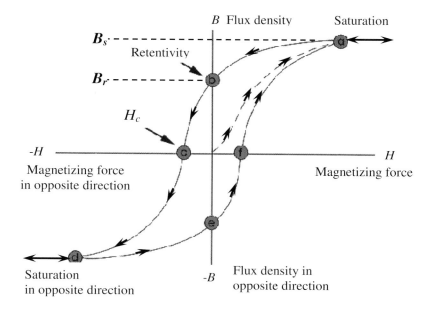

Fig. 6.20 Hysteresis loop for a ferromagnetic material.

B_s: **saturation induction** - maximum value of induction; how high the magnetisation can be;

B_r: **remanent induction** - value of B when H is decreased to zero; how high the induction remaining is after the external field is removed; and

H_c: **coercive force** - applied magnetic field required to decrease the magnetic induction to zero; how magnetically hard the magnetic material is.

The internal area is a measure of energy lost (or the work done) during the magnetising and demagnetising cycle.

6.2.11 *Energy losses in magnetic materials*

Energy losses for soft magnetic materials consist of two parts: hysteresis losses (also called magnetic losses or iron losses) and eddy current losses (also called electrical losses).

6.2.11.1 *Hysteresis losses*

For the transformers and inductors used in industrial alternative current (AC) systems, the current produces the entire hysteresis loop 50 (or 60) times per second. During each cycle there is energy lost due to the domain wall movement. This is hysteresis losses.

Hysteresis losses can be expressed using the areas within the hysteresis loop. It is smaller than $4H_cB_s$. In order to reduce hysteresis losses, H_c has to be reduced because a high B_s is required. Increasing the ac frequency will increase the hysteresis losses.

6.2.11.2 *Material factors*

Impurities, defects, such as voids, dislocations and grain boundaries, etc, all act as barriers to impede domain wall movement and thus increase the energy losses. Mechanical strain causes magnetostriction and consumes energy. Mechanical deformation increases the dislocation density, which also increases the energy losses.

When there is a spherical non-magnetic inclusion in the middle of a domain. The dipoles are separated by the inclusion. Energy is increased.

There are two ways to reduce the energy:

1. the inclusion lies in a domain wall, this reduces part of the wall energy and brings the poles closer together; and
2. the inclusion may have its own domain. This domain tends to be blade-shape so poles can be closer.

6.2.11.3 *Eddy current energy losses*

AC current produces a fluctuating magnetic field. It results in transient voltage gradients in a conducting magnetic core. This creates stray electric current called eddy current. It causes energy losses and core heating.

Eddy current losses can be expressed as:

$$W_e = \rho \cdot J^2, \tag{6.11}$$

where ρ = resistivity of the core material, and J = eddy current density. W_e is in W/m^3, i.e. energy losses in unit volume.

In transformer cores, W_e can be given by

$$W_e = (\pi \cdot b \cdot f \cdot B_s)^2 \cdot \sigma/6 = (\pi \cdot b \cdot f \cdot B_s)^2/(6\rho), \tag{6.12}$$

where b = thickness of the laminations, f = frequency of the a.c. current, B_s or B_m = maximum flux density, σ = conductivity, and ρ = resistivity.

Therefore, the energy losses are proportional to the conductivity of the core material ($W_e \propto \sigma$).

To reduce eddy current losses, thin sheets of the transformer material are used. It is also beneficial to increase the resistivity of the core material. Eddy current losses will be very high in high frequency applications since $W_e \propto f^2$. Ferrimagnetic oxides (ceramic magnets) have been used because of their high resistivity.

6.3 Soft Magnetic Materials

Soft magnetic materials are easily magnetised and demagnetised (typical hysteresis loop shown in Fig. 6.21a). The most important characteristics include:

1. high saturation induction B_s,
2. low coercivity H_c,
3. low hysteresis losses, and sometimes
4. high initial permeability, μ_i.

Soft magnetic materials are used in a variety of applications that involve changing electromagnetic induction, such as transformers, motors, generators, inductors, solenoids, lamp ballast units, direct current relays and magnetic shielding. The largest volume of soft magnetic materials used is for fractional horsepower motors, heavy rotating equipment, and power frequency transformers.

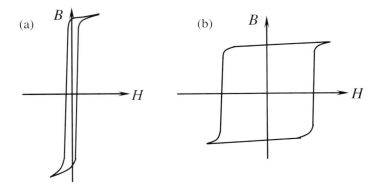

Fig. 6.21 Hysteresis loops for (a) a soft and (b) a hard magnetic material.

Materials used in these industries include high-purity iron, low-carbon steels (cold-rolled lamination steels), silicon steels (electrical steels), iron-nickel alloys, iron-cobalt alloys, ferritic iron-chrome alloys (solenoid-quality stainless steels), soft ferrites (ceramic magnets), and soft magnetic amorphous alloys.

6.3.1 *Iron-silicon alloys*

Fe-Si alloys are the most extensively used soft magnetic materials. They usually contain 3–4 wt.% Si and were first used in 1905. Non-oriented Fe-Si alloys have an approximately isotropic grain texture. Its Si concentration varies from 1 to 3.7 wt.%. Al of 0.2–0.8 wt.% and Mn of 0.1–0.3 wt.% is usually added. Non-oriented Fe-Si alloys are preferentially used in medium and high power rotating machines whereas low carbon steel laminations are preferred. The development of improved non-oriented alloys is the control of impurities, grain size, crystallographic texture, surface state, residual and applied stresses.

In 1934 a process was developed to produce grain oriented Fe-Si alloys in which the crystallites have their [001] easy axis close to the rolling direction and their (110) plane nearly parallel to the lamination surface. The easy directions for magnetisation in Fe and Fe-Si are the <100> directions. Fig. 6.22 shows the magnetisation curves of Fe-Si with or without grain texture. The Fe-Si grains can be oriented in the sheet plane by special texturing processes, which usually includes hot rolling pass, two cold rolling passes and some annealing treatments.

Two types of textured structure are produced as shown in Fig. 6.23:

1. "cube on edge" - (110) and [001]
2. "cubex" - (100) and [001]

When the applied field is parallel to the rolling direction of a textured sheet, the permeability μ is much higher and the hysteresis losses are much lower compared with those randomly oriented Fe-Si alloys.

Fe-6.5 wt.% Si alloys are promising low loss materials and provide a potentially excellent soft magnetic alloy for applications at power and medium frequencies. These alloys are hard and brittle and then cannot be prepared by cold rolling due to the heterogeneous formation of ordered FeSi and Fe_3Si phases during cooling. Ductile Fe-6.5% Si ribbons however can be obtained by planar flow casting, a method where a molten metal stream is ejected onto a rotating metallic drum.

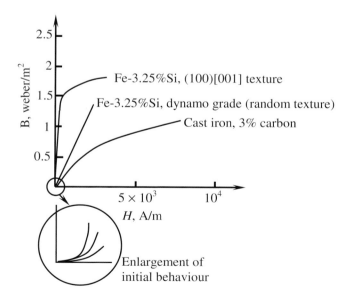

Fig. 6.22 The effect of Si on the magnetisation behaviour of Fe (after Rose, Shepard and Wulff, "Electronic Properties", John Wiley and Sons, 1966).

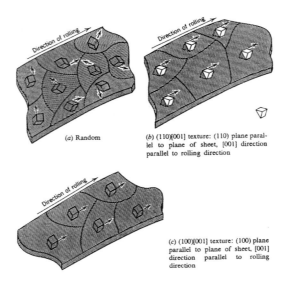

(a) Random

(b) (110)[001] texture: (110) plane parallel to plane of sheet, [001] direction parallel to rolling direction

(c) (100)[001] texture: (100) plane parallel to plane of sheet, [001] direction parallel to rolling direction

Fig. 6.23 Random and preferred orientations in polycrystalline Fe-Si sheet (after Rose, Shepard and Wulff, "Electronic Properties", John Wiley and Sons, 1966).

6.3.2 *Effect of the addition of 3-6%Si to Fe*

The positive effects of Si in Fe may include:

1. Si increases resistivity (ρ), thus reducing the eddy current losses;
2. Si decreases the magnetocrystalline anisotropic energy, thus increasing the magnetic permeability (μ), and decreasing the hysteresis losses; and
3. Si decreases the magnetostriction and therefore the hysteresis losses and mechanical vibration (the transformer noise "hum").

However, Si in steel also has some negative effects:

4. Si decreases the ductility of Fe. Usually, only up to 4% Si can be added without too much mechanical degradation. Fe-6% Si steels can only be produced by a more complicated rolling-heat treatment process; and

5. Si also decreases the saturation induction and Curie temperature.

6.3.3 *Ni-Fe alloys* (*permalloy*)

The magnetic permeabilities of pure Fe or Fe-Si alloys are relatively low. Low initial permeability is not a problem for power applications such as transformer cores because they are operated at high magnetisation. For communication equipment of high sensitivity, used to detect or transmit small signals, a high initial permeability is very important. Ni-Fe alloys, with a composition of 35-80% Ni and a well-defined structure, display a broad range of magnetic properties and have much higher permeabilities at low fields and are commonly used for these applications.

The high permeability of the Ni-Fe alloys is due to the low magnetocrystalline and magnetostriction energies.

Several groups of Ni-Fe alloys are commercially produced:

1. ~50%Ni alloys: they have a moderate μ (μ_i = 2,700 and μ_{max} = 25,000), and a very high B_s = 1.6 T. They have a higher permeability and better corrosion resistance than silicon-iron alloys. These alloys can be prepared as strongly (100)[001] textured sheets by severe cold rolling and primary re-crystallization annealing around 1000°C;
2. 36%Ni alloys: they have higher resistivity than the 50%Ni alloys but the saturation magnetisation and permeability are lower;
3. ~75-80%Ni alloys: these alloys variously called Mumetal and Permalloy have the highest permeability (μ_i = 100,000 and μ_{max} = 1,000,000) due to the almost zero crystalline anisotropy and magnetostriction. However they have a lower B_s = 0.8 T.

The initial permeability of Ni-Fe alloys can be further increased three to four times by annealing in a magnetic field. Field annealing causes directional ordering of the atoms of the Ni-Fe lattice and therefore increases μ.

The magnetic properties of some soft magnetic materials are listed in Table 6.5.

Table 6.5 Magnetic property of some soft magnetic materials.

Materials	Saturation induction (B_s, T)	Coercive force (H_c, A/cm)	Initial relative permeability (μ_i)
Magnetic iron, 0.2cm sheet	2.15	0.88	250
M36 cold-rolled Si-Fe (random)	2.04	0.36	500
M6 (110) [001], 3.2% Si-Fe (oriented)	2.03	0.06	1,500
45Ni-55Fe (45 Permalloy)	1.6	0.024	2,700
75Ni-5Cu-2Cr-18Fe (Mumetal)	0.8	0.012	30,000
79Ni-5Mo-15Fe-0.5Mn (Supermalloy)	0.78	0.004	100,000
48%MnO-Fe$_2$O$_3$, 52%ZnO-Fe$_2$O$_3$ (soft ferrite)	0.36		1,000
36%NiO-Fe$_2$O$_3$, 64%ZnO-Fe$_2$O$_3$ (soft ferrite)	0.29		650

6.3.4 *Amorphous alloys (metallic glasses)*

Amorphous alloys are alloys without crystal structure (without long range atomic ordering). The arrangement of the atoms is like those in a liquid. Glasses usually have amorphous structures. Amorphous alloys have many attractive properties that crystal alloys do not possess, including excellent magnetic, mechanical and corrosion properties.

Amorphous metallic alloys can be produced by a number of different methods, including high energy beam radiation, vapour deposition, electroplating etc. These methods produce bulk or thin film of amorphous structure. The most widely used method to produce amorphous alloy ribbons or sheets is the melt spinning process (rapid solidification or rapid quenching) of liquid metals. During this quenching process, the cooling speed can reach $\sim 10^6$ K/s. The atoms do not have enough time to rearrange their positions to form a crystal structure. Thus an alloy with liquid type of atomic structure, an amorphous structure, is formed. This process can produce a continuous ribbon of about 0.03 mm thick and up to 200 mm wide.

Table 6.6 Property of some soft magnetic amorphous alloys.

Composition	J_s [a]	H_c [b]	σ [c]	T_c [d]	W_T [e]
$Fe_{81}B_{13.5}Si_{3.5}C_2$	1.6	3.2	0.77	643	0.3
$Fe_{67}Co_{18}B_{14}Si_1$	1.8	4.0	0.77	688	0.5
$Fe_{40}Ni_{38}Mo_4B_{18}$	0.9	1.2	0.62	626	0.4
$Fe_{78}B_{13}Si_9$	1.5	1.5	0.8	600	0.3
Fe-6%Si	1.9	8.0	1.3	900	0.8

[a] J_s in T at 300 K
[b] Induction coercivity in A/m
[c] Conductivity in MS/m at 300 K
[d] Curie temperature in K
[e] Total losses in W/kg at $B_m = 1.4$ T
(Source: L.A.A. Warnes, "Electronic Materials", Macmillan Education Ltd. 1990)

Some of the amorphous alloys have very attractive soft magnetic properties, including low coercive force, low hysteresis loss and high electrical resistivity. Domain walls in these materials are able to move with exceptional ease, mainly because there are no grain boundaries and no long-range crystal anisotropy.

Fe, Co and Ni based amorphous alloys containing Si, B, C and/or P are used as the core materials in transformers with very small hysteresis losses and eddy currents. They are also used to make magnetic sensors and recording heads. To show the energy saving, a 25 kVA distribution transformer with a grain-oriented Si-steel has ~85 W of core loss. Replacing the Fe-Si with an amorphous Fe-Si-B alloy core can reduce this loss to 25–30%. Because a transformer is expected to work for 20–40 years, the energy saving during the service life time is about ~3×10^7 Wh. The cost increase of the core material is relatively smaller. Table 6.6 lists the properties of some soft ferromagnetic amorphous alloys.

6.3.4 *Nanocrystalline materials*

Development of nanocrystalline magnetic materials is generally based on iron alloys and has grains typically ranging from 10 to 15 nm. These materials normally have coercivities below 1 A/m and high relative permeabilities (~10^5) and relatively high saturation magnetization (10^6

A/m). The material system most widely investigated is Fe-Cu-Nb-Si-B (e.g., Fe73.5Si13.5B9Nb3Cu1). This alloy is produced by rapid solidification and then annealed above its crystallization temperature to produce the nanocrystalline structure. Cu, insoluble in Fe, promotes massive nucleation, while Nb retards grain growth. The nanocrystalline phase in these alloys is α-Fe_3Si, which occupies ~70–80 vol% and 10 – 15 nm in size.

Other magnetically soft nanostructured materials may include: (1) amorphous ribbons consisting of nanocrystalline α-Fe/Co in an amorphous matrix produced by partial recrystallization, (2) nanocomposites of Fe and Fe/Ni dispersed in a Mn-Zn ferrite, and (3) magnetic nanowires that were electrodeposited into the pores of membranes.

6.4 Hard Magnetic Materials

Hard magnetic materials are also called permanent magnetic materials. They have: (1) a high coercive force H_c (>10 kA/m) and (2) a high remanent magnetic induction B_r. The hysteresis loops are wide and the areas inside the loop are large.

When a hard magnetic material is placed in a magnetic field (strong enough to orient the domains), some of the energy is converted into potential energy and stored in the magnet. Therefore a permanent magnet in the magnetised state holds higher energy than in the demagnetised state.

Permanent magnets can be broadly classified into sintered or bonded magnets. The material can be in the form of alloys, intermetallics and ceramics. Permanent magnets are made of hard ferrites, Alnico alloys (Fe-Al-Ni-Co), rare-earth alloys (Nd-Fe-B and Sm-Co), Cunife alloys (Cu-Ni-Fe), Fe-Co alloys containing Nb or V, certain steels and Pt-Co alloys. Cobalt is a major element used, but neodymium (Nd), which is cheaper than Co, is rapidly becoming an important alloying element that provides very high energy product.

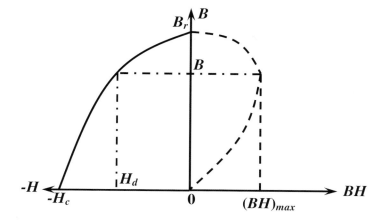

Fig. 6.24 A schematic of the energy product curve of hard magnetic material (after W.F. Smith, "Foundations of Materials Science and Engineering", McGraw-Hill, 1993).

1: Sm(Co,Cu)$_{7.4}$
2: SmCo$_5$
3: Bonded SmCo$_5$
4: Alnico 5
5: Mn-Al-C
6: Alnico 8
7: Cr-Co-Fe
8: Ferrite
9: Bonded ferrite

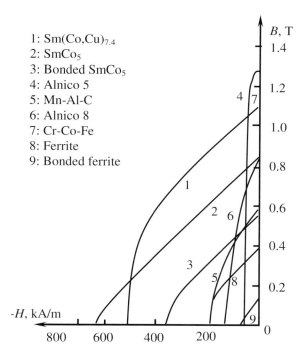

Fig. 6.25 Demagnetisation curves of various hard magnetic materials (after G.Y. Chin and J.H. Wernic, "Magnetic Materials", Wiley, 1981).

6.4.1 *Maximum energy product, (BH)$_{max}$*

Hard magnetic materials are difficult to demagnetise once magnetised. The demagnetising curves can be used to compare the strength of permanent magnets (Fig. 6.25).

The potential magnetic energy is measured by its maximum energy product, $(BH)_{max}$. The maximum energy product is the area occupied by the largest rectangle that can be fitted in the second quadrant of the hysteresis loop (Fig. 6.24). From Fig. 6.24, we can see that the magnetic energy of a hard magnetic material depends not only on the value of the B_r and H_c, but also on the shape of the demagnetisation curve. The more convex the curve, the greater the $(BH)_{max}$.

6.4.2 *Alnico alloys*

Alnico alloys consist of Fe, Al, Ni, Co and ~3% Cu. They possess a wide range of useful magnetic properties and are popular and most commonly used commercial hard magnetic materials. They have reasonably high energy product, $(BH)_{max}$ = 40–70 kJ/m^3, high remanent induction, B_r = 0.7–1.35 T, but a moderate coercivity, H_c = 40–160 kA/m. They have a very wide useful temperature range but are mechanically hard, impossible to forge, and difficult to machine.

Fig. 6.26 gives the compositions of some Alnico alloys. Alnico 1-4 contain a relatively high amount of Fe and are isotropic. Alnico 5-9 contain relatively high amounts of Co and Ni, and are anisotropic due to heat treatment in a magnetic field while the precipitates form.

Alnico alloys are two-phase alloys. Above 1250°C they have a single phase with *bcc* structure. During cooling to 850–750°C, the single phase decomposes into two phases, α and α'. Both of these have a *bcc* structure: α is the matrix phase, rich in Ni and Al but magnetically weak. α' is the precipitate phase, rich in Fe and Co. It is magnetically strong.

With a heat treatment carried out at ~800°C in a magnetic field, the α' phase forms fine, long rods aligned in the applied field direction. The high coercive force H_c is due to the difficulty of rotating these single domain α'-particles. The larger the length-to-width ratio of the rod, the more difficult the rod rotation becomes and the greater the H_c. Field treatment and Ti addition increase H_c even further.

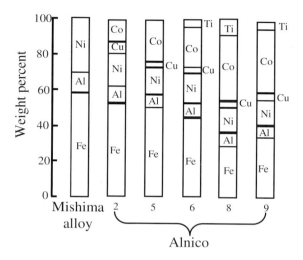

Fig. 6.26 Chemical compositions of the alnico alloys (after B.D. Cullity, "Introduction to Magnetic Materials", Addison-Wesley, 1972).

6.4.3 *Rare earth alloys*

Rare earth hard magnetic alloys have the best magnetic properties of the commercially available magnetic materials. The origin of magnetism in the rare earth transition elements is due to their unpaired 4f electrons. For instance, Nd has the electron structure of 1s2 2s2 2p6 3s2 3p6 3d10 4s2 4p6 4d10 5s2 5p6 4f4 6s2, and Sm has the electron structure of 1s2 2s2 2p6 3s2 3p6 3d10 4s2 4p6 4d10 5s2 5p6 4f6 6s2.

6.4.3.1 *Nd-Fe-B alloys*

Nd-Fe-B magnets were developed in 1983 and commercialized in 1985. The maximum energy product $(BH)_{max}$ can reach ~450 kJ/m^3, and the coercive force H_c = ~3200 kA/m. They have made significant contributions to the permanent magnets industry. These powerful magnets greatly reduce the size of electronic devices and instruments.

Two disadvantages of these neodymium alloys are low resistance to corrosion and a relatively low Curie temperature resulting in instability with a rapid decrease in flux with increasing temperature. A breakthrough for the Nd-Fe-B type magnetic materials with desired alloying elements and defined microstructures would be achieved for Curie temperatures of around 450°C. A higher Curie temperature would promote the acceptance of these magnets in the electro motor industry.

Nd-Fe-B magnets can be produced by two methods: conventional powder metallurgy followed by sintering or by rapid solidification melt-spinning processes. The major content of Nd-Fe-B alloys is iron with a typical composition of $Nd_2Fe_{14}B$. The microstructure of $Nd_2Fe_{14}B$ contains highly ferromagnetic matrix grains surrounded by a non-ferromagnetic Nd-rich thin intergranular phase. The high H_c and $(BH)_{max}$ of this material are resulted from the difficulty of nucleating reverse magnetic domains, which usually start at the grain boundaries. Fig. 6.27 demonstrates this process.

Recently, nanocrystalline-composite hard magnetic materials, such as $Nd_2Fe_{14}B$ + α-Fe and $Pr_2Fe_{14}B$ + α-Fe, are also being under development. α-Fe has a large magnetic moment so that the composite magnetic material could achieve a relatively high exchange coupling maximum energy products of $(BH)_{max} > 100$ kJ/m^3. These values are 40% larger than the values of conventional bonded magnets.

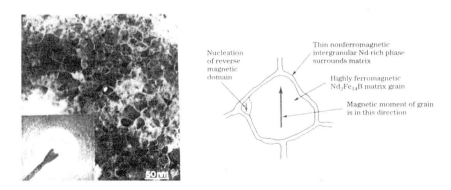

Fig. 6.27 Microstructure of a quenched Nd-Fe-B sample (after J.J. Croat and J.F. Herbst, MRS Bull., June 1988).

6.4.3.2 *Sm-Co alloys*

Sm alloys have very good permanent magnetic properties, but they are more expensive than Nd alloys. Sm alloys can be divided into two groups: single-phase $SmCo_5$ and precipitation-hardened $Sm(Co,Cu)_{7.5}$. $SmCo_5$ alloys are magnetically hardened by grain boundary/surface pinning mechanism. They are produced by powder metallurgy using alloy powder produced either by reducing the rare earth powder together with cobalt powder or by preparing the alloy and powdering it down. The powder is compacted and pressed to produce an anisotropic compact in a magnetic field. This is then sintered at $1155°C$ and followed by a tempering treatment at about $900°C$.

Precipitation-hardened $Sm(Co,Cu)_{7.5}$ can also be produced by powder metallurgy. Fine particles (~10 nm) are produced with low ageing temperatures that pin the domain walls. It is the most stable rare earth alloys and has a higher remanence and $(BH)_{max}$ than $SmCo_5$ alloys.

6.4.4 *Other materials*

Cunife (Cu-Ni-Fe) and Cr-Fe-Co are alloys have properties similar to those of anisotropic Alnico. Pt-Co alloys have a very high coercivity. Lodex, consisted of fine particles of iron cobalt, are now obsolete because of cost, lack of demand and production difficulties.

6.4.5 *Applications*

Applications of permanent magnetic materials include electric motors, especially those like automotive starting motors where reduction in weight and volume are important. Other applications include small devices needing a high efficiency, e.g. medical devices such as small motors in implantable pumps, electronic wristwatches, and DC motors.

6.5 Ferrites – Ceramic Magnetic Materials

Ferrites are ceramic magnetic materials. The magnetic moment is due to the two sets of the unpaired inner electron spin moments which are in opposite directions, but do not cancel each other. They are produced by ceramic powder processing: mixing a powder oxide (e.g. Fe_2O_3) with other oxides or carbonates, grinding, pressing and sintering a number of times. Texturing of the easy magnetising direction can be achieved by "wet press" in a magnetic field. Finishing machining may be needed to produce the final products. Their magnetic saturations are high and electrical conductivities are low. They are also mechanically hard and wear resistant.

6.5.1 *Soft ferrites*

Most soft ferrites have the composition of $MO \cdot Fe_2O_3$, where M is a divalent metal such as Fe, Mn, Ni, Zn, Co, Cu, Li or Al. Their crystal structure is the inverse spinel structure. Iron, nickel and cobalt ferrites all have this crystal structure. They are ferrimagnetic due to a net magnetic moment of their ionic structures. A mixture of ferrites has a higher saturation magnetisation than a single ferrite. The two examples of the mixtures of ferrites are the nickel-zinc-ferrite ($Ni_{1-x}Zn_xFe_{2-y}O_4$) and the manganese-zinc-ferrite ($Mn_{1-x}Zn_xFe_{2+y}O_4$).

These soft magnetic oxides are insulators and have high electrical resistivity. They are used in high frequency applications such as core materials in transformers where eddy-current losses would be very high if a conducting core is used.

Applications of soft ferrites also include those used in low signal levels such as memory cores, transformers and inductors, audio-visual recording heads, and convergence coils for TV receivers. Mn-Zn and Ni-Zn ferrites are used for recording heads because the operating frequencies (100 kHz to 2.5 GHz) are too high for metallic magnets.

6.5.2 *Hard ferrites*

A group of hard ferrites used for permanent magnets have general formula of $MO \cdot 6Fe_2O_3$ where M is either barium or strontium. It has a hexagonal crystal structure. Barium ferrite, $BaO \cdot 6Fe_2O_3$, was introduced by Phillips in 1952 and still widely used. Another example, $SrO \cdot 6Fe_2O_3$, has even better magnetic properties. The high magnetic strength is due to the high magnetocrystalline anisotropy.

Table 6.7 Magnetic properties of some hard magnetic materials.

Materials	Remanent induction (B_r, T)	Coercive force (H_c, kA/m)	Maximum energy product (($BH)_{max}$, kJ/m^3)
Alnico 1, 12Al, 21Ni, 5Co, 2Cu, bal Fe	0.72	37	11.0
Alnico 5, 8Al, 14Ni, 25Co, 3Cu, bal Fe	1.28	51	44.1
Alnico 8, 7Al, 15Ni, 24Co, 3Cu, bal Fe	0.72	150	40.0
Rare earth – Co, 35Sm, 65Co	0.90	675-1200	160
Rare earth – Co, 25.5Sm, 8Cu, 15Fe, 1.5Zr, 50Co	1.10	510-520	240
Fe-Cr-Co, 30Cr, 10Co, 1Si, 59Fe	1.17	46	34.0
$Mo \cdot Fe_2O_3$ (M = Bs, Sr) (hard ferrite)	0.38	235-240	28.0

Table 6.8 Applications of permanent magnetic materials.

Application	Recommended/Alternative material
Auto d.c. motors	Ferrite/bonded Nd-Fe-B
Auto cranking motors	Ferrite/bonded Nd-Fe-B
Aircraft magnetos	Sm-Co/cast Alnico 5
Alternators	Sm-Co/ferrite, Alnico
Magnetos for lawn equipment	
Small d.c. motors	Bonded ferrite/bonded Nd-Fe-B, sintered
Large d.c. motors	ferrite
	Sm-Co/Nd-Fe-B
Voice coil motor	Nd-Fe-B/Sm-Co
Acoustic transducer	Ferrite/Nd-Fe-B
Small-gap magnetic couplings	Ferrite/bonded Nd-Fe-B
Large-gap magnetic couplings	Nd-Fe-B/Sm-Co
Transport systems	Nd-Fe-B/Sm-Co
Separators	Ferrite/Nd-Fe-B
Magnetic focusing system	Nd-Fe-B/Sm-Co
Holding devices	Ferrite/Alnico
Ammeters and voltmeters	Alnico
Watt-hour meters	Alnico 5 & 6

These hard ferrites have a high coercive force. The magnetisation takes place by domain wall nucleation and motion. The grain size is very large. These materials have relatively low density and are low cost materials.

Applications include permanent magnets for loudspeakers and telephones, generators, motors. They are also used for door closers, latches, seals, holding devices and in many toys. Table 6.8 lists the applications of different types of hard magnetic materials.

6.5.3 *A summary - "hard" and "soft" materials*

There are similarities between mechanically hardness and magnetically hardness. What does the "hard" mean? Hard generally means "not easily yielding to physical pressure" (from Webster's Dictionary). Table 6.9 is a comparison of mechanical hardness and magnetic hardness.

Table 6.9 Mechanical and magnetic hardness (After Livingstone's Seminar at MIT).

	Mechanical Hardness	Magnetic Hardness
Physical pressure	Stress (σ)	Reverse field (H_r)
Yielding	Plastic strain	Magnetisation change
Moving entities	Dislocations	Domain walls
Hardness measure	Yield stress (σ_y)	Coercivity (H_c)

6.6 Dilute Magnetic Semiconductor (DMS)

Currently, extensive researches have been carried out to create diluted magnetic semiconductor materials with new or enhanced functionalities. Typical magnetic semiconductors are the dilute systems where magnetic transition metal ions partially substitute main group cations in traditional zinc blende or wurtzite semiconductors. In such material systems, electrical manipulation of magnetism or magnetic manipulation of electrical signals is highly possible.

Substantial progress has been made with $Ga_{1-x}Mn_xAs$ and $In_{1-x}Mn_xAs$ in understanding spintronic concepts. However, their ferromagnetic critical temperatures are quite low, $T_c < 170$ K for $Ga_{1-x}Mn_xAs$ and even lower for $In_{1-x}Mn_xAs$. GaN and ZnO are the most promising candidates for ferromagnetic DMS. They are expected to have high Curie temperatures, approaching or exceeding the room temperature. It is also expected that magnetic ZnO will have an enormous potential to be a highly multifunctional material with its existing electronic, photonic, piezoelectric properties, and also its richest family of low-dimensional nanostructures. To date, transition metal (Cr, Co, Fe, Mn, Ni, Sc, Ti, or V) has been doped into ZnO polycrystalline powders, single crystals and thin films to achieve a T_c ranging from 30–550 K.

The push for semiconductor spintronics is motivated by the materials compatibility with traditional semiconductor electronics and by the desire to produce true three-terminal spintronic devices with potential applications such as nonvolatile programmable logic, spin based optoelectronics and quantum computation. For example, spin injection from (Ga,Mn)As to an (In,Ga)As quantum well has been demonstrated as a means for generating polarized light. A persisting bottleneck for semiconductor based spintronics has been spin injection from the ferromagnet into the semiconductor, which is a critical step in the implementation of any spin logic devices. Traditionally, the spintronics community has tackled this problem from two fronts: materials synthesis and interface engineering. The work on materials has focused on synthesis of magnetic semiconductors and other novel materials with high spin polarization.

6.7 Case Study – Materials in Magnetic Recording

Magnetic recording is the storage of signals on a magnetic medium for subsequent use. In cost (not in tonnage), magnetic recording is the most important application for magnetic materials. Over the last few decades, a huge industry has been built up to produce magnetic recording materials. This industry is closely linked with telecommunication, computer and audio-video manufacturers. The main applications for magnetic materials in recording include recording heads, magnetic tapes, hard and floppy discs, and computer cores.

6.7.1 *Magnetic recording principle*

Fig. 6.28 shows a recording system. A gapped core head is commonly used to write tapes and discs. The current in the coil produces a magnetic field in the core and in the gap. The core is made from a soft magnetic material and is much easier to magnetise than the air in the gap. A small amount of the field "leaks out" from the gap and creates a fringe field. This field is used to write a tape or disc.

According to the required properties, the applications can be classified into two groups, i.e., recording heads and recording medium.

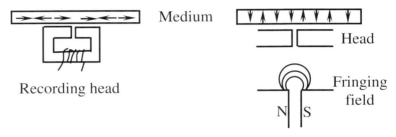

Fig. 6.28 (a) A longitudinal recording medium; (b) a perpendicular recording medium (source: L.A.A. Warnes, "Electronic Materials", Macmillan Education Ltd. 1990).

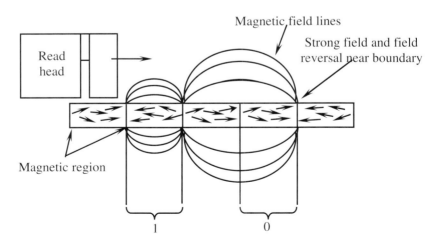

Fig. 6.29 A cross section of the magnetic surface in action.

6.7.2 *Recording heads*

The read-write recording heads are highly permeable magnetic rings with a very small gap of ~1 μm. The distances between the recording heads and medium are also very small, less than 25 nm have been achieved to date with disc flies while contact is normal with tapes. Small gaps, small fringing fields and small distances between the head and tape result in accurate recording and reading. The properties required for recording heads include:

1. High magnetisation B_s,
2. High initial magnetic permeabilities μ_i,
3. Low coercive force H_c, and
4. Good wear resistance.

Soft magnetic materials such as Permalloy (80Ni-20Fe), 45Ni-55Fe, Alfesil (Al-Fe-Si) and soft ferrites (MnZn ferrites) are often used for magnetic heads. The most commonly used material in disk drives is electroplated 45Ni-50Fe, with a saturation magnetization of ~1.6 T. For high density recording, it is also important to consider eddy current losses, so metallic cores must be laminated, or soft ferrites should be used.

Spin valve, as shown schematically in Fig. 6.30, is the technology that has evolved for reading from media. It is based on the giant-magnetoresistive (GMR) effect discovered in 1988. A typical spin valve consists of a sequence of thin films of MnFe/Co/Ru/Co/Cu/Co/$Ni_{80}Fe_{20}$, where MnFe/Co/Ru/Co films form a synthetic antiferromagnetic film.

Tapes and floppy discs require the recording head to contact recording medium to ensure satisfactory read-write operations. Therefore, the mechanical properties (in particular, wear resistance) are important for the head materials. For computer hard discs, the wear problem is greatly reduced by allowing the head to fly on an air cushion above the spinning disc surface. The wear is then restricted to the start and stop operation when the head lands. "Crash" may take place when the flight control is lost and a very bad head-disc collision occurs.

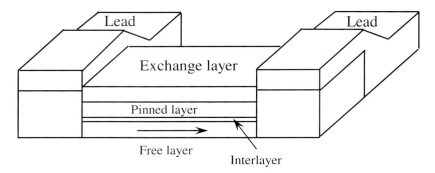

Fig. 6.30 A schematic of spin valve read head.

6.7.3 *Recording medium*

Recording media includes magnetic tapes, floppy discs and hard discs. They require hard magnetic properties (reasonably high H_c) to keep the written signals stable, so that the information stored will not be accidentally erased if it is unexpectedly exposed to a small magnetic field. But H_c should not be too high because it must be allowed to write onto and to delete off. The medium also needs a relatively high saturation magnetisation and remanence so that the leakage field from the surface of the material is large enough to be picked up by the reading head. The H_c largely depends on the recording technique: how strong the recording signals and how close the tape and head. Materials with a higher H_c can be used in a stronger recording system. Coercive forces of the recording media are typically in the range of 20–150 kA/m while saturation magnetisations are in the range of 0.3–2.0 MA/m.

Recording media materials include Cr_2O_3, γ-Fe_2O_3, Co-modified γ-Fe_2O_3, ferromagnetic powders, metallic films etc. Particulate-type media are generally composed of single-domain particles suspended in a polymer matrix. Particulates are suitable materials when soft substrates such as tape or polyester floppy disks are used. γ-Fe_2O_3 has long been used in recording tapes and still is the most popular recording material. Cr_2O_3 has higher H_c and B_s but lower Currie temperature (T_c) than γ-Fe_2O_3. Higher H_c and B_s give better recording performance including higher recording densities and better signal-to-noise ratio. However, the low Curie temperature may result in poor thermal stability and consequent loss of recording information. Co-modified γ-Fe_2O_3 consists of a 3 nm thick cobalt layer on the surface of γ-Fe_2O_3. This increases its coercivity. Ferromagnetic powders have very high H_c and B_s. They are used in high performance applications. Metallic films are widely used in hard discs. They can be easily deposited on the surface of the rigid discs. Tables 6.10 and 6.11 list the properties of these recording materials.

Table 6.10 Properties of magnetic medium materials.

Materials	H_c (kA/m)	T_c (°C)	B_s (kA/m)
γ-Fe_2O_3	25	600	370
Cr_2O_3	60	128	500
Co modified γ-Fe_2O_3	50		370
Ferromagnetic powders	120		1700
Metallic films	80		1000

Table 6.11 Properties of magnetic thin films.

Material	H_c (Oe)	M_s (G)	Method of preparation
Co	1,000	-	Sputtering
CoP	1,000	-	Plating
CoCrTa	1,400	-	Sputtering
CoNiCr	2,000	-	Sputtering
CoNiPt	900	800	Sputtering

Note that the H_c of the medium materials is smaller than those of hard magnetic alloys.

6.7.4 *Tape and disc processing*

The following steps are used to produce tapes and discs:

1. Powder making: a chemical presses to produce fine oxide particles;
2. Slurry making: mixing with resin, coat the slurry on a film; and
3. Magnetic field treatment to orientate the particles, then drying and surface finishing.

The typical chemical process for the powder preparation is:

$$FeSO_4 + NH_4OH \rightarrow Fe(OH)_2 \downarrow, \qquad (6.13)$$

$$Fe(OH)_2 \rightarrow \alpha\text{-}Fe_2O_3 \cdot H_2O \quad \text{(reaction in oxygen)}, \qquad (6.14)$$

$$\alpha\text{-}Fe_2O_3 \cdot H_2O \rightarrow \alpha\text{-}Fe_2O_3 \quad \text{(red oxide)}, \qquad (6.15)$$

$$\alpha\text{-}Fe_2O_3 \rightarrow Fe_3O_4 \quad \text{(reaction in hydrogen)}, \qquad (6.16)$$

$$Fe_3O_4 \rightarrow \gamma\text{-}Fe_2O_3 \quad \text{(reaction in oxygen)}. \qquad (6.17)$$

6.7.5 *Storage density*

Storage density is an important property for recording media used in various information technology (IT) devices, especially for computer disc systems (such as floppy disks and hard disk drives), since it means more data can be stored in a given space. The storage density depends on the magnetic properties of the material and the writing technology.

On magnetic discs, linear bit density is measured by bits per inch. The number of bits per inch times the number of tracks per inch gives the storage density in bits per square inch (psi). Fig. 6.31 shows the improvement of the storage density of hard disk drives (HDDs).

Storage density is related to the type of recording system. Longitudinal recording systems have a medium recording layer of a high anisotropy in the lateral direction to the film surface. The magnetic film has such a microstructure in which a phase of small magnetic grains is isolated by a nonmagnetic phase. Magnetic recording media used in modern disk drives are all thin films of cobalt alloys prepared by sputtering deposition. The structure is typically composed of Cr alloy underlayers, CoPt-based magnetic alloy layers (CoPtCrX, where X is Ta or B or both) and amorphous carbon overcoat film for protection and offering a mechanically rigid surface on which the slider can fly at less than 25 nm. Aluminium plated with electroless Ni-P, plastic or glass is the material of substrate.

Longitudinal recording systems are widely used in commercial hard disk devices. However, the storage density of is generally limited to 200 Gbpsi, therefore cannot be used for ultra-high density recording. Increased areal density requires decreased mean grain size but, equally important, tighter distributions of grain size. A problem with this is that the ratio of magnetic to nonmagnetic volume in the film is decreasing faster than the average grain size, which results in the number of grains per bit decreasing faster than the average grain size.

To overcome these problems, perpendicular recording technology is being actively developed and is now starting to supplant longitudinal recording. In perpendicular recording the magnetization in the recording media is held perpendicular to the surface and higher information density stabilizes the bit against demagnetization. Thus it is no longer necessary to prepare thinner recording media for improvements in recording density in longitudinal recording.

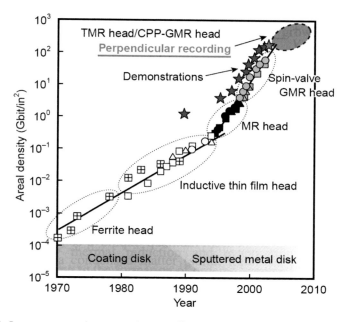

Fig. 6.31 Improvements in magnetic recording densities for hard disk drives (after I. Kaitsu, R. Inamura, J. Toda and T. Morita, Ultra High Density Perpendicular Magnetic Recording Technologies, Fujitsu Scientific & Technical Journal, 42, 2006, pp.122–130).

Perpendicular recording system is composed of a film perpendicularly magnetized and a single-pole magnetic head. It means that the single-domain particles are aligned vertical to the tape plane. It has advantages over the longitudinal recording including higher storage density and higher and more sharply-defined leakage field. However, there are problems that the mechanical properties and stabilities of the medium are relatively poor, and it requires a very small head-to-medium distance. Thus there are still considerable technical and practical challenges in migrating to perpendicular recording.

Media under consideration for perpendicular recording include: (1) Co-Cr and Co-Cr-Pt based granular alloys, (2) Co/Pd and Fe/Pt based multilayered systems, and (3) oxides such as CoCrPtO and $CoCrPt-SiO_2$.

Hard disk drives (HDD) using the perpendicular recording technology was commercialized in 2005. Hitachi Global Storage Technologies has then reported an areal density of 230 Gb/in^2 using perpendicular magnetic recording and a tunnelling current-perpendicular-to-the-plane (CPP) GMR read head. It is believed that perpendicular recording will become the mainstream of HDD in future.

Summary:

Ferromagnetism comes from unpaired electron spin and orbital rotation. In ferromagnetic materials such as iron, there are regions called magnetic domains in which atomic magnetic dipole moments are aligned parallel to each other. The domain structure is determined by minimisation of five energies: exchange, magnetostatic, magnetocrystalline anisotropy, domain wall, and magnetostrictive energies. When the ferromagnetic domains are randomly orientated, the sample is in a demagnetised state. When a magnetic field is applied, the domains are aligned and the material becomes magnetised and remains magnetised to some extent when the field is removed.

Hysteresis loops are used to describe the magnetisation behaviour of magnetic materials. Soft magnetic materials are those which are easy to be magnetised and demagnetised. Important properties are high permeability, high saturation induction, and low coercive force. Hard magnetic materials are those difficult to be magnetised. They remain magnetised to a great extent after the field is removed. Important properties are high coercive force and high saturation induction. The power of a hard magnetic material is measured by its maximum energy product.

Ferrites are ceramic compounds and they are ferrimagnetic due to a net magnetic moment produced by their ionic structure. Due to the high electrical resistivity, they have found special applications in electronic devices such as those involving high frequency.

Important Concepts:

Ferromagnetic Material: The material which is capable of being highly magnetised. Elemental Fe, Co and Ni are ferromagnetic materials.

Magnetisation (M): A measure of the increase in magnetic flux due to the insertion of a given material into a magnetic field.

Magnetic Induction, B: The sum of the applied field H and the magnetisation M. In SI units, $B = \mu_0 \times (H + M)$.

Magnetic Permeability, μ: The ratio of the magnetic induction B to the applied magnetic field H for a material; $\mu = B/H$.

Relative Permeability, μ_r: The ratio of the permeability of a material to the permeability of the vacuum; $\mu_r = \mu/\mu_0$.

Magnetic Susceptibility, χ_m: The ratio of M to H; $\chi_m = M/H$.

Diamagnetism: A weak, negative, repulsive reaction of a material to an applied magnetic field.

Paramagnetism: A weak, positive, attractive reaction of a material to an applied field.

Ferromagnetism: The creation of a very large magnetisation in a material when subjected to an applied field. After the applied field is removed, the ferromagnetic material may retain much of the magnetisation.

Antiferromagnetism: A type of magnetism in which magnetic dipoles of atoms are aligned in opposite directions by an applied field so that there is no net magnetisation.

Ferrimagnetism: A type of magnetism in which the magnetic dipole moments of different ions are aligned by a field in an antiparallel manner so that there is a net magnetic moment.

Curie Temperature: The temperature at which a ferromagnetic material completely loses its ferromagnetism and becomes paramagnetic.

Magnetic Domain: A region in a ferro- or ferrimagnetic material in which all magnetic dipole moments are aligned due to the magnetic exchange energy.

Exchange Energy: The energy associated with the coupling of individual magnetic dipoles into a single magnetic domain. The exchange energy can be positive or negative.

Magnetostatic Energy: The magnetic potential energy due to the external field surrounding a sample of a ferromagnetic material.

Magnetocrystalline Anisotropic Energy: The energy required during the magnetisation to rotate the magnetic domains because of crystalline anisotropy. For instance, the difference in magnetising energy between the hard [111] and easy [100] directions in Fe is $\sim 1.4 \times 10^4$ J/m^3.

Domain Wall Energy: The potential energy associated with the disorder of dipole moments in the wall volume between magnetic domains.

Magnetostriction: The change in length of a ferromagnetic material in the direction of magnetisation due to an applied field.

Hysteresis Loop: The B vs. H or M vs. H graph traced out by the magnetisation and demagnetisation of a ferro- or ferrimagnetic material.

Saturation Induction, B_s: The maximum value of induction for a ferro- or ferrimagnetic material.

Saturation Magnetisation, M_s: The maximum value of magnetisation for a ferro- or ferrimagnetic material.

Remanent Induction, B_r: The value of B in a ferromagnetic material when H is decreased to zero.

Coercive Force, H_c: The applied magnetic field required to decrease the magnetic induction to zero.

Hysteresis Energy Loss: The work or energy lost in tracing out a *B-H* hysteresis loop.

Eddy-Current Energy Loss: Energy losses in magnetic materials while using alternating fields. The losses are due to induced current in the material.

Energy Product, $(BH)_{max}$: The maximum value of *B* times *H* in the demagnetisation curve of a hard magnetic material (J/m^3 in SI units).

Soft Magnetic Material: A magnetic material with a high permeability and low coercive force.

Hard Magnetic Material: A magnetic material with a high coercive force and high saturation induction.

Amorphous Alloy (Metallic Glass): A metallic alloy whose atomic structure has no long-range order.

Questions:

1. What property represents the (magnetic) hardness of a magnet? How is the potential magnetic energy of a hard magnetic material measured?
2. What property represents the softness of a magnetic material? How can the energy loss be expressed graphically during a magnetising and demagnetising cycle? What are the magnetic loss and electrical loss for a core material in a transformer? Express these two losses with equations.
3. What is the fundamental difference between hard and soft magnetic materials?
4. List the most commonly used soft and hard magnetic materials. Discuss their characteristics in magnetic properties and typical applications.

5. What is mechanism behind magnetostriction?
6. A magnetic "bit" (binary digit) can be represented by the direction of magnetization in a certain region of material.

    (a) What is the smallest size (physically) could one isolated bit be? Choose from: subatomic, atomic, 1000 atoms, μm, and mm etc.
    (b) What factors might limit the size of a bit in practice?
    (c) Suppose we have a disc coated with single-domain bit of ~0.1 μm in size. Estimate the bit-capacity of a disc with 0.01 m^2 of useful area.
    (d) The disc in (c) spins at a linear speed of ~36 m/s. At what rate could data be transferred to or from the disc?
    (e) What magnetic and other properties should be considered for the recording disc application?

7. What are the potential energies that are related to magnetic domains? Briefly explain how these energies influence the size and shape of the domains.
8. Understand the effect of temperature on ferromagnetism and Curie temperature. Plot out the maximum magnetisation of Fe, Co and Ni versus temperature
9. A magnetic field of 1,000 A/m is applied to an Fe-Si alloy which has a relative permeability of 30,000. Calculate (a) the magnetization, (b) the permeability, and (c) the magnetic induction. The magnetic constant in vacuum $\mu_0 = 4\pi \times 10^{-7}$ H/m.
10. A supermalloy is surrounded by a 20 m long, 30 turn coil of a conductor through which a current of 5 A is passed. Calculate (a) the magnetic field H, (b) the magnetization M, and (c) the induction B. The relative permeability of this alloy is 800,000.
11. A Ni-Cu alloy is prepared which has a lattice constant of 0.354 nm. The maximum magnetization of the alloy is measured as 1.46×10^6 A/m. Assuming no interactions between Ni and Cu atoms, estimate the wt.%Cu in the alloy. Both Ni and Cu have a *fcc* structure with four atoms in a unit cell.

Chapter 7

Dielectrics

7.1 Introduction

Dielectric materials are insulators. They have a large energy gap between the valance and conduction bands, the electrons in the valance bands cannot jump to the conduction band. Thus the resistivities of these materials are high. Most ceramics dielectric materials have a mixture of ionic and covalent bonding. Although these materials do not conduct electric current when an electric field is applied, they are not inert to the electric field. The field may cause a slight shift in the balance of charge within the material to form an electrical dipole, therefore the material is called "dielectric" material.

The two important applications of dielectric materials are insulators for preventing electricity transfer and capacitors for the storage of electrical charges. The most important properties for dielectrics are:

1. Relative permittivity, ε_r,
2. Tangent of loss angle, tan δ, and
3. Dielectric strength.

Other important properties of dielectrics include:

1. Ferroelectricity
2. Piezoelectricity
3. Electrostriction
4. Pyroelectricity

7.2 Fundamentals

7.2.1 *Permittivity or dielectric constant ε*

The capacitance of a parallel plate capacitor (Fig. 7.1), C_0, is proportional to the plate area A and inversely proportional to the distance between the two plates, d:

$$C_0 \propto A, \text{ and } C_0 \propto 1/d.$$

In vacuum, C_0 can be expressed as:

$$C_0 = \varepsilon_0 \cdot A/d, \tag{7.1}$$

where ε_0 is the *permittivity* of vacuum, and is a constant.

$$\varepsilon_0 = 8.85 \times 10^{-12} \text{ F/m.}$$

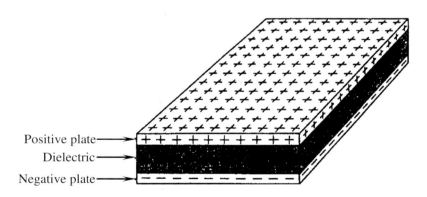

Positive plate

Dielectric

Negative plate

Fig. 7.1 A parallel plate capacitor with dielectric between the plates.

If a dielectric material is inserted into the plate, Eq. (7.1) should be rewritten as:

$$C = \varepsilon \cdot A/d, \tag{7.2}$$

where ε is defined as the permittivity of the material, an indication of how much more charges can be stored in the capacitor due to inserting of this dielectric material.

Table 7.1 Relative permittivity of a variety of materials at room temperature.

Materials	Relative permittivity, ε_r
NaCl	5.9
LiF	9.0
KBr	4.9
Mica	2.5-7.3
MgO	9.6
BaO	34
BeO	6.5
Diamond	5.7
Al_2O_3	8.6-10.6
Mullite	6.6
TiO_2	15-170
Cordierite	4.5-5.4
Porcelain	6.0-8.0
Forsterite (Mg_2SiO_4)	6.2
Fused SiO_2	3.8
Steatite	5.5-7.5
High-lead glass	19.0
Soda-lime-silica glass	6.9
Zircon	8.8
$BaTiO_3$	1600
$BaTiO_3$ + 10% $CaZrO_3$ + 1% $MgZrO_3$	5000
$BaTiO_3$ + 10% $CaZrO_3$ + 10% $SrTiO_3$	9500
Paraffin	2.0-2.5
Beeswax	2.7-3.0
Rubber, polystyrene, polyacrylates, polyethylene	2.0-3.5
Phenolic	7.5

(Source: D. Richerson, "Modern Ceramic Engineering", Marcel Dekker Inc. 1992, p256).

Relative permittivity ε_r:

Relative permittivity, ε_r, is the ratio of the permittivity of the dielectric material to the permittivity of vacuum.

$$\varepsilon_r = \varepsilon/\varepsilon_0, \tag{7.3}$$

$$C = \varepsilon_r \cdot \varepsilon_0 \cdot A/d = \varepsilon_r \cdot C_0. \qquad (7.4)$$

This means that the inserted dielectric material has increased the capacity by a factor of ε_r. Table 7.1 lists the relative permittivity for a variety of materials at room temperature.

For a capacitor containing n parallel conductor plates,

$$C = \varepsilon_r \cdot \varepsilon_0 \cdot (n\text{-}1) \cdot A/d. \qquad (7.5)$$

From Eq. (7.5) we see that in order to build a capacitor with high capacity, many plates are needed with a large surface area. A small separation between the plates, and an dielectric material with high permittivity and high dielectric strength are also required.

7.2.2 Tangent of loss angle

In a capacitor, the voltage, V, and electrical charge, Q, have the following relation:

$$V = Q/C = 1/C \int I dt. \qquad (7.6)$$

Therefore, the current through the capacitor is:

$$I = C \times (dV/dt). \qquad (7.7)$$

An AC voltage can be expressed as:

$$V = V_0 \times sin(\omega \cdot t), \qquad (7.8)$$

$$I = C \cdot V_0 \cdot \omega \times cos(\omega \cdot t). \qquad (7.9)$$

This means that the current leads the voltage by $90°$.

For a real dielectric, the current leads the voltage by $90° - \delta$, where the angle δ is a measure of the dielectric power loss.

$$\text{Power loss} \propto f \cdot V_0^2 \cdot \varepsilon \times \tan \delta, \qquad (7.10)$$

where f is the frequency and $\tan \delta$ is called the tangent of loss angle.
Tangent of loss angle, *tan δ*, can also be expressed as:

$$\tan \delta = F_L / \varepsilon_r, \qquad (7.11)$$

where ε_r is the relative permittivity and F_L is defined as the relative loss factor.

Table 7.2 Loss tangents of ceramics and glasses.

Materials	Loss Tangent (tan δ)
Al_2O_3	0.001
BaO	0.001
LiF	0.0002
KBr	0.0002
KCl	0.0001
MgO	0.0003
NaCl	0.0002
TiO_2	0.0002 ($\perp$ *c*-axis); 0.0016 ($\parallel$ *c*-axis)
Diamond	0.0002
Mg_2SiO_4	0.0003
Fused silica glass	0.0001
Vycor glass	0.0008
High lead glass	0.0057
Soda-lime-silica glass	0.01

(Source: David W. Richerson, Modern Ceramic Engineering: Properties, Processing, and Use in Design, CRC Press, 2006).

Table 7.2 lists tangent of loss angle for some ceramics and glasses at 25°C and 10^6 Hz. Dielectric loss results from several mechanisms including ion migration, ion vibration, ion deformation and electronic polarisation. The most important mechanism is ion migration, which is strongly affected by the temperature and frequency.

7.2.3 *Dielectric strength*

Dielectric strength is the maximum voltage gradient which a dielectric material will withstand before failure occurs. It is also called breakdown strength, and is the most important property for dielectric materials used as insulators.

In a plate capacitor, dielectric strength can be expressed as:

$$\xi_{max} = (V/d)_{max}, \tag{7.12}$$

where V is the voltage and d is the distance between the two plates.

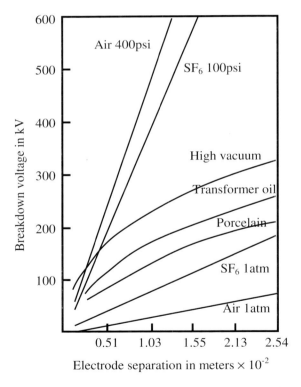

Fig. 7.2 DC breakdown strength of various dielectric in uniform fields (after Rose, Shepard and Wulff, "Electronic Properties", John Wiley and Sons, 1966, p254).

Table 7.3 Dielectric constant of various materials.

Material	Dielectric Constant (Relative to air)
Al_2O_3	8.6-10.6
BaO	34
BeO	6.5-6.7
$BaTiO_3$	1600
Gallium arsenide (GaAs)	13.1
Germanium	16
LiF	9.0
KBr	4.9
MgO	9.6
Mica	2.5-8.0
Mullite	6.6
NaCl	5.9
Silicon	11.7-12.9
$SrTiO_3$	233
TiO_2	15-170
Fused quartz	3.8
Fused silica	3.8
Glass	4-10
Glass (Corning 7059)	5.8
High lead glass	19.0
Soda lime silica glass	6.9
Paper	3.0
Nylon	3.2-5.0
Phenolic	4.0-15.0
Polyethylene	2.3
Polyamide	2.5-2.6
Polypropylene	2.2
Polystyrene	2.5-2.6
Polyvinychloride (PVC)	3.0
Rubber	2.0-4.0
Teflon	2.0-2.1
Distilled water	76.7-78.2
Wood	1.2-2.1

There are three different mechanisms of dielectric breakdown: intrinsic, thermal and discharge. Intrinsic breakdown begins with the appearance of a number of electrons in the conduction band. These electrons are accelerated rapidly by the high field in the dielectric and obtain high kinetic energy. As a large number of electrons initiates this process, it multiplies itself. The current increases rapidly and finally results in a breakdown.

Thermal breakdown occurs when dielectric losses cause heating which lowers the breakdown strength. Each dielectric has a temperature limit over which thermal breakdown may take place. Discharge breakdown takes place when the gas in the dielectric becomes ionised by the field. The gaseous ions are accelerated by the field and impact the side of the cavity causing damage and more ionisation. Dielectric breakdown may also result in local melting, burning or vaporising.

Impurity atoms can donate electrons to the conduction band. Voids and interconnected pores can provide direct breakdown channels as a result of electrical gas discharge. They reduce the breakdown strength, and must be avoided at all costs.

Fig. 7.2 shows the breakdown voltage versus dielectric thickness for various materials. Table 7.3 lists the dielectric strength for some ceramics, glasses and organic materials.

7.2.4 *Dipole and polarisation*

In both dielectric and magnetic materials, the application of a field causes the formation and movement of dipoles. This is called polarisation. For magnetic materials, the applied field is a magnetic field and the dipoles are magnetic dipoles. For dielectric materials, the applied field is an electric field and the dipoles are electrical dipoles.

7.2.4.1 *Dipole*

When an electric field is applied to a material, dipoles are induced within the atomic or molecular structure and become aligned with the direction of the applied field. In addition, any permanent dipoles already present in the material are aligned with the field. The material is then said to be polarised. The polarisation P (C/m^2) can be expressed as:

$$P = Z \cdot e \cdot d \qquad (7.13)$$

where Z is the number of the charge centres displaced per cubic meter, e is the electronic charge, and d is the displacement between the positive and negative ends of the dipole.

7.2.4.2 *Polarisation*

Electrical polarisation in dielectric materials has the following distinguishable mechanisms:

1. Electronic polarisation,
2. Ionic polarisation,
3. Molecular polarisation, and
4. Space charge polarisation.

7.2.4.2.1 Electronic polarisation

When an electric field is applied to an atom, the electron structure is distorted, with the electrons concentrating on the side of the nucleus near the positive end of the field. The atom acts as a temporarily induced dipole. This effect is small and occurs in all materials.

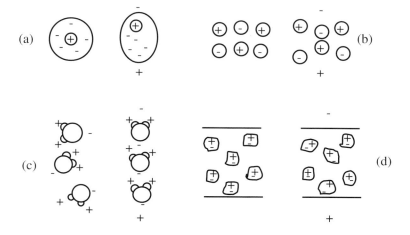

Fig. 7.3 Polarisation mechanisms in materials: (a) electronic polarisation, (b) ionic polarisation, (c) molecular polarisation and (d) space charges (after D.R. Askeland, "The Science and Engineering of Materials", Chapman and Hall, 1990, p683).

7.2.4.2.2 Ionic polarisation

When an ionically bonded material is placed in an electric field, the bonds between the ions are elastically deformed. Consequently, the charge is slightly redistributed within the material. Depending on the direction of the field, cations and anions move either closer together or further apart. Fig. 7.3(b) shows a schematic drawing of the ionic polarisation. The temporarily induced dipoles may also change the overall dimensions of the material.

7.2.4.2.3 Molecular polarisation

Some materials contain natural dipoles. When a field is applied, the dipoles rotate to line up with the applied field, as shown in Fig. 7.3(c). This is also a type of temporary polarisation.

In some materials, polarisation occurs in the same manner but when the field is removed, the dipoles remain in alignment, causing permanent polarisation. For example, there is displacement of O^{-2} and Ti^{+4} ions in $BaTiO_3$. The unit cell is a permanent dipole and produces excellent polarisation. When an ac voltage is applied to $BaTiO_3$, the Ti ions move back and forth between the two allowable positions. In this type of material, polarisation is highly anisotropic. Thus single crystals should be used, and the crystals must be properly aligned in the applied field to obtain the maximum polarisation.

7.2.4.2.4 Space charge polarisation

This type of polarisation is caused by the accumulation of charges at phase interfaces in the multiphase dielectrics. When one of the phases has a much higher resistivity than the other, the charge moves on the surface when the material is placed in an electric field. This polarisation is often found in ferrites or semiconductors at elevated temperatures.

7.2.5 *Frequency, temperature dependence of ε_r and energy loss*

7.2.5.1 *Relaxation frequency*

When a parallel plate capacitor works in an ac field, the total polarisation P and the relative permittivity ε_r depend on how easily the dipoles can reverse alignment with the field change. Some polarisation mechanisms do not permit rapid reversal of the dipole alignment. The time required to reach the equilibrium orientation is called the relaxation time, and its reciprocal is called relaxation frequency.

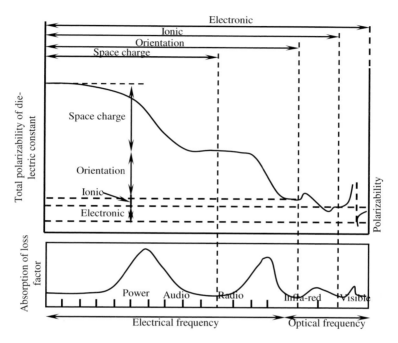

Fig. 7.4 Variation of the total polarisation and dielectric absorption as a function of frequency (after E.J. Murphy and S.D. Morgan, Bell System Tech. Il., 16, 1937, p493).

The relaxation frequencies of the four polarisation processes, described above, are different, so they can be separated experimentally. Fig. 7.4 displays the plots of total polarisation and absorption versus the frequency, showing that the different polarisation mechanisms have different frequency behaviours. Each contribution to the polarisation decays as its characteristic resonant frequency is exceeded.

7.2.5.2 *Temperature effect*

The effect of temperature on the dielectric constant of ionic materials is generally not high at low temperatures if there are no structural changes. It increases with temperature. At elevated temperatures, ion mobility is much higher than at lower temperatures. However, some materials may show large and sudden changes in ε_r when temperature increases.

Fig. 7.5 shows the effects of frequency and temperature on the permittivity of a silica glass. It can be seen that ε_r generally increases with increasing temperature under a fixed frequency, and decreases with increasing frequency at a fixed temperature. However, ε_r varies very little when the temperature is relatively low (< 100 °C).

If there are changes in crystal structure, temperature will have a strong effect on the polarisation characteristics. As shown in Fig. 7.6, for example $BaTiO_3$ has a cubic structure above 120°C. The permittivity, ε_r, is high around this temperature. The thermal vibration is high enough to produce the random orientation of the Ti ions, resulting in the same ε_r along a and c-axis. Below 120°C, the structure changes to tetragonal with the Ti ion in an off-centre position. This results in a large permanent dipole in the a-axis.

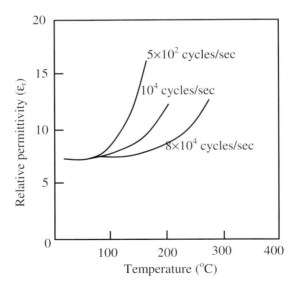

Fig. 7.5 Effect of frequency and temperature on the permittivity of a soda-lime-silica glass (after Rose, Shepard and Wulff, "Electronic Properties", John Wiley and Sons, 1966, p254).

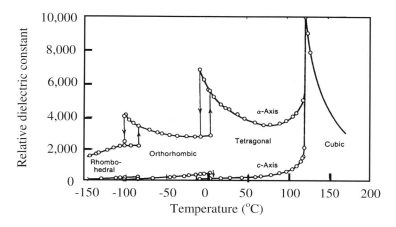

Fig. 7.6 Relative permittivity of $BaTiO_3$ as a function of temperature and crystallographic form (after W.J. Merz, Phys. Rev. 76, 1949, p1221).

7.2.5.3 *Energy loss in dielectrics*

The energy loss is due to the DC conductivity and dipole relaxation. It is proportional to the AC frequency, *f*, applied voltage, V_0, and the loss factor, *tan δ*. The energy loss is in the form of heat.

For a piece of dielectric material inserted in a capacitor, the power losses are due to the DC conductivity and dipole relaxation. The following equation is used to calculate the power loss:

$$P_L = 5.556 \times 10^{-11} \times \varepsilon_r \times (tan\ \delta) \times E^2 \cdot f \cdot V, \qquad (7.14)$$

where *E* is the electric field, *f* is the frequency, and *V* is the volume of the dielectric material. The losses can be minimised even with a large dielectric constant if the loss angle is small enough.

Because ε_r is related to *f*, the energy loss is also related to *f*. There are certain frequencies corresponding to the energy loss peaks for different types of polarisation mechanisms, as shown in Fig. 7.4. Dielectrics can be divided into low and high loss materials according to their applications.

7.3 Dielectric Materials

7.3.1 *Classification of dielectric materials*

Dielectrics can be generally classified according to their properties:

1. $\varepsilon_r < 12$ - used mainly for insulating materials,
2. $\varepsilon_r > 12$ - often used for capacitor materials (electricity storage), and
3. Ferroelectrics, piezoelectrics and pyroelectrics.

From the nature of the material, dielectrics can be classified into:

1. Fibrous material: papers, press-board, yarns, cloths, tapes, wood etc,
2. Filling and bonding materials: resins, paraffin wax, polymer materials etc,
3. Rubber and rubber based materials,
4. Mica and mica based insulating materials,
5. Glass,
6. Ceramic insulators, and
7. Electronic ceramics.

7.3.2 *General applications*

Application of dielectric materials depends on their properties and natures. The dielectric materials with $\varepsilon_r < 12$ are not usually used for energy storage, they are used for insulators. High breakdown voltage is the essential requirement for insulation. Application in capacitors requires high ε_r, low tan δ and high breakdown voltage.

The operating temperatures of the dielectrics are often limited by the nature of the materials:

1. Cotton, silk, paper and many polymers are used mainly at temperatures up to ~90°C.
2. Inorganic fillers like mica or asbestos bonded organic materials are used up to 130°C; inorganic bonded silicone is used up to 180°C.

3. High temperature applications such as in high temperature furnaces require insulation consisting entirely mica, porcelain, or similar inorganic materials.

Mechanical strength and oxidation resistance are also important considerations. For high voltage applications, mica and porcelain are suitable because of the high breakdown voltages.

7.4 Case Study – Capacitors

7.4.1 *Functions of capacitors*

Capacitors are important elements in an electric circuit and can be used for a number of different functions including energy storage, blocking, coupling and decoupling, bypassing, filtering, transient voltage suppression and arc suppression. For energy storage, a large charge is built up in the capacitor for release at a later time. This can be used, for instance, for photoflash light or arc welding. Blocking involves the interaction of DC and AC currents. A DC current results in polarisation in the capacitor and blocks the current, while AC results in charge and discharge of the capacitor, which is equivalent to current flowing. This characteristic is used to "couple" one circuit to another. Bypass is achieved by placing a capacitor in parallel with a circuit device. The AC signals pass through the capacitor and DC signals pass the circuit device. Filtering involves the use of a capacitor to separate AC signals with different frequencies.

7.4.2 *Capacitor materials*

Capacitors have been used for tuning in various broadcast stations on a radio receiver since the early 1900's. The materials available at that time had a relative permittivity < 10. In the late 1930's, people discovered that some ceramics containing rutile (TiO_2) had relative permittivity of 80 to 100. Barium titanate ($BaTiO_3$, relative permittivity $= 1200–1500$) was discovered in 1943, leading to the new age of dielectric materials.

Property improvement can be achieved by controlled substitutions into $BaTiO_3$. For example, the relative permittivity can be increased to above 5000 by adding 10 wt.% $CaZrO_3$ plus 1 wt.% $MgZrO_3$. Additions of 10% $CaZrO_3$ and 10% $SrZrO_4$ resulted in a room temperature relative permittivity of 9500. The relative permittivity of $BaTiO_3$ based dielectrics now reaches 18000. Recently, new ceramics have been developed with even higher permittivity. Examples include lead magnesium niobate and lead iron tungstate with $\varepsilon_r = 25000$. Internal boundary-layer (IBL) capacitors based on $SrTiO_3$ have $\varepsilon_r = 100,000$.

Like the ultra-large-integrated-circuits technology, this new dielectric materials has been an important factor in miniaturisation of electric and electronic devices. A typical transistor radio contains at least 10 capacitors with capacitance ranging from 0.005 to 0.05 mF. Each is about 1 cm in diameter with a permittivity of 5000 to 7000. A 0.01 mF capacitor made with rutile ($\varepsilon_r = 80$) in 1942 would be about 10 cm, and made in 1930 with mica ($\varepsilon_r = 7$) would be about 40 cm in diameter.

7.4.3 *Type of capacitors*

There are two types of capacitor constructions: single layer and multi-layer. Single layer capacitors have relatively low capacitance because of the small area-to-thickness ratio. Higher capacitance is possible in multi-layer capacitors because of the high area-to-thickness ratio, as shown in Fig. 7.7. Table 7.4 summarise the types of capacitors.

Thin film technology has been used to fabricate multilayer capacitors. These capacitors consist of metal sheets on which a thin layer of oxide is formed by electrolytically anodising the surface of the metal. Aluminium and tantalum sheets are used. Aluminium oxide (Al_2O_3) and tantalum oxide (Ta_2O_5) have relative permittivities of 8 and 27, respectively. The high capacitance is achieved due to the very thin dielectric layers. Special etching techniques have also been developed to make the surface area of Al sheets larger, resulting in a larger capacitance.

Table 7.4 Types of capacitors, general characteristics and applications.

Type	Permittivity	Characteristics	Applications
Electrolyte			
(a) wet Al	8	Polar, large capacitance-volume ration	Smoothing, decoupling, bypass, etc.
(Al and Ta) Ta	27	Limited temperature and frequency range	Ta for higher reliability and lower leakage current. Al for intermitten use in ac circuits
(b) solid (Ta)		Generally as for wet, improved performance at higher frequencies and over wider temperature range	As above
(c) thin film (Ta)		Largely nonpolar, high capacitance per unit area	Microelectronic circuits
evaporated silicon monoxide	5	Lower capacitance than Ta types but superior dissipation factor	Microelectronic circuits
Paper	~5	Cheap, generally good performance, insulation resistance falls rapidly with increasing temperature	General purpose: power factor correction, blocking, audio, and high frequency bypass
Plastic			
(a) polystyrene	2.3	High insulation resistance, low dielectric absorption and dissipation factor	Charge storage, filter circuits
(b)	3.1	Capable of working at higher temperatures than paper or polystyrene	Replacement from paper in higher temperature systems
Mica	2.5-7	High stability and low dissipation factor	Filter circuits, standard capacitors
Ceramic			
(a) low permittivity	5-20	Low dissipation factor	May be used instead of mica in some circuits
(b) medium permittivity	20-200	Moderately high capacitance-volume ratios. Controlled temperature coefficient of capacitance	LC resonance circuits
(c) high permittivity	1,000-25,000	Large capacitance-volume ratio.	Bypass, blocking, and smoothing circuits
Glaze	10-10,000	Relatively cheap, printed, RC networks. Encapsulation not required	Microelectronic circuits

Source: C.E. Jowett, Materials in Electronics, Business Book, Ltd., London, 1971, p174.

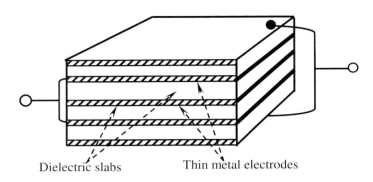

Dielectric slabs   Thin metal electrodes

Fig. 7.7 Schematic of a multilayer ceramic capacitor.

Polymer capacitors are of low cost and relatively easy to fabricate. Together with ceramic capacitors, they are mainly used in small-capacitance applications such as consumer electronics, personal computers and microprocessors. Aluminium and tantalum capacitors are used in high-capacitance applications such as mainframe computers, military systems and telecommunications. Their applications have been greatly increased in the last 30 years.

7.5 Ferroelectric Materials

7.5.1 *Fundamentals and materials*

Some dielectrics exhibit a polarisation versus applied electric field behaviour (*P-E* curve) very similar to the *B-H* curve of a ferromagnetic material. This type of *P-E* plots is called ferroelectric hysteresis loop, and is shown in Fig. 7.8. Ferroelectricity was discovered in 1920 in Rochelle salt by Valasek.

Like ferromagnetism, this type of spontaneous electric polarisation takes place below a certain phase transition temperature, which is also called Curie temperature, T_c. Above the Curie temperature, the material is in the paraelectric state.

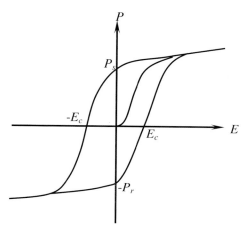

Fig. 7.8 A ferroelectric hysteresis curve (source: D. Richerson, "Modern Ceramic Engineering", Marcel Dekker Inc. 1992, p273).

There are two types of ferroelectrics:

1. The order-disorder type: Its polarisation process involves the movement of hydrogen atoms. For example, the polarisation process of KH_2PO_4 and $NaNO_2$ has this type of mechanism; and
2. The displacive type: For instance, $BaTiO_3$ is this type of ferroelectrics which has a perovskite crystal structure.

The displacive type of polarisation is dependent on the crystal structure. The crystal structure must be non-centric and contain alternate atom positions or molecular orientations that permit the reversal of the dipoles and the retention of polarisation after the field is removed.

For instance, $BaTiO_3$ has a perovskite crystal structure as shown in Fig. 7.9. Above 120°C, it has a cubic structure and does not show ferroelectric. The Ti^{4+} ion lies in the centre of the unit cell. When the temperature is below 120°C (its Curie point), a displacive transformation takes place. The structure of $BaTiO_3$ changes from cubic to tetragonal. When an electric field is applied, the anions (Ba^{2+} and Ti^{4+}) all move in one direction and the cations (O^{2+}) move in the other, creating a net dipole moment in the unit cell. Dipoles also form domains in which the dipoles are aligned in a common direction, similar as for the magnetic domains.

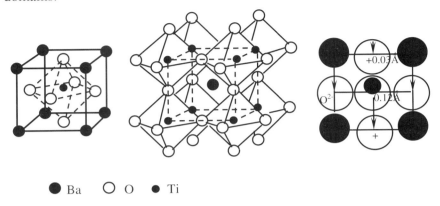

● Ba   ○ O   ● Ti

Fig. 7.9 Crystal structure and ferroelectricity in barium titanate (after Rose, Shepard and Wulff, "Electronic Properties", John Wiley and Sons, 1966, p264).

When the field is removed, a remanent polarisation is retained in the ferroelectric material. This can only be removed by applying a field, E_c, of opposite direction (coercive field) and a ferroelectric hysteresis loop is generated as shown in Fig. 7.8. There are also hard and soft ferroelectric materials.

In an order-disorder ferroelectric, there is a dipole moment in each unit cell, but at high temperatures they are pointing in random directions. Upon lowering the temperature and going through the phase transition, the dipoles order, all pointing in the same direction within a domain.

7.5.2 *Applications*

The nonlinear nature of ferroelectric materials can be used to make capacitors with tuneable capacitance. The permittivity of ferroelectrics is commonly very high in absolute value, especially when close to the phase transition temperature, thus ferroelectric capacitors can be smaller compared to dielectric capacitors of similar capacitance.

The hysteresis effect resulted from the spontaneous polarization of ferroelectric materials can be used as a memory function. Ferroelectric capacitors have been used to fabricate ferroelectric RAM for computers and radio frequency identification (RFID) cards. Ferroelectric capacitors are also finding their applications in medical ultrasound machines, high quality infrared cameras, fire sensors, sonar, vibration sensors, and even fuel injectors on diesel engines. Also, the electro-optic modulators that form the backbone of the Internet are made with ferroelectric materials.

One new development is the ferroelectric tunnel junction (FTJ) in which a contact made up by ferroelectric film placed between metal electrodes. The thickness of the ferroelectric layer is thin (nano-meter level) enough to allow tunnelling of electrons. The piezoelectric and interface effects as well as the depolarization field may lead to a giant electroresistance (GER) switching effect.

7.6 Piezoelectric Materials

When an electric field is applied to a dielectric material, the polarisation may change its dimensions. Atoms act as egg-shaped particles, or the bonds between ions change its length during polarisation. This is called electrostriction.

On the other hand, when a dimensional change (strain) is imposed on some dielectrics, polarisation occurs and a voltage or field is created. Dielectric materials that display this type of behaviour are called piezoelectrics, which means "pressure electricity". A number of natural and laboratory made crystals show this behaviour, including quartz, zinc blends, boracite, tourmalite, topaz and sugar. All these crystals have one thing in common: they are structurally anisotropic and do not have a center of symmetry.

The electric field ξ produced by a stress σ can be written as:

$$\xi = g \cdot \sigma, \qquad (7.15)$$

and the strain ε produced by a field ξ is:

$$\varepsilon = d \cdot \xi, \qquad (7.16)$$

where g and d are constants depending on the material and related to the modulus of elasticity E (Young's modulus):

$$E = \sigma/\varepsilon. \tag{7.17}$$

As the stress and strain are within elastic deformation range, we can combine (7.17), (7.16) and (7.15):

$$E = 1/(g \cdot d). \tag{7.18}$$

Although many materials were found to be piezoelectric, relatively few of them have been optimised for practical application. Most commonly used piezoelectric materials include quartz, barium titanate ($BaTiO_3$), lead titanate ($PbTiO_3$), lead zirconate ($PbZrO_3$), CdS and ZnO.

7.7 Pyroelectric Materials

Pyroelectric materials are a special class of piezoelectrics. Their crystal structure contains at least one crystallographic direction along which spontaneous polarisation exists. Heating of the material results in mechanical deformation due to thermal expansion and the change in temperature slightly modifies the positions of the atoms within the crystal structure, such that the polarization of the material changes. This polarization change gives rise to a temporary electric potential.

Pyroelectric property of a material is measured by pyroelectric temperature coefficient dP/dT:

$$\Delta q/A = \Delta T \times (dP/dT), \tag{7.19}$$

where $\Delta q/A$ is the charge released on area A when temperature changes by ΔT.

Therefore, temperature signals can be transferred into electric signals.

All pyroelectric materials are also piezoelectric since these two properties are closely related. However, some piezoelectric materials have a crystal symmetry that does not allow pyroelectricity. Typical examples of pyroelectric materials include hexagonal ZnS, tourmaline, triglycine sulfate, $BaTiO_3$, $Pb(Zr,Ti)O_3$, lithium sulfate and $LiTaO_3$.

Two example materials are discussed below. The first one is $Pb(Zr,Ti)O_3$, short as PZT. PZT ceramics are solid solutions of PbO, ZrO_2, and TiO_2. By manipulating the composition and microstructure of PZT, it is possible to control their properties to suit nearly all technical needs. The variation of composition is mainly between Zr and Ti, therefore, the general formula can be written as $Pb(Zr_zTi_{1-z})O_3$, with $z = 0-1$. Small amounts of other oxides such as oxides of La, K, Ba, Cr, Co and Nb can be added into the $Pb(Zr,Ti)O_3$ solution. The properties such as conductivity, Curie temperature, coercivity, compliance and pyroelectric coefficient can be controlled by doping and heat treatments.

Most pyroelectric materials lose their pyroelectric properties at high temperatures. $LiTaO_3$ retains its pyroelectricity to 609°C. This material can therefore be used over a large temperature range for accurate temperature measurement. Temperature changes in the order of 10^{-6} K can be detected. $LiTaO_3$ has been developed to be used as the detector for scanning microcalorimeter (DSC) with a very high sensitivity.

Many new pyroelectric materials are being developed, such as gallium nitride (GaN), caesium nitrate ($CsNO_3$), polyvinyl fluorides, derivatives of phenylpyrazine, and cobalt phthalocyanine. In general, these materials are in the form of thin film.

Pyroelectric materials can be repeatedly heated and cooled to generate usable electrical power. It had been reported that a pyroelectric could reach 84-92% of Carnot efficiency. Advantages of pyroelectric generators in comparison with conventional heat engines and electrical generators may include potentially lower operating temperatures, less bulky equipment, and fewer moving parts.

7.8 Case Study – Materials for Transducers

Ferroelectric, piezoelectric and pyroelectric materials are often used to make transducers, which can convert mechanical or temperature information into electrical signals.

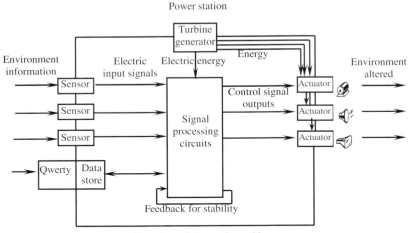

Fig. 7.10 Transducers in the electronic world (after N. Braithwate and G. Weaver, "Electronic materials", Butterworths, 1990, p19).

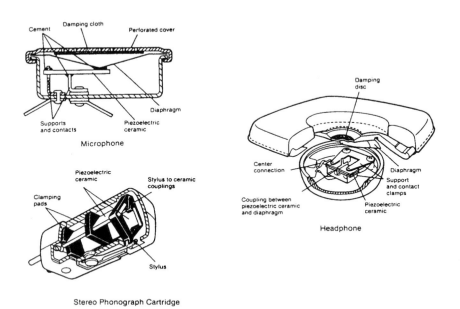

Fig. 7.11 Examples of applications of piezoelectric materials (after Piezoelectric Technology Data for Designers brochure, Vernitron Piezoelectric Division).

Transducers are the devices that are used to translate various environmental information into electrical signals. The environmental information includes temperature or heat flow, pressure, light, sound (acoustical waves), smoke, humidity, chemical compositions (oxygen, acidity or pollution) and electromagnetic waves. This information can be processed and analysed and the measured signals can be used for display, recording or automatic control. Alternatively, these materials can also convert electronic signals into other forms of energy. Fig. 7.10 shows the position of transducers in the modern electronic world.

Transducers are widely used in modern technology. Three typical examples, as shown in Fig. 7.11, are presented to demonstrate the different types of translation and their application.

7.8.1 *Piezoelectric transducers for acoustical systems*

Applications of piezoelectric materials in acoustical systems are based on the conversion of a mechanical force to an electric signal. One example is the phonograph pickup. The stylus is caused to vibrate by the contours in the groove of the record. This vibration is converted into an electrical signal by a piezoelectric ceramic. The signal is then amplified and converted into audible sound waves. Modern stereo phonographs use polycrystalline ferroelectric ceramics such as $Pb(Zr,Ti)O_3$.

Microphones and headphones also use piezoelectric materials as the transducers to receive vibrations from sound pressure through a diaphragm. Piezoelectric ceramics convert the mechanical vibration to voltage signals. Headphones work in an opposite way that the piezoelectric elements vibrate according to the voltage signals. These vibrations are passed to a diaphragm and are converted to sound. Fig. 7.11 shows the schematic drawings for a stereo phonograph cartridge, microphone and headphone.

7.8.2 *Pyroelectric transducer for a burglar detector*

Fig. 7.12 shows the principle of a pyroelectric burglar detector. A small piece of pre-polarised dielectric material is used as a capacitor. The polarisation controls the charge held in the dielectric detector and hence the voltage across it. The radiant energy from an intruder warms the transducer, changes its polarisation, and alters the voltage. A low noise, high impedance amplifier conveys the signal into the system and an alarm will be activated to shatter the peace of the neighbourhood.

In order to detect a small heat signal, the pyroelectric element must be a very thin sheet, so that it heats up quickly. It also needs to be thermally insulated from the surroundings so that the temperature change from the air near by will not activate the alarm. The temperature change caused by the signal is not high, perhaps in the order of 10^{-6} K. Therefore, it is not an easy task to pick up the signal from the background noise.

Lead zirconate, $PbZrO_3$, possesses the required properties and is selected to produce the sensor in burglar detectors. It has a Curie temperature of ~200°C, high enough for working at room temperature or slightly above. It has a high relative permittivity of ~250, a high pyroelectric coefficient of 350 μC/(m²·K). Its electric resistivity is between 10^8 to 10^{11} Ω·m, and can be altered by adjusting the chemical composition.

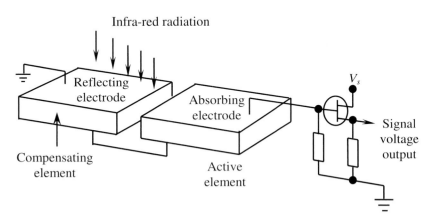

Fig. 7.12 Schematic of pyroelectric detector (after N. Braithwate and G. Weaver, "Electronic materials", Butterworths, 1990, p173).

Summary:

Dielectrics are used in capacitors and as electrical insulation. The dielectric constant or relative permittivity of a material can vary with temperature and frequency, the bonding, crystal structure and phase constitution. Because the polarisation always lags the applied field, it leads to an electrical energy loss which appears as heat and is proportional to the product of the relative permittivity and the tangent of the loss angle δ. Therefore, most dielectrics are rated by three factors:

1. relative permittivity,
2. tangent of loss angle, and
3. dielectric strength.

Some other dielectric properties such as piezoelectricity, pyroelectricity and ferroelectricity are also important. They are belonging to the same group of materials as shown below in Fig. 7.13. Examples of materials with these special properties are $BaTiO_3$ and $Pb(Zr,Ti)O_3$. They are used to produce transducers and other devices for measurements, communication and automatic control in modern industries and our everyday life.

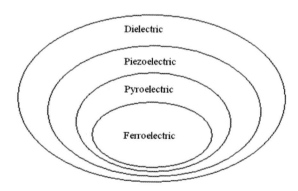

Fig. 7.13 Dielectric materials family.

Important Concepts:

Capacitor: An electric device, constructed from alternating layers of a dielectric and a conductor, which is capable of storing an electrical charge.

Polarisation: Alignment of dipoles so that a charge can be permanently stored. It is expressed as the total dipole moment per unit volume in a dielectric material.

Permittivity: The ability of the material to polarise and store a charge within the material. The permittivity of a vacuum $\varepsilon_o = 8.85 \times 10^{-12}$ F/m.

Relative Permittivity, ε: The ratio of the permittivity of a material to the vacuum, thus describing the relative ability of a material to polarise and store a charge.

Energy Loss in Dielectrics: The fraction of energy lost during each time an electric field in a material is reversed. This loss is due to the dc resistivity and dipole relaxation.

Dielectric Strength (dielectric breakdown): The maximum electric field that can be maintained between two conductor plates.

Ferroelectricity: Alignment of dipoles and dielectric domains under an electric field so that a net polarisation remains after the field is removed.

Electrostriction: The dimensional change that occurs in a material when an electric field is acting on it.

Piezoelectricity: When a dimensional change is imposed on a material, polarisation occurs and an electric field is created.

Pyroelectricity: When a temperature change occurs in a material, polarisation takes place and an electric field is created. Pyroelectric materials are a special type of piezoelectric materials.

Transducer: Devices which can translate various environmental information into electrical signals for measurement or control purposes.

Questions:

1. What are the three most important dielectric properties? Understand the analogues of the magnetic and dielectric properties.
2. Discuss the similarity and difference of crystal structure and characteristics between ferroelectric and piezoelectric materials.
3. Discuss the mechanism behind the formation of hysteresis loop in ferroelectric materials and the related applications.
4. To construct a multilayer plate capacitor having a capacitance of 0.05 μF, fused silica sheets of 1 cm × 1 cm × 0.001 cm are used. Calculate how many layers of (a) the silica sheets and (b) conductors are needed. The permittivity in vacuum is 8.854×10^{-12} F/m and the relative permittivity of fused silica is 4.5.
5. A force is applied to a 0.5 cm × 0.5 cm wafer of $BaTiO_3$ that is 1 mm thick. The piezoelectric voltage across the thickness of the wafer was measured to be 400 V. Calculate (a) the electrical field created by the force, (b) the strain produced by the force, and (c) the value of the force. The Young's modulus and piezoelectric constant for $BaTiO_3$ are 69 GPa (GPa = 10^9 Pa) and 1×10^{-10} m/V, respectively.
6. $BaTiO_3$ has a perovskite structure. Below its Curie temperature, the ions shifted from its original cubic positions, resulting in a tetragonal structure with $a = b = 0.399$ nm and $c = 0.403$ nm. Calculate (a) the total dipole moments for a unit cell, and (b) the polarisation as the density of dipole moments.
7. Suppose that the average displacement of the electrons relative to the nucleus in a Cu atom is 1×10^{-8} A when an electric field is imposed on a Cu plate. Calculate the polarization. Cu has an atomic number of 29, a *fcc* structure and $a_0 = 3.6151$ Å.
8. The ionic polarization observed in a NaCl crystal is 4.3×10^{-8} C/m². Calculate the displacement between Na^+ and Cl^- ions. There are four Na^+ in a unit NaCl cell, which has a *fcc* structure and $a_0 = 0.55$ nm.

Chapter 8

Optical Properties and Materials

8.1 Introduction

Optical materials and instruments are used in a wide range of industries and our everyday life. Optoelectronic devices, which transfer signals from electrical to optical or from optical to electrical, have rapid development in recent decades. Laser and optical-fiber communication systems are two examples that have tremendous impact on our society. The science of "photonics" studies the generation, emission, transmission, modulation, signal processing, switching, amplification, detection and sensing of light. The reactions between light (radiation) and materials are the basis of photonics.

8.1.1 *Radiation and resources*

Optical properties of materials are related to the interaction of material with radiation. The energy of the radiation (frequency or wavelength) is determined by the source (Table 8.1). Fig. 8.1 also shows the spectrum of electromagnetic radiation.

Table 8.1 Radiation and its source.

Radiation	Source
γ-rays	Change in nuclear structure
X-rays, ultraviolet & visible light	Change in electronic structure
Infrared, microwave & radio waves	Vibration of atoms in crystal structure

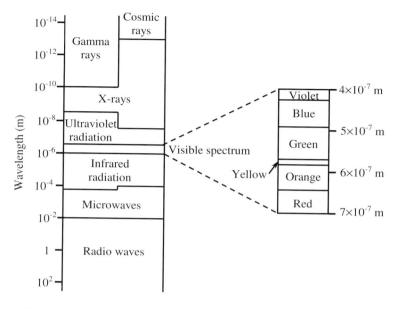

Fig. 8.1 The electromagnetic spectrum of radiation (after D. Askeland, "The Science and Engineering of Materials", Chapman and Hall, 2nd SI Ed.1990, p732).

8.1.2 *Reaction between radiation and materials*

Interactions between radiation and materials include:

1. Absorption,
2. Photo electron emission,
3. Colours and characteristic spectrum,
4. Fluorescence,
5. Heat conduction,
6. Refraction, reflection and transmission, and
7. Some other electronic behaviours.

The study of these phenomena allows an understanding of the optical properties of materials and how to use them in important applications such as:

1. Optical imaging systems,
2. Lasers,
3. Optical-fibre telecommunication systems,
4. Light emitting and liquid crystal display systems,
5. Solar cells and energy absorption/storage materials,
6. Analytical instruments for determination of crystal structure and micro-composition, and
7. Aircraft that cannot be detected by radar.

8.2 Emission of Continuous and Characteristic Radiation

8.2.1 *Continuous spectrum*

When a high energy electron beam strikes a material, it will be decelerated, and energy will be given up as emitted photons.

Each time an electron strikes an atom, some of its energy is given up (Fig. 8.2). Each interaction may be more or less severe, so the electron gives up a different fraction of its energy in each collision, producing photons with different wavelength. This is called "continuous spectrum", or "white radiation".

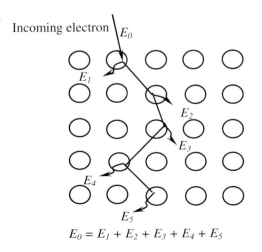

$$E_0 = E_1 + E_2 + E_3 + E_4 + E_5$$

Fig. 8.2 Production of continuos spectra of radiation emitted from a material.

Some of the electrons may lose all their energy in one impact. The wavelength of the emitted photons would be equivalent to the original energy of the incident electron. We call this "short wavelength limit", λ_{swl}, because no shorter wavelength can be emitted by this incident electron beam.

When the energy of the stimulus increases, λ_{swl} decreases and the number and energy of the emitted photons increase, which gives a more intense continuous spectrum, as shown in Fig. 8.3.

8.2.2 *Characteristic spectrum*

If the incident stimulus has a sufficiently high energy, electrons from an inner energy level can be excited into an outer energy level. To restore equilibrium, electrons from the higher energy level will fill the inner empty level.

There are discrete differences between any two energy levels. When an electron drops from one level to another, a photon with that particular energy and wavelength is emitted. This results in "characteristic spectrum".

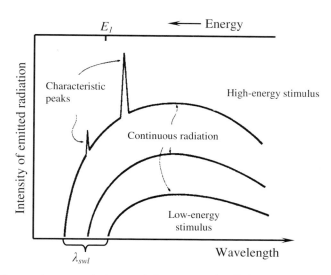

Fig. 8.3 The continuous and characteristic spectra of radiation emitted from a material (after D. Askeland, "The Science and Engineering of Materials", Chapman and Hall, 2nd SI Ed.1990, p730).

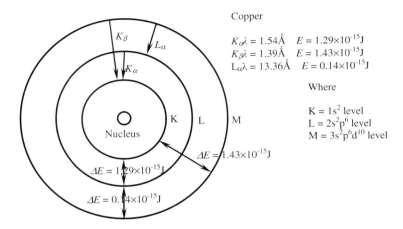

Fig. 8.4 Characteristic X-rays are produced when electrons change from one energy level to a lower level, as shown here for Cu (after D. Askeland, "The Science and Engineering of Materials", Chapman and Hall, 2nd SI Ed.1990, p731).

Table 8.2 Characteristic emission lines for selected elements.

Elements	K_α (nm)	K_β (nm)	L_α (nm)
Al	0.8337	0.7981	
Si	0.7125	0.6768	
S	0.5372	0.5032	
Cr	0.2291	0.2084	
Mn	0.2104	0.1910	
Fe	0.1937	0.1757	
Co	0.1790	0.1621	
Ni	0.1660	0.1500	
Cu	0.1542	0.1392	1.3357
Mo	0.0711	0.0632	0.5724
W	0.0211	0.0184	0.1476

(Source: B. Cullity, "Elements of X-ray Diffraction, 2nd Ed., Addison-Wesley, 1978).

The characteristic spectrum usually appears as a series of peaks added on the continuous spectrum as shown in Fig. 8.3. They are in X-ray range. We use K, L, M ... to represent the energy levels where K = $1s^2$, L = $2s^22p^6$, M = $3s^23p^63d^{10}$ .... Different ways to fill the inner energy level will emit photons with different energies (and wavelengths). This is the characteristic X-ray emission of materials (see Fig. 8.4). Table 8.2 lists the characteristic emissions for some elements. This principle is used for materials compositional analysis including microanalysis.

Electron reactions in the outer energy levels of certain materials may produce visible light.

8.3 Applications of Photon Emission

8.3.1 *Electron probe microanalysis*

Electron probe microanalysis (EPMA) is a powerful technique which permits the characterisation of materials down to a very small scale (< 1 μm). A thin electron beam is used to excite photons from materials in order to measure the composition in a small area under a scanning electron microscope (SEM) or transmission electron microscope (TEM). This instrument is called electron probe.

When the inner shell electrons are stimulated by the incident electrons, X-rays are produced. The most useful X-rays are the most energetic photons produced by filling the K and L levels. These X-rays are used to determine the composition of the material.

8.3.1.1 *Qualitative analysis*

Use a high energy beam to bombard an unknown material, X-rays of both the characteristic and continuous spectra will be excited. The continuous spectra can be subtracted. The wavelength or energy for each peak in the characteristic spectra can then be recorded and matched with those known for certain elements. We can determine what elements are in the unknown material as the spectra for every element occurs at different wavelengths – it is characteristic of the element.

8.3.1.2 *Quantitative analysis*

The intensities of the characteristic peaks can also be measured and calculated with a ZAF calibration programme (Z represents the atomic number correction, A is the absorption correction, and F is the fluorescence correction), or compared with the intensities from a standard specimen with a similar composition. Thus the amount or percentage of each element in the unknown material can be determined.

The size of the electron beam used in a modern microscope can be very small, a few nm in diameter. This allows us to identify small features such as individual phases, inclusions (impurities) or precipitates in a material, and explain many material properties which are important but could not be explained before.

Scanning electron microscopes (SEM) are now widely used in microelectronics for quality control, research and development. For example, SEM and EPMA are very powerful and popular tools for observing, analysing and studying the extremely small features on a semiconductor chip, thin film coatings, tiny metal leads, the tracks and holes on a compact disc (CD) and the grain size and alignment in magnetic, dielectric and superconductor materials.

8.3.2 *Luminescence*

Luminescence is the conversion of radiation or other forms of energy to visible light. It is caused by reactions in the outer energy levels of an atom. The incident radiation excites electrons from the valence band into the conduction band. When the electrons drop back to the valence band, photons are emitted. If the wavelength of these photons is in the visible light range (400–700 nm), luminescence occurs.

In metals, the energy of emitted photon is very small, since the valence and conduction bands overlap. The wavelength of the emitted photons is longer than visible light spectrum. Therefore, luminescence does not occur.

In certain ceramics, the energy gap between the valence and conduction bands is relatively large. When an electron drops back through the gap, it may produce a photon in the visible light range. The wavelength of the photon corresponds to the energy gap E_g (Fig. 8.5).

The band gap energy that corresponds to the visible light with wavelength between 400–700 nm can be calculated as follows:

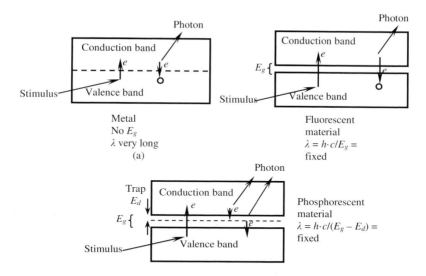

Fig. 8.5 Luminescence occurs when photons have a wavelength in the visible spectrum: (a) luminescence does not occur in metals, (b) fluorescence occurs when there is an energy gap, (c) phosphorescence occurs due to new level traps in the energy gap (after D. Askeland, "The Science and Engineering of Materials", Chapman and Hall, 2nd SI Ed.1990, p731).

$E_{g,400} = h \cdot v = h \cdot c/\lambda = 6.63 \times 10^{-34} \times 3 \times 10^{8}/(400 \times 10^{-9} \times 1.6 \times 10^{-19}) =$ 3.11 eV, and
$E_{g,700} = 1.78$ eV.

Therefore, an energy gap ranging from 1.8 to 3.1 eV corresponds to the visible light.

Two effects may be observed in luminescent materials:

Fluorescence - Luminescence stops when the stimulus removed. All excited electrons drop back to the valence band and the photons are emitted within a very short time ~ 10^{-8} s.

Phosphorescence - Materials have impurities which introduce a new energy level within the energy gap. The stimulated electrons first drop into the new energy level and trapped there. The electrons must escape the trap before returning to the valence band (Fig. 8.5c). There is a delay ($> 10^{-8}$ s) before the photons are emitted. Therefore, when the source is removed, electrons in the traps gradually escape and emit light over some additional period of time.

The intensity of the luminescence, I, can be expressed as:

$$ln\ (I/I_0) = -\ t/\tau, \qquad\qquad (8.1)$$

where τ is the relaxation time and I_0 is a constant for the material.

When $t = 0$, $I = I_0$; when $t = \tau$, ln $(I/I_0) = -1$, $I/I_0 = 1/2.718$, $I = 0.368\ I_0$.

Phosphorescent materials are used in fluorescent lights, oscilloscope screens, TV screens, and photocopy lamps. The fluorescent light consists of a sealed glass tube coated on the inside with a halogen phosphate such as $Ca_5(PO_4)_3(Cl,F)$ or $Sr_5(PO_4)_3(Cl,F)$ doped with Sb and Mn. The tube is filled with mercury (Hg) vapour and argon. A power source provides electric charges that stimulate radiation of the Hg vapour at a wavelength of 254 nm, corresponding to ultraviolet radiation. This UV radiation in turn excites a broad band of radiation from the phosphorescence materials in the visible range, producing light.

In TV screens, photons are excited by an electron beam that sweeps across the phosphor-coated screen. The relaxation time is in the range of 0.1–0.01 s. In colour TV, three types of phosphorescence materials are used. The energy gaps are designed to be different so that the red, green and blue colours are produced.

8.4   Laser and Laser Materials

8.4.1   *What is a laser?*

Laser is the short form of "Light Amplification by Stimulated Emission of Radiation". In certain materials, electrons that are excited by a stimulus produce photons which in turn excite additional photons of same wavelength. Consequently, a large number of photons are emitted from the material. By the proper choice of stimulant and luminescence material, the wavelength of the photons can be in the visible light range. This is called laser. The output of the laser is a beam of photons that are parallel, coherent and of same wavelength. No destructive interference occurs in a coherent beam, so that the beam can be very strong.

8.4.2 *Laser materials*

The first visible light (694 nm) laser was a ruby (Al_2O_3) laser, developed by Maiman in May 1960. A wide range of laser materials and equipments have been developed since then. The most important ones are listed below:

1. Ruby laser - Al_2O_3 doped with a small amount of Cr_2O_3,
2. YAG laser - yttrium aluminium garnet doped with Nd, Er, Tm or Yb,
3. CO_2 gas laser,
4. Inert gas lasers (He and Ne),
5. Vapour lasers (Hg and Cd),
6. Liquid lasers (dye lasers), and
7. Semiconductor lasers (GaAs, GaN, or GaAlAs).

CO_2 lasers have high penetrating ability. They are widely used in high power applications such as melting, welding and surface treatment and surgery. Semiconductor materials such as GaAs have an energy gap corresponding to visible light (E_g for GaAs = 1.42 eV).

8.4.2.1 *Solid-state laser*

Cr^{3+}:Al_2O_3 (chromium-doped corundum) is a typical single-crystalline material for solid-state laser. Al^{3+} in Al_2O_3 crystal is partially substituted by Cr^{3+}. Cr^{3+} has been the most successful transition metal ion used for laser applications. It has a relatively wide absorption band, but only has two radiative transition paths. Thus this material has a relatively high efficiency.

The most commonly used Nd-based laser materials may include Nd:YAG (yttrium aluminium garnet), Nd:YIG (yttrium iron garnet),

Nd:YLF (yttrium lithium fluoride) and Nd:YVO$_4$ (yttrium orthovana-date). These materials can be either crystalline or amorphous. These la-sers can produce high powers at 1064 nm (the most popular wavelength) and 1340 nm. These lasers can also be frequency doubled, tripled or quadrupled to produce 532 nm (green), 355 nm (UV-A) and 266 nm (UV-C) light when those wavelengths are needed.

8.4.2.2 *Semiconductor laser*

GaAs semiconductor laser was first invented in 1962. It has the ad-vantages of high efficiency, simple structure/configuration, and rigidity, so attracted extensive attention. In 1968, semiconductor laser based on GaAlAs-GaAs heterojunction was developed and operated at ambient temperature with high output powder density. After this, semiconductor laser gained rapid development and found more and more applications in optical disks and optical fiber communication. Semiconductors used in laser mainly include GaAs, GaN, GaAlAs, CdS, InP and ZnSe.

Lasing from a semiconductor is sourced from the radiative recombi-nation of charge carriers, i.e., electrons (e^-) and holes (h^+), in the active area of the device. There are three basic requirements for this special optical emission:

1. Formation of reverse states: a sufficient number of electrons must be excited from the valence band to the conduction band, so that stimu-lated emission would be dominant (as shown in Fig. 8.6).
2. Have optical resonator: a positive feedback process is necessary for the amplified stimulated emission (Fig. 8.7); and
3. Multiplication must be greater than loss, i.e., the probability of light absorption should be lower than that of emission.

From Fig. 8.6, it can be also seen that the wavelength (or frequency) of the laser light is mainly determined by the bandgap energy of the semi-conducting material used. However, the type of the bandgap must be clarified. Semiconductors that are suitable for lasing operation should have a direct bandgap. In this respect, elemental semiconductors of Si and Ge therefore are not ideal candidates for semiconductor lasers.

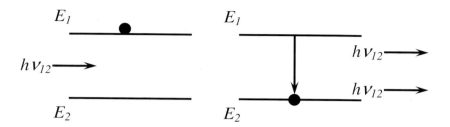

Fig. 8.6 A schematic of stimulated emission in semiconductor.

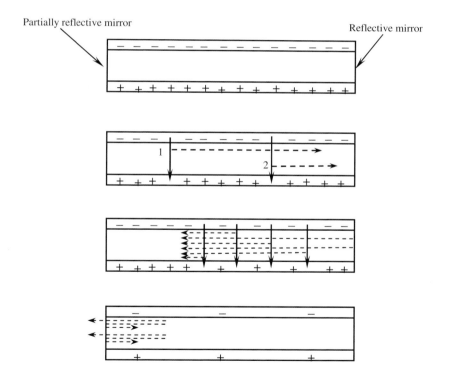

Fig. 8.7 Creation of a laser beam from a semiconductor (after D. Askeland, "The Science and Engineering of Materials", Chapman and Hall, 2nd Ed, 1990, p740).

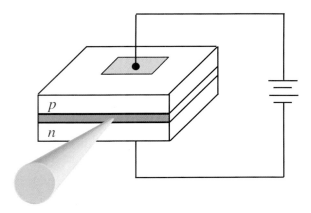

Fig. 8.8 A schematic of a simple semiconductor laser device.

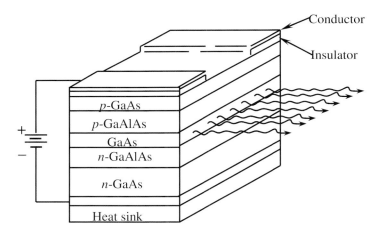

Fig. 8.9 Schematic show of a GaAs laser (after D. Askeland, "The Science and Engineering of Materials", Chapman and Hall, 2nd SI Ed.1990, p741).

A simple semiconductor laser device is shown in Fig. 8.8. It is consisted of consists of a *p-n* junction which is biased in the forward direction. Electrons are injected from the *n*-type region and holes from the *p*-type region, respectively.

However, the device using one semiconductor that is *p*- and *n*-doped, e.g., *p*-GaAs and *n*-GaAs, to form a so-called homojunction, is showing a relatively low efficiency due to the low carrier density in the active area and also the weak overlap between the inverted region and the optical mode. Double heterostructure has then developed to achieve room-temperature continuous wave (CW) operation with high efficiency. Such a device is shown in Fig. 8.9 schematically. The active material, GaAs, is embedded in GaAlAs, a semiconductor with a slightly larger bandgap energy. This causes a potential well in which electrons and holes are confined in order to achieve high carrier densities in the active area.

8.4.3 *Laser applications*

Laser has a long list of applications. They are almost everywhere in our industries, research labs, offices, homes, hospitals and entertainment centres. Some of them are only in the research-testing stage.

1. Laser communication: used in optical fibre signal transmission,
2. Materials surface treatment: fast melting, surface defects removal and surface alloying etc,
3. Medical applications: surgery (accurate operations) and diagnostics,
4. Machining: cutting, drilling and welding,
5. Materials processing: zone melting of metals, ceramics, superconductors – a highly concentrated melting-solidification process,
6. Measurement instrumentation and accurate long-distance measurement,
7. Sensors: fibre-optical sensors, laser sensors etc,
8. Laser induced mass analysis: used for micro-compositional analysis,
9. Optical data storage systems: including computer CD ROM, and noise-free stereo play-back/recording performance,
10. Data acquisition systems: including barcode readers and document readers,
11. Office automation: including fax and copy machines, and laser printers etc,
12. Holography,

13. Entertainment: including coloured laser beams, CD players, video players, erasable disc systems, camcorders etc,
14. Missile guidance systems and smart bombs,
15. Laser weapons (for the "Star Wars" defence system), and
16. Nuclear fusion (potentially).

8.5   Case Study – A Compact Disc (CD) System

Compact discs (CD) and digital optical recording were first developed to achieve noise-free stereo play-back/recording performance. Audio CDs have been commercially available since October 1982. This technology is now widely used for multimedia systems in computers. There is already a large variety of material available on CD ROM including encyclopaedias, atlases and books. Multimedia computers provide text, graphics, images, sound, full motion video, games, films and interactive video. People believe that interactive multimedia will speed up the learning process because one-on-one instruction means a student can master the material before moving on, although nobody knows the percentage of time our teenagers spend in front of a computer to learn something or to play games. Nevertheless, multimedia systems provide professionals with a powerful business and presentation tool. No doubt the impact to our society from them is tremendous.

8.5.1   *Principle of digital optical storage*

The CD optical recording is digital. All information including pictures, sound and text is reduced to strings of binary zeros and ones. Writing is operated by ablation of the surface layer of CD with a laser beam to produce small holes, which represent "1", while shiny surface areas represent "0". The holes have a diameter of 0.5–1 μm, the distance between tracks ~1.4 μm, and the recording density is 0.5–1.0×10^5 bits/mm². A disc having a surface area ~10,000 mm² will hold >5 Gb memory, which is equivalent to the information typed in "single spacing" on 3 million A4 pages!

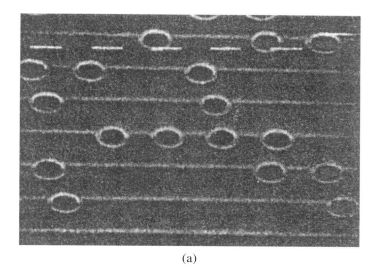

(a)

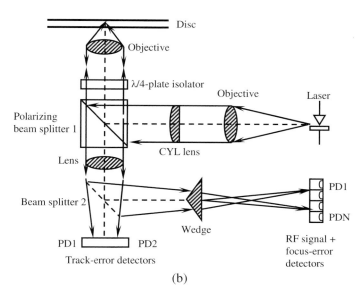

(b)

Fig. 8.10 (a) Micrograph of pre-addressed and recorded tracks on a compact disc and (b) Diagram of an electro-optic recording system (after J.C.A. Chaimowicz, "Lightwave Technology, Butterworth & Co.1989, p246).

Each bit is checked immediately after writing (a DRAW system – direct read after write). If an error is detected, the relevant data will be re-written elsewhere on the disc. Data is recorded on a continuous, tightly wound spiral, pre-grooved surface, which provides an optical "hand-rail" to the write/read head (Fig. 8.10a).

The electro-optic head has four jobs to do: to keep itself on track, to keep itself in focus, to write and to read. Fig. 8.10(b) is a schematic drawing of the recording system, showing how the tracking, signal and focus errors can be detected. The accurate control in beam focus makes possible to produce holes smaller than 1 μm.

8.5.2   *CD materials*

A GaAlAs laser is used for both recording and reading (with different power). The compact discs are made from polymer based materials (such as polycarbonate), which are light and flexible, and are strengthened to have a relatively high bending strength. A very shiny, thin layer of Bi compound or other metallic compounds of ~30 nm thick is coated on the surface. The recording holes are deeper than the coating layer, ~200 nm. The pre-grooving and pre-addressing are done on an 8 μm layer of photopolymer. Finally, there are two "giant" sizes: the total thickness and diameter of a disc are ~1 mm and 12 cm, respectively. The total mass of a compact disc is ~15–20 g.

8.6   **Thermal Photoemission**

When a material is heated, electrons are thermally excited to higher energy levels, particularly the electrons in the outer levels where they are less tightly bounded to the nucleus. The excited electrons then drop back to their normal levels immediately and release photons with relatively low energy. This is called "thermal photoemission".

When we heat a piece of steel in a workshop, we see the colour changes gradually as follows:

black - dark red - red - orange - yellow - white - white with blue

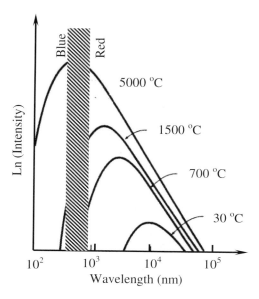

Fig. 8.11 The intensity versus wavelength of photons emitted thermally from a material (after D. Askeland, "The Science and Engineering of Materials", Chapman and Hall, 2nd SI Ed.1990, p742).

As the temperature rises, the thermal energy increases, and the maximum energy of the emitted photons increases. A continuous spectrum of radiation is emitted, with a minimum wavelength and intensity distribution depending on the temperature.

By measuring the intensity of a narrow band of the emitted wavelength with a pyrometer, one can measure the temperature of the material (Fig. 8.11). This is an optical temperature measurement method, which is widely used in metallurgical and energy production industries (for instance in smelting and casting plants) where the high temperature objects are difficult to approach.

8.7   Interaction of Photons with Materials

Photons, either in characteristic or continuous spectra, cause a number of optical phenomena when they interact with the electronic or crystal structure of a material. They are:

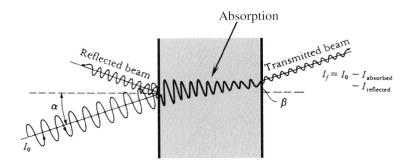

Fig. 8.12 Interaction of photons with a material (after D. Askeland, "The Science and Engineering of Materials", Chapman and Hall, 2nd SI Ed.1990, p742).

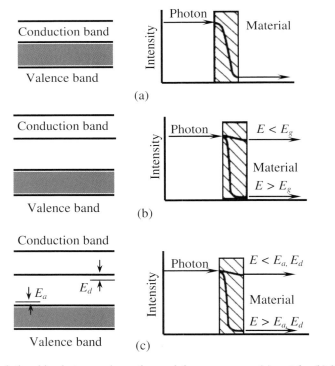

Fig. 8.13 Relationships between absorption and the energy gap: (a) metals, (b) insulators and intrinsic semiconductors, (c) extrinsic semiconductors (after D. Askeland, "The Science and Engineering of Materials", Chapman and Hall, 2nd SI Ed.1990, p743).

1. Reflection,
2. Refraction,
3. Absorption, and
4. Transmission.

The relation between them is:

$$I_{trans} = I_0 - I_{ab} - I_{refl}, \qquad (8.2)$$

where I_{trans}, I_0, I_{ab} and I_{refl} are the intensities of the transmitted, incident, absorbed and reflected radiations, respectively (Fig. 8.12).

The most important factor to decide the interaction is the energy required to cause an electron to jump from the valence band to the conduction band in the material, i.e. the bandgap energy.

As a first approximation,

- If $E_p > E_g$, E_p will be absorbed and absorption or reflection will take place;

- If $E_p < E_g$, E_p will not be absorbed and transmission will take place.

where E_p is the energy of the incident photons, and E_g is the band gap energy of the material. Fig. 8.13 illustrates the general relationships between the energy gap and absorption.

8.7.1 *Absorption and transmission*

In metals, the valance and conduction bands overlap so there is no energy gap. Therefore, radiation of almost any wavelength will be absorbed, and metals are considered to be opaque.

If the material is very thin, and the wavelength of the radiation is very short, not all the incident photons are absorbed. Some of the radiation can penetrate the thin foil. In this situation, the amount of transmitted intensity, I, can be calculated by:

$$ln\ (I/I_0) = -\mu \cdot x, \tag{8.3}$$

where I_0 is the original intensity, x is the path through which the photon move, usually this is the thickness of the foil. μ is the linear absorption coefficient of the material for photons and is related to the density of the material and the energy required to stimulate an electron either from the valence band to conduction band or from one energy level to another.

Therefore, $I_0 - I$ is the absorbed intensity, which increases with the increasing μ and x.

Fig. 8.14 shows the plot of μ vs. λ for Mo, Ni, and Cr. μ changes abruptly at a particular wavelength corresponding to the energy required to excite an electron from the K shell of the atom. This wavelength is called absorption edge. Note the absorption coefficient μ increases dramatically as the wavelength increases beyond the absorption edge. By reaching the visible light wavelength, absorption is virtually complete unless the specimen is extremely thin.

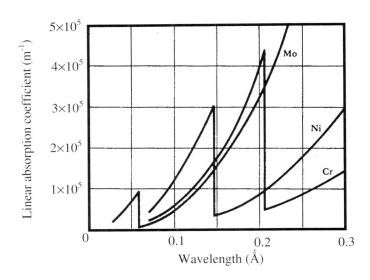

Fig. 8.14 The linear absorption coefficient vs. wavelength for Mo, Ni and Cr (after D. Askeland, "The Science and Engineering of Materials", Chapman and Hall, 2nd SI Ed.1990, p744).

In insulators, the energy gap between the valence and conduction bands is large. If the energy of the incident photons is less than the gap energy, no electron will gain enough energy to escape the valence band, and therefore, no absorption will occur, unless the photons interact with imperfections in the material. These materials should be transparent. This is the case for glass, many high purity crystalline ceramics, and amorphous polymers etc.

In semiconductors, the energy gap is smaller than in insulators. In intrinsic semiconductors, absorption will occur when the photons have energy greater than E_g, while transmission will take place when photons have less energy than E_g. Extrinsic semiconductors have been doped to produce donor or acceptor levels. The energy criterion will be E_d or E_a. This means that extrinsic semiconductors tend to absorb light. Therefore, semiconductors are often opaque to short-wavelength radiation but may be transparent to long-wavelength radiation.

8.7.2 *Case study – Colour*

Colour is an important optical property of materials. It results from absorption of a relatively narrow wavelength of radiation within the visible light range (400–700 nm). The absorption is caused by electron transitions. There are four types of electron transitions:

1. Intrinsic band gap electron transition,
2. Internal electron transitions within $3d$ (transition metals V, Cr, Mn, Fe, Co, Ni, Cu) and $4f$ (rare earth elements),
3. Electron transfer from one ion to another, and
4. Electron transitions associated with crystal imperfections.

The first type of transition is decided by the intrinsic property of the material (bandgap energy). The last three transitions are related to the small amounts of impurities and crystal defects.

The visible colour is the complementary colour of the absorbed colour. A beam of white light contains the complete spectrum of colours. When one wavelength is absorbed by a material, the rest of the wavelengths pass through. Our eyes see the colour of the combined wavelengths that pass through. For example, red light is absorbed by Fe^{2+} in olivine [$(Mg,Fe)_2SO_4$], olivine appears to be a greenish colour.

The intrinsic band gap energies decide the electron transition. For instance, CdS has a bandgap energy of 2.45 eV. The blue and violet wavelengths have enough energy to promote electrons into the conduction band. These wavelengths are absorbed. The light with longer wavelengths are transmitted through the CdS, thus we see the complementary colour of yellow. Si has a band gap of ~1.1 eV and it absorbs all the wavelengths from the visible light. Therefore, Si appears opaque and black.

In some compounds, replacement of ions by transition metal or rare earth elements may create new energy levels. For example, when Cr^{+3} ions replace Al^{+3} in Al_2O_3, the new energy levels absorb the light in the violet and green/yellow portions. Red wavelengths are transmitted. This produces the beautiful reddish colour in ruby. Ceramic colorants are widely used as pigments in paints and for decorations. They have the advantage of high temperature resistance. For example, porcelain enamels that are fired at ~800 °C require ceramic colorants where other types of pigments will be destroyed. Zr compounds can be used at even higher temperatures (up to 1800 °C).

Table 8.3 Effect of doping on colour produced in glasses.

Ion	Colour
Cr^{2+}	Blue
Cr^{3+}	Green
Cu^{2+}	Blue-green
Mn^{2+}	Orange
Fe^{2+}	Blue-green
U^{6+}	Yellow

Glasses can also be doped with ions that produce selective absorption and transmission, and obtain certain colours. Table 8.3 shows the relations of doping and colour. Photosensitive glasses are based on SiO_2-Na_2O-Al_2O_3-ZnO containing halides and sensitisers (Ag^+, Ce^{3+}, Sn^{2+} and Sb^{3+}). Precipitates will form during cooling stage. The morphology and crystal structure of the precipitates can be controlled by ultraviolet light radiation and subsequent heat treatment. The different precipitates have different absorption bands, resulting in the ability of producing a broad range of colours. This type of glass is called photochromatic glass.

The colours of metals are related to their ability to absorb and reflect light with certain range of wavelengths. For example, Au and Cu absorb the shorter wavelength photons on the blue and violet end and reflect photons with longer wavelength, therefore, they appear yellow-reddish. Al and Ag reflect photons of all wavelengths in the visible spectrum, they appear white.

8.7.3 *Refraction*

Even if the photons are transmitted by the material, they lose some of their energy. The wavelength of the transmitted photons becomes slightly longer. The photons behave as if the speed of the light in the material has been reduced, and the photon beam changes the directions. In the case of a light beam travelling in a vacuum to a material:

$$n = c/v = \lambda_{vac}/\lambda = (sin\ \alpha)/(sin\ \beta), \qquad (8.4)$$

where n is the index of refraction, c is the speed of light in vacuum, v is the speed of the light in the material, and α and β are the incident and refracted angles. Table 8.4 lists the refraction indexes of some materials.

Table 8.4 Index of refraction of selected materials for photons of wavelength 589 nm.

Material	Index of refraction
Ice	1.309
NaCl	1.544
Quartz	1.544
Diamond	2.417
TiO_2	2.7
Water	1.333
Plastics	1.5
Glasses	1.5
Leaded glass	2.5
Air	1.0

If the photons travel from material 1 to material 2 with speeds of v_1 and v_2, Eq. (8.4) will be:

$$v_1/v_2 = n_2/n_1 = (sin\ \alpha)/(sin\ \beta). \tag{8.5}$$

This means that the ratio of the index of refraction depends on the ratio of the light travelling speed in the two materials. Eq. (8.5) can also be used to determine whether the beam will be transmitted or not. The light will be reflected if the angle β becomes greater than 90 degrees. This is called total internal reflection.

More interaction and refraction of the photons occur when the electrons in the material are more easily polarised. There is a relationship between the index of refraction n and the dielectric constant (permittivity) ε of the material:

$$n = \sqrt{\varepsilon_r}\ . \tag{8.6}$$

This means that a material that is polarised easily has a higher index of refraction.

Note that n is not a constant for a material, it is related to the frequency of the incident radiation. n is also anisotropic in certain crystalline materials.

8.7.4 *Reflection*

When a beam of photons strikes a metal, electrons are excited into a higher energy level in the conduction band. We would expect that the photons would be absorbed and the metal would appear black.

However, most metals are not black. This is because that as soon as the electrons are excited into the higher energy level, they immediately drop back to their stable levels, emitting photons of the same wavelength as the incident photons. This is reflection.

Metals are reflective materials. Even in materials which are not opaque, some reflection of the incident beam may occur.

The reflectivity R is defined as the percentage of the beam that is reflected. R is also related to the index of refraction n. In vacuum, we have:

$$R = [(n - 1)/(n + 1)]^2 \times 100. \tag{8.7}$$

This means that R increases with the increasing n. If $n = 1$, $R = 0$, which means that the beam is totally transmitted and no reflection occurs.

8.8 Case Study – Optical Fibers and Photonic Systems

Alexander Bell transmitted a telephone signal over a distance of 200 m using a beam of sunlight as the carrier in 1880. This is the first time of "optical communication". However, this "photophone" did not have real application due to the lack of intense light source and low-loss transmission medium. About 100 years later, people started to replace electrical cable system with optical fiber system. From 1988 to now, more than 20 lightwave undersea systems have been established, including Transatlantic, Transpacific and Pacific Basin Systems. Tasman-2 system was established in 1992, linking New Zealand with Australia. Nowadays big companies such as AT&T, MCI and Sprint all use fibre optical cables in their long distance networks almost exclusively.

8.8.1 *Advantages of optical telecommunications*

Optical telecommunications are rapidly growing as they have many advantages over conventional transmission using Cu cables, including:

1. Much greater bandwidth: this means larger numbers of massages can be transmitted at the same time. It has been estimated that a single optical fibre could simultaneously carry 500 million telephone conversations;
2. Immunity from electromagnetic interference and safer;

3. Smaller in size and lighter, thus a lower cost of building the system;
4. Stable or declining price: unlike Cu, which is a strategic material, raw materials for optical fibres are cheap and plenty. Sand is the basic material for optical glass; and
5. Better corrosion resistance and more friendly to our environments.

8.8.2   *Material properties for optical fibres*

There are two basic requirements for the optical fibres: high transparency and no light leakage.

Silica glass (SiO_2 doped with GeO_2, P_2O_5, and F) is the basic material for optical fibres. The transparency of glass is mainly determined by its purity. The transparency was slowly improved from 3000 B.C. in ancient Egypt to 1966, when optical fibres were needed for telecommunication. Today's silica fibres are 100 orders of magnitude ($\times 10^{100}$) more transparent than those in 1966 (Fig. 8.15).

The index of refraction n is the major factor which is related to the leakage. n of the fibre should be greater than n of the surroundings, to reflect the light and keep it inside the fibre, provided the angle between the beam and the length of the fibre is small enough.

Composite fibres are often used, which have an outer layer with a lower n than the inner. The light beam will travel in the inner fibre core. Another way to make n of the core higher or n of the outside fibre lower is doping. The core is usually doped with germanium and phosphorous pentoxide to increase its n, while the cladding should be pure silica or silica doped with B_2O_3 or fluoride. The doping method has an advantage that the roundabout of the light path will shorten the total light travelling distance. Fig. 8.16 illustrates the two methods of controlling the leakage from an optical fibre.

The wavelength of the photon signal is also a factor that may affect the losses. For instance, ultraviolet photons can excite electrons in the atoms of the glass fibre and produce new photons with different wavelength. Inferred beams can cause vibrational excitation of the bonds between the atoms. Therefore, appropriate wavelength should be used to minimise the internal losses.

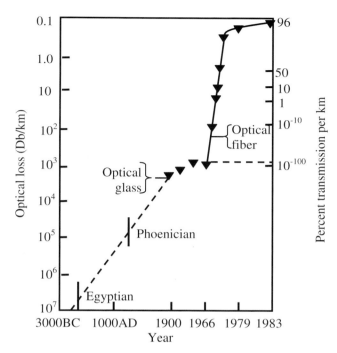

Fig. 8.15 Historical improvement in glass transparency (After National Research Council, "Materials Science and Engineering For The 1990s", National Academy Press, DC, 1989, 24).

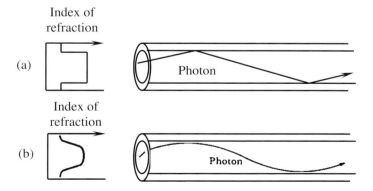

Fig. 8.16 Two methods for controlling leakage from an optical fibre: (a) a composite glass fibre (b) a glass fibre doped at the surface to lower the index of refraction.

Preform Manufacture

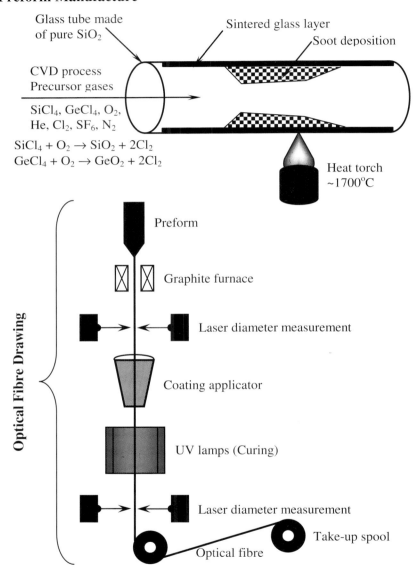

Fig. 8.17 Schematic drawings of a typical manufacturing process of optical fibres.

There are two main steps in the process of converting raw materials into optical fibres: manufacturing of the pure glass preform and drawing of the preform. Fig. 8.17 is a schematic drawing of a typical manufacturing process of optical fibres.

8.8.3 *Structure of optical fibre cables*

There are different types of cables for all conventional installations. These include underground conduit, direct buried, aerial, underwater, and in-building. The structure of a cable usually consists of outside sheath and wrapping, central strength member, buffer materials and optical fibre. Figs. 8.17 and 8.18 show the structures of an underground cable and an aerial cable, respectively.

The transmission properties of a cable are also strongly affected by the stresses it encounters, because any stress-induced density fluctuation in the glass can change the light's direction or scatter light out of the fibre. Therefore, cables are designed to minimise stresses during construction and installation, and over a wide range of environmental temperatures.

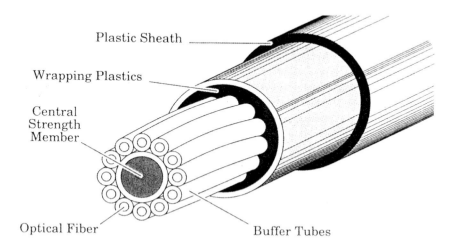

Fig. 8.18 Underground duct type cable (after "Tomorrow's Technology").

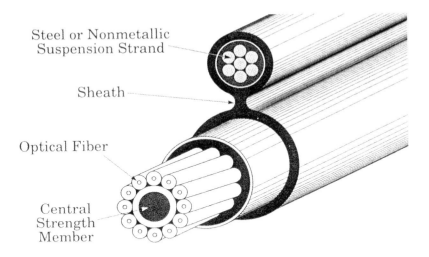

Steel or Nonmetallic
Suspension Strand

Sheath

Optical Fiber

Central
Strength
Member

Fig. 8.19 Aerial type optical fibre cable (after "Tomorrow's Technology").

8.8.4 *Photonic communication system*

A photonic system consists of a laser generator (emitter) as a light source, an optical fibre system for signal transmission, and a receiver for signal processing and conversion.

Emitter: Group III-V semiconductors such as GaAs, GaAlAs have energy gaps that provide emitted photons in the visible spectrum. They are used as the laser generator materials. Laser is generated by a voltage. By varying the voltage applied to the laser, the intensity of the laser beam can be varied. This beam can then be used to carry information.

Transmitter: Silica glass is used to produce optical fibres to transmit the light signals over long distances.

Receiver: A receiver consists of light-emitting diodes or Si semiconductor diodes. When a photon reaches the *p-n* diode, an electron is excited into the conduction band, leaving behind a hole. If a voltage applied to the diode, the electron-hole pair will create a current that can be amplified and further processed. Fig. 8.20 shows a schematic drawing of a photonic communication system.

Normally the received signals are immediately converted into electronic signals and processed using conventional Si based devices. Certain materials (so called "non-linear optical materials", or "acousto-optical materials") such as $LiNbO_3$, $LiTaO_3$, $PbMoO_4$ and $PbMoO_5$ can act as transistors and amplify the optical signals. The indexes of refraction of these crystals change when a pressure or acoustic wave is applied. Development of photonic transistors may lead to integrated optics and optically based computers.

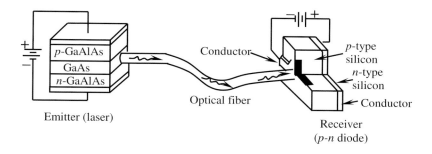

Fig.8.20 A photonic system for transmitting information (after D. Askeland, "The Science and Engineering of Materials", Chapman and Hall, 2nd SI Ed.1990, p758).

Summary:

The optical properties of a material depend on the interactions between the radiation and atomic structure of the material. The most important factor to decide the interaction is the energy required to cause an electron to jump from the valence band to the conduction band. If the energy of the light is greater than the band gap energy in the material, the light will be absorbed or reflected. If the energy is lower than the band gap, the light is likely to be transmitted through the material. Optoelectronic materials are experiencing rapid development in recent decades. We can influence the optical properties by altering the atomic arrangement of the materials. This permits us to create materials with special photonic properties, which are used for such as light-emission/display, laser, optical telecommunication, digital recording and photonic computing systems.

Important Concepts:

Characteristic spectrum: The spectrum of radiation that emitted from a material which occurs at fixed wavelengths corresponding to particular energy level differences within the atomic structure of the material.

Continuous spectrum: Radiation emitted from a material having all wavelengths longer than a critical short wavelength limit.

Short wavelength limit: The shortest wavelengths or highest energy radiation that emitted from a material under particular conditions.

Luminescence: Conversion of radiation to visible light.

Fluorescence: Emission of radiation from a material only when the material is actually being stimulated.

Phosphorescence: Emission of radiation from a material after the material is stimulated.

Laser: A beam of monochromatic, coherent radiation produced by the controlled emission of photons.

Thermal emission: Emission of photons from a material due to excitation of the material by heat.

Linear absorption coefficient: Describes the ability of a material to absorb radiation.

Absorption edge: The wavelength at which the absorption characteristics of a material abruptly change.

Index of refraction: Relates the change in velocity and direction of radiation as it passes through a transparent material.

Reflectivity: The percent of the incident radiation that is reflected.

Questions:

1. Discuss the basic property and fabrication process of silica based optical fibers.
2. Discuss the operation principle of semiconductor laser.
3. Determine the difference in energy between electrons in (a) the K and L shells, (b) the K and M shells, and (c) the L and M shells in J in tungsten. The K_α, K_β and L_α radiations in W are 0.0211, 0.0184 and 0.1476 nm, respectively.
4. The intensity of a phosphorescent material is reduced to 95% of its original intensity after 2.56×10^{-6} s. Calculate the time required for the intensity to decrease to 0.1% of the original intensity.
5. An optical fibre ($n = 1.4581$) is to be clad with a second glass to ensure internal reflection that will contain all light travelling within 3 of the fibre axis. What is the maximum index of refraction of glass which could be used for the cladding?
6. A composite glass fibre has the inner portion of the fibre with an index of refraction of 1.5 and outer sheath of the fibre with an index of refraction of 1.45. Calculate the maximum angle that a light beam can deviate from the axis of the fibre without escaping from the inner potion of the fibre.
7. A beam of X-ray having a wavelength of 0.2291 nm is passed through a platinum (Pt) foil. If only 1% of the original intensity of the beam is transmitted, what is the thickness of the foil? The linear absorption coefficient of Pt is 6.54×10^5/m.
8. Suppose that 25% of the intensity of a beam of photons entering a material at $90°$ to the surface is transmitted through a 1 cm thick material with a dielectric constant of 1.44. Determine the fraction of the beam that is (a) reflected, and (b) absorbed, (c) calculate the linear absorption coefficient of the photon in this material.

Chapter 9

Thermal and Thermoelectric Properties

Thermal properties are the responses of a material to the application of heat. These properties include heat capacity, thermal conductivity and thermal expansion. They are strongly influenced by the atomic vibration in the materials and in the case of thermal conductivity, also affected by the electronic properties. Thermoelectric properties deal with thermally induced electrical properties. They are also decided by the electronic structures of the materials.

9.1 Heat Capacity

9.1.1 *Phonon*

The atomic vibration may be expressed as an energy or the wavelike nature of the energy.

At 0 K, the atoms have a minimum energy. When heat is supplied to the material, the atoms gain thermal energy and vibrate at a particular amplitude and frequency. This vibration produces an elastic wave called "phonon".

The energy of the phonon can be expressed in terms of wavelength or frequency in the same way as for a photon with Planck's relation:

$$E = h \cdot v = h \cdot c / \lambda. \tag{9.1}$$

Therefore, the material gains or loses heat by gaining or losing phonons.

9.1.2 *Heat capacity and specific heat*

Heat capacity is the energy, or the number of phonons, that are required to change the temperature by 1 K for one mole of certain material.

Specific heat is the energy required to raise the temperature by 1 K per unit mass (1 kg in SI system) of a material. Table 9.1 lists the specific heat of selected materials at 300 K.

Table 9.1 Specific heat of selected materials at 300 K.

Material	Specific heat ($J \cdot kg^{-1} \cdot K^{-1}$)
Al	913
Cu	385
B	1030
Fe	440
Pb	126
Mg	1017
Ni	460
Si	700
Ti	523
W	130
Zn	385
Water	4187
He	5240
N	1040
Silica	1110
Al_2O_3	837
SiC	1050
Si_3N_4	720
Polymers	840-1470
	525

The relation between specific heat and heat capacity is:

$$\text{Specific heat} = (\text{heat capacity/atomic weight}) \times 1000. \quad (9.2)$$

Specific heat is used for most engineering applications.

The heat capacity can be expressed either at constant pressure, C_p, or at constant volume, C_v. At "relatively high temperatures", the internal energy per mole of a solid, E, can be expressed as:

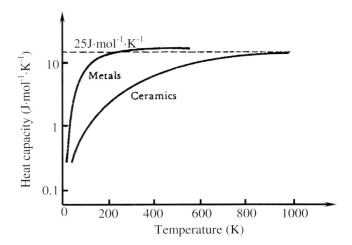

Fig. 9.1 Heat capacity as a function of temperature for metals and ceramics (after D. Askeland, "The Science and Engineering of Materials", Chapman and Hall, 2nd SI Ed.1990, p762).

$$E = 3R \cdot T \text{ cal/mole,} \qquad (9.3)$$

and $\qquad C_v = (dE/dT)_v = 3R = 5.96 \text{ cal/(mol·K)} = 25 \text{ J/(mol·K).} \qquad (9.4)$

Eq. (9.4) means that the heat capacity at constant volume, C_v, is a constant at relatively high temperatures for different materials.

The term of "relatively high temperatures", however, is not an accurate one. For metals, it means above or around room temperature; for ceramics, it means ~1000 °C. Fig. 9.1 shows the different heat capacities between metals and ceramics.

9.2 Thermal Expansion

When an atom gains thermal energy, it begins to vibrate as it has a larger atomic radius. Therefore, the average distance between atoms and the overall dimensions of the material increase. This causes thermal expansion.

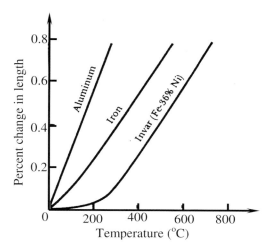

Fig. 9.2 Thermal expansion of Al, Fe and Invar alloy (after D. Askeland, "The Science and Engineering of Materials", Chapman and Hall, 2nd SI Ed.1990, p766).

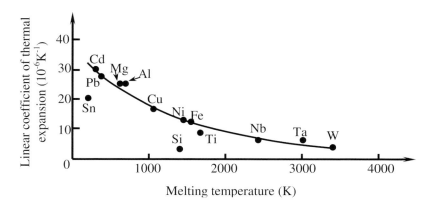

Fig. 9.3 Relationship between the linear coefficient of thermal expansion at 25 °C and the melting point (after D. Askeland, "The Science and Engineering of Materials", Chapman and Hall, 2nd SI Ed.1990, p765).

The linear coefficient of thermal expansion α is used to express the expansion:

$$\alpha = 1/l \times \Delta l / \Delta T, \qquad (9.5)$$

where l is the initial length, Δl is the change of the length, and ΔT is the increase in temperature.

Fig. 9.2 shows the thermal expansion of Al, Fe and a Fe-Ni alloy. Linear coefficient of thermal expansion, α, is related to the strength of the atomic bonds. Stronger atomic bonding results in higher melting points, and also smaller thermal expansion α (Fig. 9.3).

Several factors should be noted when considering the thermal expansion:

1. The expansion behaviour of some single crystals or textured materials may be anisotropic, i.e. the thermal expansion in one crystal orientation may be greater than in others.
2. Abrupt dimensional change takes place when the material undergoes a phase transformation. This is important in materials thermal processing such as heat treatment because the abrupt dimensional change may cause parts cracking and fracture failures.
3. The linear coefficient of thermal expansion, α, is not a constant at all temperatures. Normally, α can either be expressed as a complicated temperature dependent function, or listed as a constant for a particular temperature range.
4. Interaction of materials with electric or magnetic fields may prevent normal expansion. For instance, Invar alloy (Fe-36%Ni) has almost no dimensional change below 200 °C (its Curie temperature). This makes Invar alloy have special application as bimetallics.

9.3 Thermal Conductivity

The thermal conductivity is a measure of the rate at which heat is transferred through a material. Table 9.2 lists the thermal conductivity of selected materials at 300 K. Thermal conductivity, K, is defined as follows:

$$Q/A = K \times (\Delta T/\Delta x), \qquad (9.6)$$

where Q is the heat transferred, A is the area, and $\Delta T/\Delta x$ is temperature gradient (see Fig. 9.4). K is similar to diffusion coefficient D where D is a measure of the rate of material transfer.

Table 9.2 Thermal conductivity of selected materials at 300 K.

Material	Thermal conductivity $(\mathbf{W \cdot m^{-1} \cdot K^{-1}})$	Material	Thermal conductivity $(\mathbf{W \cdot m^{-1} \cdot K^{-1}})$
Al	238	60/40 brass	126
Cu	400	Cu-30%Ni	42
Fe	78	Argon	0.018
Mg	156	Carbon (graphite)	335
Pb	35	Carbon (diamond)	2320
Si	139	Soda-lime glass	0.96
Ti	22	Vitreous silica	1.34
W	174	Vycor glass	1.25
Zn	117	Fireclay	0.27
Zr	23	Al_2O_3	16
Low carbon steel	52-65	ZrO_2	5.0
Ferrite	75	Si_3N_4	15
Cementite	50	Silicon carbide	88
Stainless steel	15-30	Polyimide	2.1
Grey cast iron	80	6,6-Nylon	121
Al-1.2%Mn alloy	180	Polyethylene	188

(After D Askeland, "The Science and Engineering of Materials", Chapman and Hall, 2nd SI Ed.1990, p768).

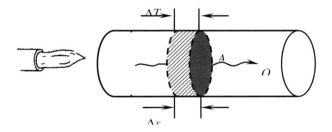

Fig. 9.4 Thermal conduction.

9.3.1 *Thermal conductivity in metals*

The electronic contributions are the dominant factor in the thermal conduction in metals, since the valence band is not completely filled. They are easily excited into the conduction band, thermal energy can be transferred by the electrons. The amount of energy transferred depends on the number of free electrons and their mobility. Thus the relationship between thermal and electrical conductivity can be expressed as:

$$K/(\sigma \cdot T) = L. \tag{9.7}$$

This means that the ratio of thermal and electrical conductivities is a constant at a certain temperature. This is called Wiedeman–Franz relationship (see Chapter 2), and L is the Lorentz constant.

The lattice and microstructure defects also affect the thermal conductivity. For instance, cold-worked metals (with a high concentration of defects) show lower thermal conductivity.

9.3.2 *Effect of temperature on thermal conductivity*

The effect of temperature on the thermal conductivity of metals is rather complicated. Fig. 9.5 shows the plots of thermal conductivity vs. temperature for a few metals and ceramics. When temperature increases, there are two factors that influence the thermal conductivity in an opposite way:

1. As temperature increases, the mobility of electrons decreases, therefore, the thermal conductivity decreases.
2. As temperature increases, the energy of electrons increases, therefore, the thermal conductivity increases.
3. As temperature increases, the lattice vibration increases, therefore, the thermal conductivity increases.

Therefore, the effect of temperature on the thermal conductivity depends on the dominant factor(s). Usually, when temperature increases, thermal conductivity first decreases due to the decrease of the electron mobility, then stays constant or increases slowly due to the increase of the electron energy and lattice vibration.

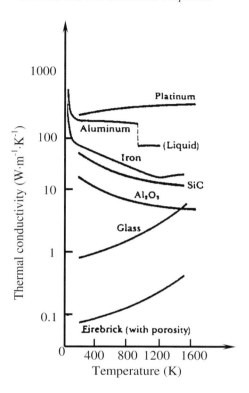

Fig. 9.5 The effect of temperature on the thermal conductivity (after D. Askeland, "The Science and Engineering of Materials", Chapman and Hall, 2nd SI Ed.1990, p769).

9.3.3 *Thermal conductivity in ceramics and semiconductors*

In ceramics, the energy gap is too large for electrons to be excited into the conduction band except at very high temperatures. Lattice vibration (phonos) is responsible for the heat transfer. Therefore, most ceramics and glasses have a higher thermal conductivity at high temperatures due to the higher energy phonos and some electronic contributions. However, some other ceramics (e.g. Al_2O_3 and SiC) have lower K at higher temperature because of the complicated conduction mechanism.

Microstructure also influences the thermal conductivity. Materials with a close-packed structure and high modulus of elasticity produce high energy phonos that encourage high thermal conductivity.

Porosity reduces thermal conductivity. Materials with high porosity are used for thermal insulation.

In semiconductors, both phonons and electrons may contribute to the thermal conductivity. At relatively low temperatures, phonons are the principal carriers of energy. At higher temperatures, electrons are excited through the small energy gap into conduction band, and the thermal conductivity increases significantly. The study of thermal conductivity of semiconductors is important for releasing heat from very closely packed chips.

9.4 Thermoelectricity in Metals

9.4.1 *Thermoelectricity*

Thermoelectricity deals with the phenomena of thermally induced electricity. The main applications include:

1. thermocouple,
2. thermoelectric heating,
3. refrigeration, and
4. electricity generation.

The thermocouple is the most important applications of thermoelectricity in industries. In a thermocouple, a voltage is created between the hot and cold junctions of two dissimilar metals. The temperature differences change the distribution of electrons in the energy states at the hot and cold ends, leading to a flow of electricity. This is called Seebeck Effect (Thomas Johann Seebeck, 1821).

When a piece of metal is heated in one end, more electrons will be excited in this end than in the cold end. The excited electrons then move toward the cold end, create a voltage V_1.

We cannot measure V_1 with a single wire, therefore, another wire is needed.

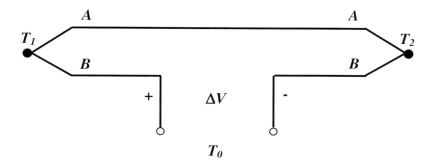

Fig. 9.6 Schematic drawing of Seebeck effect in a thermoelectric material.

If the second wire was assumed to have a thermally induced voltage of V_2, the voltage difference will be:

$$\Delta V = V_1 - V_2. \tag{9.8}$$

If the two wires are made of the same material, then $V_1 = V_2$ and $\Delta V = 0$. So the material for the two wires needs to be different. When one junction of the thermocouple is held at a known temperature (e.g. ice water), ΔV is then a function of the temperature in the other end.

$$S_{12} = \Delta V/\Delta T = \Delta V_1/\Delta T - \Delta V_2/\Delta T = S_1 - S_2, \tag{9.9}$$

where S_{12} is thermoelectric power of the junction, which depends on the two materials. It is the voltage produced by the two materials with 1 K temperature difference.

S_1 and S_2 are the thermoelectric power (or Seebeck coefficient) for the two metals. Thermoelectric powers (Seebeck coefficients) of selected metals are given in Table 9.3.

The relation between the measured voltage ΔV and thermoelectric power is:

$$\Delta V = (S_1 - S_2) \times \Delta T. \tag{9.10}$$

Table 9.3 Thermoelectric power of selected metals.

Material	S at 273K (μV/K)	S at 300K (μV/K)
Ag	+1.38	+1.51
Al	-1.6	-1.8
Au	+1.79	+1.94
Cu	+1.70	+1.84
Mo	+4.71	+5.57
Pb	-1.15	-1.3
Pd	-9.00	-9.99
Pt	-4.45	-5.28

9.4.2 *Thermocouple materials*

Constantan (60Cu-40Ni) and Fe thermocouples (Type-J) are widely used from low temperatures up to ~850 °C. They have very good accuracy and stability. Because of the iron component, these thermocouples are susceptible to oxidation and rusting.

Table 9.4 Thermoelectric EMFs of typical thermocouple materials with respect to Pt (Reference at °C).

Material	EMF at 100°C (mV)	EMF at 200°C (mV)
Alumel	-1.29	-2.17
Chromel	+2.81	+5.96
Constantan	-3.51	-7.45
Copper	+0.76	+1.83
90%Pt-10%Rh	+0.64	+1.44

Chromel (90Ni-10Cr) and Alumel (94Ni-2Al-3Mn-1Si) can be used up to 1200 °C. They have good thermal stability and excellent oxidation resistance, and are inexpensive. Chromel-Alumel thermocouples (Type-K) offer a wide measurement range with good temperature precision and are widely used for furnace temperature measurement and control.

Pt (platinum) and Pt-10%Rh (rhodium) (Type-S) and Pt and Pt-13%Rh (Type-R) thermocouples can be used up to ~1600 °C. It is very stable and can be used in oxidising and corrosive atmospheres, but they are expensive and can also be easily contaminated.

W–Re (tungsten-rhenium) alloy thermocouples can be used at temperatures up to 2400 °C. But these thermocouples have a poor oxidation resistance and are very brittle. Special sheaths are needed to provide additional protection from oxidation and mechanical stress. The most commonly used W–Re pairs are: W–W-26%Re, W-3%Re–W-25%Re and W-5%Re–W-26%Re.

B_4C–C can be used up to 2200 °C. The output of this type of thermocouple is very high, ~290μV/K. They can only be used in vacuum or non-oxidising atmospheres.

Table 9.4 is tabulating the electromotive forces (EMFs) of typical thermocouple materials with respect to Pt.

9.4.3 *Peltier effect*

Peltier effect was discovered by the French physicist Jean Peltier (1785–1845) in 1834. It describes a phenomenon that when a current I flows from material a to material b, heat will be released or absorbed in the junction between a and b.

$$Q_p = (\Pi_a - \Pi_b) \times J, \qquad (9.11)$$

where Q_p is the heat released or absorbed; Π is the Peltier coefficient; and J is the current density. The Peltier coefficients represent how much heat current is carried per unit charge through a given material. This effect is the basis of Peltier cooler/heater (or thermoelectric heat pump) which transfers heat from one side of the device to the other side.

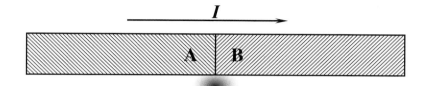

Fig. 9.7 Schematic of Peltier effect.

9.4.4 *Thomson effect*

Thomson effect was discovered by William Thomson in 1851. It says that heat is absorbed or produced when current flows in a material with a temperature gradient. The heat is proportional to the electric current and the temperature gradient. The constant, known as the Thomson coefficient is related by thermodynamics to the Seebeck coefficient.

$$Q_t = \mu_A \cdot J(T_1 - T_2), \qquad (9.12)$$

where Q_t is the heated produced or absorbed, μ_A is the Thomson coefficient, and J is the current density.

Antimony (Sb), cadmium (Cd), copper (Cu) and silver (Ag), display a positive Thomson effect, while bismuth (Bi), cobalt (Co), iron (Fe), nickel (Ni) and platinum (Pt) exhibit a negative Thomson effect. In one metal, lead (Pb), the Thomson effect is zero. In certain metals, the Thomson effect may reverse sign as the temperature is increased or as the crystal structure is changed.

9.5 Figure-of-merit

The figure-of-merit, Z, of a thermoelectric material is defined as

$$Z = S^2 \cdot \sigma/(\kappa_L + \kappa_E), \qquad (9.13)$$

where S is the Seebeck coefficient, σ is the electrical conductivity, κ_E is the electronic thermal conductivity and κ_L is the lattice thermal conductivity of the material.

Since Z varies with temperature, a useful dimensionless figure-of-merit can be defined as ZT. Values of $ZT = 1$ are considered good and ZT values of $2 - 3$ had been reported to date. Thermoelectric materials with a ZT value in the range of 3–4 would be a good candidate for thermoelectric devices that can compete with mechanical generation and refrigeration in efficiency. Recent research in thermoelectric materials is focusing on increasing the Seebeck coefficient and reducing the thermal conductivity, especially through nanostructural engineering of the materials (quantum wells, superlattices, nanowires and quantum dots).

9.6 Thermoelectricity in Semiconductors

In semiconductors, thermoelectric effects are much larger than in metals. Thermoelectric power S is usually two orders of magnitude larger than that of metals, and S can be positive or negative. In n-type semiconductors, electron is the majority carrier, S is negative. In p-type semiconductors, hole is the majority carrier. S is positive.

Because more electrons are excited into the acceptor levels at the hot end, more holes will be available there. Electrons near the top of the valence band at the cold end can lower their energies by moving into holes at the hot end. The hot end becomes negative charged, and the cold end becomes positively charged, resulting in a positive S.

9.6.1 *Typical materials*

A large number of semiconductor materials had been investigated. Some of them emerged with Z values significantly higher than in metals or alloys. The much higher thermoelectric effects are related to their temperature sensitive carrier density. Typical bulk thermoelectric materials can be categorised depending on their operation temperature.

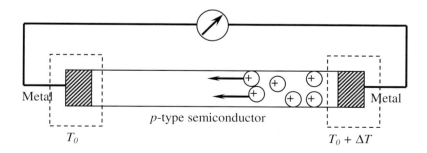

Fig. 9.8 Schematic of Seebeck effect in a *p*-type semiconductor.

1. Low temperature thermoelectric materials based on bismuth (Bi) with antimony (Sb), tellurium (Te), and selenium (Se). They can be used up to ~450 K;
2. Intermediate temperature thermoelectric materials based on lead telluride which can be used up to ~850 K; and
3. High temperature thermoelectric materials based on silicon germanium alloys for operation up to 1300 K.

9.6.2 *Applications*

The thermoelectricity of semiconductors has been used to provide thermoelectric heating, refrigeration and electricity generation. Typical thermoelectric devices are made from alternating *p*-type and *n*-type semiconductor elements that are connected by metallic interconnectors.

Fig. 9.9 depicts the simplest thermoelectric generator consisting of *p*-type and *n*-type thermoelement connected electrically in series and thermally in parallel.

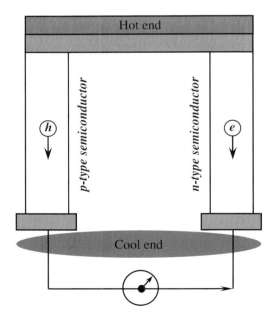

Fig. 9.9 Schematic of thermoelectric generator.

Heat is absorbed at one side of the couple and rejected from the opposite side. Charge flows through the *n*-type element, crosses the metallic interconnect and then passes into the *p*-type element. As a result, an electrical current is produced, proportional to the temperature gradient between the hot and cold ends.

For waste heat recovery up to 500 K, Bi_2Te_3 alloys have been proved to possess the greatest figure of merit for both *n*- and *p*-type thermoelectric systems. For power generation in temperatures ranging from 500 to 900 K, PbTe, GeTe and SnTe are commonly used. The highest *ZT* in optimised *n*-type materials is ~0.8. Alloys, particularly with $AgSbTe_2$, have been reported to have *ZT* > 1 for both *n*-type and *p*-type materials. A *p*-type alloy, $(GeTe)_{0.85}(AgSbTe_2)_{0.15}$, commonly referred to as TAGS, has a *ZT* > 1.2. This material has been successfully used in long-life thermoelectric generators. For thermoelectric generators operating at temperatures higher than 900 K, silicon-germanium (Si-Ge) alloys have typically been used.

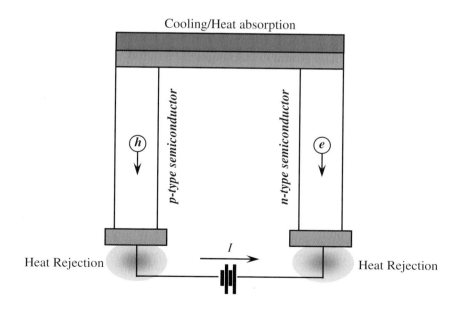

Fig. 9.10. A schematic of thermoelectric cooler.

On the other hand, if an electric current is applied, the thermoelectric device may act as a cooler with which heat is pumped from the cold end to the hot end. This is the Peltier effect. The temperature gradient between these two ends will vary according to the magnitude of current applied. For thermoelectric refrigeration, n-type BiSb has been coupled with p-type $(Bi,Sb)_2(Te,Se)_3$.

9.7 Case Study – Temperature Control

Furnaces and various heating devices are widely used in industries and our everyday life. These systems often need accurate and automatic temperature control. Here, we use a high temperature furnace as an example to demonstrate how temperature can be automatically controlled.

9.7.1 *Temperature measurement*

Thermocouples are the most commonly used temperature measuring devices (Fig. 9.11). Nowadays, ice-water is not necessary as circuits consisting of thermistors can be used to compensate room temperature and act as the cold end. A third wire can also be added into the pair without changing the voltage signal according to the law of intermediate elements (Fig. 9.12).

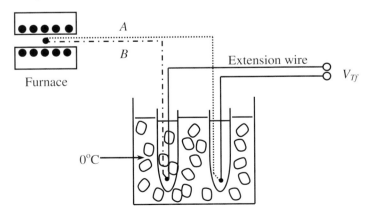

Fig. 9.11 Thermocouple ice-water cold end setting (after R. Speyer, "Thermal Analysis of Materials", Marcel Dekker Inc., 1994, 15).

Fig. 9.12 Law of intermediate elements.

Table 9.5 Thermocouple polynomial coefficients.

	Type-E -100-1000°C ±0.5°C	Type-J 0-760°C ±0.1°C	Type-K 0-1370°C ±0.7°C	Type-S 0-1750°C ±1°C
a_0	0.104967248	-0.048868252	0.226584602	0.927763167
a_1	17189.45282	19873.14503	24152.10900	169526.5150
a_2	-282639.0850	-218614.5353	67233.4248	-31568363.94
a_3	12695339.5	11569199.78	2210340.682	8990730663
a_4	-448703084.6	-264917531.4	-860963914.9	-1.63565E+12
a_5	1.10866E+10	2018441314	4.83506E+10	1.88027E+14
a_6	-1.76807E+11		-1.18452E+12	-1.37241E+16
a_7	1.71842E+12		1.38690E+13	6.17501E+17
a_8	-9.19278E+12		-6.33708E+13	-1.56105E+19
a_9	2.06132E+13			1.69535E+20
	Type-R 0-1760°C ±0.8°C	Type-T -160-400°C ±0.1°C	Type-B 0-700°C ±8.1°C	Type-B 700-1820°C ±0.9°C
a_0	2.04827	0.100860910	39.9967	169.055
a_1	167954	25727.94369	1.65406E+6	366415
a_2	-3.22234E+7	-767345.8295	-5.82049E+9	-1.14871E+8
a_3	8.60144E+9	78025595.81	1.33216E+13	3.57672E+10
a_4	-1.48143E+12	-9247486589	-1.76704E+16	-7.7762E+12
a_5	1.60727E+14	6.97688E+11	1.41E+19	1.1317E+15
a_6	-1.09762E+16	-2.66192E+13	-6.87206E+21	-1.07335E+17
a_7	4.57554E+17	3.94078E+14	2.00084E+24	6.33792E+18
a_8	-1.0622E+19		-3.19443E+26	-2.10872E+20
a_9	1.05141E+20		2.15035E+28	3.01366E+21

(After R Speyer, "Thermal Analysis of Materials", Marcel Dekker Inc., 1994, 16).

Thermocouples of types K, S, B, E, R and J are commonly used for furnaces. Table 9.5 lists the temperature coefficients with which the furnace temperature T can be calculated with the equation:

$$T = a_0 + a_1V + a_2V + \ldots \qquad (9.14)$$

9.7.2 *Furnace control*

Fig. 9.13 shows a block diagram of furnace instrumentation. SCR stands for a semiconductor-controlled rectifier. The signals received from the temperature transducer are used to control the power to the transformer. In order to obtain accurate temperature control, proportional integral derivative systems (PID) are used.

The simplest form of control is on-off type, which results temperature oscillation about the set-point. The proportional control is an important improvement, which can eliminate the temperature oscillation by applying corrective actions to the system. As shown in Fig. 9.14, a proportional band is assigned according to the deviation from the set-point. If the temperature is exceeds these outer limits, the system reverts back to the on-off control.

Integral function has been introduced to eliminate the parallel ramping shown in Fig. 9.14. This function continuously sums the difference between furnace and set-point. If the furnace temperature is persistently below the set-point, this accumulated area will increase the power output and therefore increase the temperature, till no area difference is accumulated.

Derivative function is used to minimise the overshooting and under-cooling problems in temperature control. The furnace temperature will more closely follow the programme during heating or cooling with this function (Fig. 9.14 bottom).

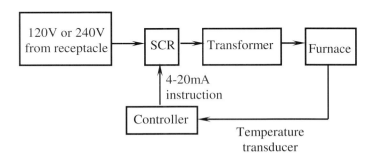

Fig. 9.13 Block diagram of furnace instrumentation (after R. Speyer, "Thermal Analysis of Materials", Marcel Dekker Inc., 1994, 23).

Proportional band, integral and derivative function can be adjusted by users. The best temperature control, however, are often resulted from repeated trial and error.

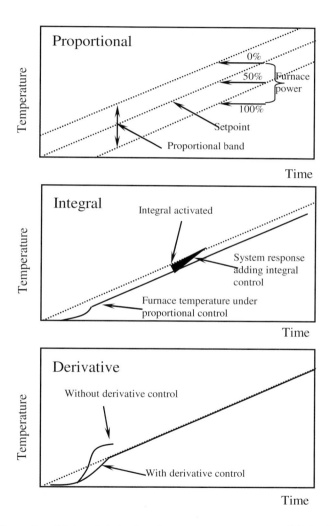

Fig. 9.14 Proportional-integral-derivative furnace temperature control (after R. Speyer, "Thermal Analysis of Materials", Marcel Dekker Inc., 1994, 31).

Summary:

The thermal activities of a material are mainly caused by the atomic vibration. Therefore, the thermal properties are influenced by the atomic structure and arrangement of the material. Thermal expansion is related to the strength of the atomic bonds. Strong atomic bonding results in both high melting points and low thermal expansion coefficients. As for the thermal conduction, the electronic contributions are the dominant factor in metals, which decreases with the increasing temperature. In ceramics and semiconductors, phonons play an important role and the thermal conductivity often increases with temperature. Thermoelectricity is used for temperature measurement and energy conversion. It may become an important topic of modern energy materials.

Important Concepts:

Heat capacity: The energy required to raise the temperature of one mole of a material one degree.

Specific heat: The energy required to raise the temperature of one kg of a material one degree.

Linear coefficient of thermal expansion: Describes the amount by which each unit length of a material changes when the temperature of the material changes by one degree.

Thermal conductivity: Measures of the rate at which heat is transferred through a material.

Thermocouple: A pair of conductor wires. The voltage produced between the wires can be used to measure temperature.

Thermoelectric power: The voltage developed in a thermocouple for a unit increase in temperature differences.

Phonon: An elastic wave that transfers energy through a material.

Questions:

1. Discuss the characteristics of Seebeck effect, Peltier effect and Thomson effect and the correlations between these thermoelectric effects.
2. How to define and improve the performance of a thermoelectric material?
3. Why do semiconductors normally have a better thermoelectric performance than metals? What are the major applications of semiconducting thermoelectric materials (discuss these based on device configuration and operation principle).
4. Suppose the temperature of 50 g of Nb increases from 50 to 125°C when heated for a period of time. Estimate the specific heat and determine the total heat required in joules.
5. An aluminium casting solidifies at 660°C. At that point, the casting is 250 mm long. Calculate the length after the casting cools to room temperature (20°C). The linear coefficient of thermal expansion of Al is 2.5×10^{-5}/K.
6. An aluminum plate $10 \times 10 \times 1$ cm thick separates a heat source at 300°C from a bath containing 1 litre of water at 25°C. Calculate (a) the heat transferred to the water each second and (b) the time required to warm the water to 26°C. The thermal conductivity of Al is 238 W/m·K over this temperature range.
7. A 10 cm long magnesium plate is coated with a thin layer of fused quartz. The composite is then heated from 20 to 400°C. Calculate the lengths of both the metal and quartz. If the coating is tightly bonded to the substrate, what is the elongation of the coating, providing the coating did not crack? The linear coefficients of thermal expansion for magnesium and fused quartz are 25×10^{-6} and 0.55×10^{-6}/K, respectively.

Chapter 10

Superconductivity and Superconducting Materials

10.1 Introduction

10.1.1 *Discovery*

Kamerlingh Onnes (1853–1926), a Dutch Physicist, discovered superconductivity. He spent most of his life studying low temperature physics. In 1908 he made liquid helium (He), which has a boiling temperature of 4.2 K. In 1911, when he (actually, his research student) passed a current through mercury with liquid He cooling, no resistance was measured ($<10^{-27}$ $\Omega \cdot$m). "When the temperature cooled to ~4.2 K," he wrote, "mercury has passed to a new state, may be called the superconductive state". Afterwards, he found Sn (~3.8 K) and Pb (~6 K) were also superconducting. They made a loop of Pb wire, started a current, and put it in liquid He. The current was still running after one year!

10.1.2 *Meissner effect*

In 1933, Walther Meissner and his collaborator, Ochsenfeld, at the Physikalisch Technische Reichsanstalt in Berlin, discovered the most fundamental property of a superconductor. A superconductor is more than a perfect electric conductor. It also has special magnetic properties.

A superconductor will not allow a magnetic field to penetrate its interior. If a superconductor is approached by a magnetic field, screen currents are set up on its surface, which create an equal but opposite field, cancelling the applied magnetic field and leaving a net zero field inside of the superconductor. This is called "Meissner Effect".

If a magnet is placed above a superconductor, an opposite magnetic field will be generated and push the magnet up. The magnet will be lifted and this is the principle of the levitating train.

10.1.3 *Theories*

Superconductivity was a mystery for many years. Einstein, Bohr, Heisenberg and many other scientists tried to explain it but were not successful.

The following people made contributions to the superconductivity theories: In 1935, Brothers Fritz and Heinz London proposed an electromagnetic theory explaining the finite penetration length of electromagnetic fields into the superconductor (London penetration depth). In 1950, Russian scientists Ginzburg and Landau used quantum mechanics to give a good macroscopic description of the superconducting state. However, these were all "phenomenological theories", which describe what experimentally observed without explaining what was occurring on a microscopic level.

In 1957, 46 years after Onnes' discovery, Bardeen, Cooper and Schrieffer of Illinois University in USA discovered a comprehensive theory later called BCS theory, which explains the mechanisms of superconductivity in a microscopic level. They won Nobel Prize for Physics in 1972.

10.1.4 *Conventional superconductors*

The superconducting materials were developed slowly. Firstly, more pure metals were found to be superconducting, including Hg, Pb, Sn, V, Nb etc, with Nb having the highest transition temperature for a pure element of 9.2 K. They have low superconducting temperatures (T_c) and low current densities (J_c). Once the current exceeds the critical current density, or if there is a magnetic field which exceeds the critical field, the material becomes non-superconducting. Very little practical applications can be made from these materials.

From 1950, alloys and metallic compounds were investigated. They are V_3Si (17 K, discovered in 1952), Nb_3Sn (18 K, discovered in 1954) and Nb_3Ge (23 K, discovered in 1973), which has the highest T_c till the discovery of the high temperature superconducting oxides in 1986.

These groups of superconductors have much higher T_c, J_c and H_c (critical magnetic field), which are essential for practical applications, especially for providing high magnetic fields. For example, Nb_3Ti and Nb_3Sn have $J_c \approx 10^6$ A/cm². It has been used for constructing superconducting magnets. "Put a million ampere current through a bar thick as your finger," J. Hulm (Director of Research at Westinghouse) wrote, "you can not do it with normal conductors such as copper unless you use a most powerful cooling system with a stream of cooling water the size of the Charles River to carry away the heat."

This development resulted in the industrial applications of superconductors, which are now called "conventional" or "low T_c" superconductors compared with high T_c superconducting oxides. Magnets wound from NbTi wires became the first commercially available applications of superconductivity and came to market in the 1960s. Nowadays these magnets are widely used in magnetic resonance imaging (MRI) and offer a profitable market, mainly in health care but also in scientific research.

10.1.5 *Discovery of high temperature superconductors*

Alex Müller and Georg Bednorz in IBM Zurich Labs were looking for superconductivity in a class of metal oxide with perovskite structure since 1984. In January 1986, they found that La-Ba-Cu-O showed superconductivity at ~30 K. They reported this with a title of "Possible High Temperature Superconductivity in Ba-La-Cu-O". People did not pay much attention to this new discovery until the Fall Meeting of Materials Research Society (MRS) in Boston, when researchers from Tokyo University not only confirmed this material had a T_c of ~30 K, but also showed it had the Meissner effect. They also gave the crystal structure of this new superconductor.

After 1986 MRS meeting, people were excited by the new high temperature oxide superconductors. Many groups spent days and nights studying and searching for new materials. Paul Chu (University of Huston) and M. Wu's (University of Alabama in Huntsville) groups raised T_c to 57 K in December 1986 and announced a new superconductor with $T_c = 93$ K in February 16th 1987, after they substituted La with Y.

People substituted La with ten other rare earth elements. Eight of the ten produced similar high temperature superconductivity (Table 10.1). Rare earth elements have different physical properties, but similar chemical properties. Therefore, people immediately thought that the formation of the Cu-O planes plays the key role of superconductivity.

Table 10.1 Superconducting temperatures in the compounds of rare earth substitutes for Y in Y123.

RE123 compound	T_c (K)
$YBa_2Cu_3O_{7-x}$ (Y123)	93.4
Nd123	95.3
Sm123	93.5
Eu123	94.9
Gd123	93.8
Dy123	72.7
Ho123	92.9
Er123	92.4
Tm123	92.5
Yb123	87.0
Lu123	89.5

(After Tarascon et al, "Novel Superconductivity", Plenum, 1987, 7050)

In 1988, H. Meada group (Japan) discovered Bi-Sr-Ca-Cu-O system which has no rare earth element in it. T_c was raised to ~110 K late with the Bi system. Shen and Hermann (Colorado University) discovered Tl-Ba-Ca-Cu-O system with a T_c around 125 K, which was the highest for several years afterwards. In 1993, about 5 years after the discoveries of Bi and Tl compounds, people found superconductivity occurs in similar Hg compounds and T_c has been raised to 130–160 K. All these superconductors have Cu-O planes in their crystal structures.

There were also reports of 200 K, 250 K and even > 0°C superconductivity, but they were too difficult to reproduce, and have not been accepted by the scientific society. The resistance drop may be caused by certain types of phase transformation other than the superconducting phase, or formation of a metallic phase, or due to measurement errors.

10.1.6 *Present situation in high T_c superconductor research*

The present high T_c superconductor systems mainly include:

Tl-Ba-Ca-Cu-O	125/115 K
Bi(Pb)-Sr-Ca-Cu-O	115/105 K
Y-Ba-Cu-O	95/85 K
Hg-Ba-Cu-O and Hg-Ba-Ca-Cu-O	130–155 K

Searching for new high temperature superconductors is still going. Most people are studying the known systems, trying to develop new compounds with higher T_c. They replace some elements in the existing systems, using doping to change the hole concentration, or apply high pressure to form unstable phases in the hope of finding new high temperature superconducting phases.

Some people refuse to predict the possible new systems. They argue that if people only follow so called "common sense" to develop superconductors, we would never find the new high temperature superconducting oxides. They want to develop totally new systems with higher T_c. However, there are so many possible chemical combinations and crystal structures. This type of searching will be very costly.

Most people in superconductor area study the existing systems, trying to develop new processing methods or optimise old processing techniques, control the microstructures and improve their properties in an effort to put these materials into industrial applications.

10.2 Superconducting Properties and Measurements

When a superconductor is cooled below a certain temperature, there is not only an abrupt loss of electrical resistance, many of the other physical properties including specific heat, thermoelectricity and thermal conductivity are also changed abruptly.

10.2.1 *DC resistivity*

When electrons flow through a metal conductor, their passage will be resisted by the vibrations and impurities of the lattice. Some kinetic energy of the electrons will lose and becomes heat. For a pure metal with a perfect crystal structure at 0 K, both residual and thermal resistivity should be zero. We called it "perfect conductor". Superconductors can be alloys and compounds. They have zero resistance at temperatures much higher than 0 K. Therefore, they are not "perfect" conductors. Their electrical conduction mechanism is different.

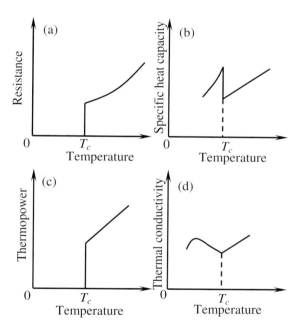

Fig. 10.1 Temperature dependence of various physical properties of a superconductor. (after D. Robins, "Introduction to Superconductivity", IBC Tech. Serv. Ltd., 1989, p15).

When a DC current is applied to a superconductor, there will still be a resistance if the temperature is higher than a certain temperature T_c. T_c is called "superconducting transition temperature", or "superconducting critical temperature". When the temperature drops below T_c, the resistance will disappear (Fig. 10.1a). We say the superconductor is now superconducting. This only applies to the dc current. If an AC current is applied to a superconductor at a temperature below T_c, there will still be energy losses due to the constantly changing direction of the moving electrons. This part of the energy loss, however, is much smaller than the dc resistance part. For example, under an ac frequency of 50 or 60 Hz, the energy loss will be reduced to less than 1/1000 when using superconducting wires compared with using Cu wires.

10.2.2 Specific heat

At T_c, the specific heat, C, increases abruptly. Below and above T_c, the temperature dependencies of specific heat are different (Fig. 10.1b).

For example, Vanadium (V) has a $T_c = 5.03$ K. In normal state (> 5.03 K):

$$C = \gamma T, \qquad (10.1)$$

where γ is the temperature coefficient of specific heat.

In superconducting state (< 5.03 K):

$$C = a \times exp(b \cdot T_c / T), \qquad (10.2)$$

where a and b are constants for the superconducting material.

10.2.3 Thermoelectric effect and thermal conductivity

The thermoelectric effect also changes abruptly at T_c. For normal conductors, a temperature difference ΔT produces an electric voltage ΔV according to the Seebeck effect.

The thermoelectric power is defined as:

$$S = \Delta V / \Delta T. \qquad (10.3)$$

Superconductors do not show the Seebeck effect at or below T_c (Fig. 10.1c), i.e.

$$\Delta V/\Delta T = 0. \tag{10.4}$$

Thermal conductivity also changes abruptly when the material is cooled into the superconducting state because of the change in electronic structure (Fig. 10.1d).

10.2.4 *Superconducting circuits*

If a conductor ring is exposed to a uniform magnetic field of flux density B, which changes with time, the current circulating in the ring at time t will obey Lenz's law:

$$A \times (dB/dt) = R \times I(t) + L \times [dI(t)/dt], \tag{10.5}$$

where A is the area enclosed by the ring, R is the resistance of the ring, and L is the inductance of the ring.

If there is no external applied field, Eq. (10.5) can be written as:

$$0 = R \times I(t) + L \times [dI(t)/dt]. \tag{10.6}$$

Eq. (10.6) has the solution:

$$I(t) = I_0 \times exp(-R{\cdot}t/L). \tag{10.7}$$

This means that any initial current I_0 circulating in a circuit will be reduced exponentially to zero.

In the superconducting state, $R = 0$, we have

$$I(t) = I_0. \tag{10.8}$$

This means that the initial current I_0 will circulate in the superconducting ring forever. Such a current is called "persistent current".

10.2.5 *Magnetic properties of superconductors - Meissner effect*

The magnetic properties of superconductors are not directly related to their electrical properties. This is important in understanding superconductivity. The resistivity of a perfect conductor may be very close to zero, but it does not possess the special magnetic properties described below. Let us inspect the following experiments as shown in Fig. 10.2:

1. When a normal conductor (a pure metal with perfect crystal structure) is cooled to 0 K, it becomes a "perfect conductor". When a magnetic field is applied externally, the flux will pass around, instead of going through the sample. When the field is switched off, the sample will be field free.

2. If the sample is first put in the same field, the flux lines will pass through the sample. When the sample is cooled down to 0 K, the flux will still pass through it. When the field is switched off, the sample will retain its internal flux with its own current and field.

3. When a superconductor is first placed in a field and then cooled below T_c, a surface current will appear and reject the magnetic field, leaving a zero field inside of the superconductor. If the field is removed, the surface current in the opposite direction will cancel the magnetically induced currents. The superconductor will remain field and current free.

This is the "Meissner Effect". It can be seen that the magnetic behaviours of a "perfect conductor" and a superconductor are different. For the normal conductor at 0 K, we say the state is "frozen in". For superconductors, the state changes at T_c.

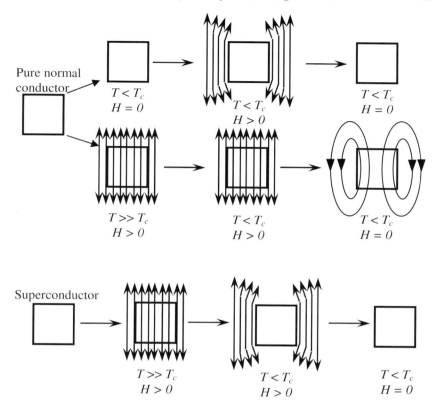

Fig. 10.2 Behaviour of conductors and superconductors in a magnetic field, Meissner Effect (after D. Robins, "Introduction to Superconductivity", IBC Tech. Serv. Ltd., 1989, p15).

The Meissner effect is limited by the strength of the flux. Strong magnetic fields will destroy superconductivity and the Meissner effect will disappear.

10.2.6 *Type I and Type II superconductors*

Superconductors have been divided into two groups according to their basic properties: Type I and Type II superconductors. The main difference of the Type I and II superconductors is in their magnetic behaviour (Fig. 10.3). For any superconductor at a temperature below T_c, there is a magnetic field above which the superconductive state cannot exist. This field is called critical magnetic field, H_c.

With the exception of Nb and V, all elemental superconductors are Type I, exhibiting a full Meissner flux expulsion before making a transition to the normal state at H_c. Type I superconductors are magnetically simple, exhibiting a diamagnetic behaviour in *B-H* plot. There is a vertical drop at the critical field H_c. This means that at the field H_c, Type I superconductors pass from superconductive state to normal state abruptly.

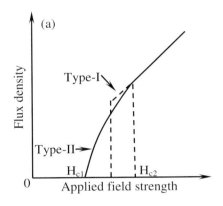

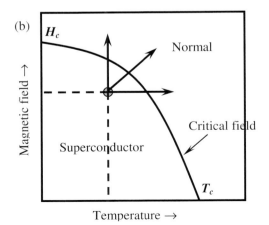

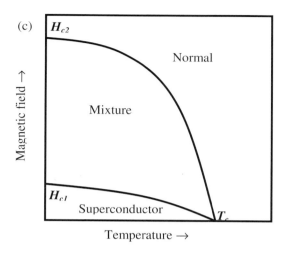

Fig. 10.3 (a) Variation of flux density with applied magnetic field for Type I and Type II superconductors, and correlation between temperature and magnetic field for Type I (b) and Type II (c) superconductors.

Type II superconductors have a more complex magnetic behaviour than Type I. Magnetic induction B begins to increase gradually at a field H_{c1} and slowly reaches the value, as in the normal state, at a field H_{c2}. In Type II superconductors, the Meissner effect is established only up to the lower critical magnetic field H_{c1}. It then becomes energetically favourable for the sample to undergo a transition in which flux lines carrying a single quantum of flux are nucleated at the surface, forming a flux line lattice in the bulk. This is known as the "mixed state", representing a mixture between the superconducting and normal phases. The mixed state extends to the upper critical magnetic field H_{c2}, above which the normal state is formed.

Type II superconductors retain a zero resistance to much higher magnetic fields (up to H_{c2}), which is very important for industrial applications.

Metals Nb, V and their alloys, intermetallic compounds (Nb_3Ti, Nb_3Sn, Nb_3Ge, Nb_3Ga, Nb_3Al, V_3Si, V_3Ga) and the high T_c superconducting oxides are Type II superconductors.

10.2.7 T_c and J_c measurement - "Four point" method

Superconducting properties can be measured electrically or magnetically. The "four point" method is an electric method. It is a direct measurement with current passing through the conductor, therefore it is called a transport measurement method.

10.2.7.1 T_c measurement

Fig. 10.4 shows the schematic drawing of the arrangement for T_c measurement. A small current is put through the sample. The specimen is lowered down into a dewar with liquid helium or liquid nitrogen. The voltage across the specimen is recorded against the reducing temperature. The voltage is then converted to resistance and the transition is measured from the resistance-temperature (R-T) plot.

Theoretically, the superconductive transition should take place at one temperature. This means that when T_c is reached, the whole conductor should be superconducting. In practice, the transition is often not so sharp, especially for polycrystalline high T_c superconductors. The whole conductor does not become superconducting at one temperature. The superconducting on-set temperature T_{on} and zero-resistance temperature $T_{R=0}$ are used to characterise T_c, as shown in Fig. 10.5.

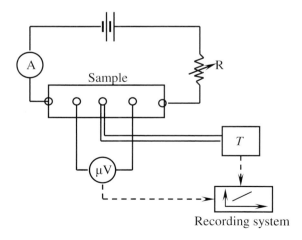

Fig. 10.4 "Four-point" measurement.

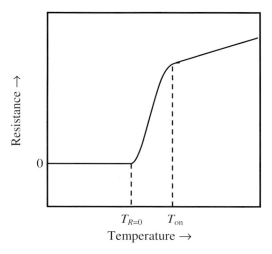

Fig. 10.5 T_c measurement: superconducting on-set temperature, T_{on}, and zero resistance temperature, $T_{R=0}$.

10.2.7.2 J_c measurement

J_c is the critical current density. When a DC current passes through a superconductor at a temperature below T_c, there will be no voltage drop in the specimen if the current is smaller than a certain value. This value is called critical current (I_c). Critical current density J_c is I_c divided by the cross section area that conducts the current:

$$J_c = I_c/A. \tag{10.9}$$

J_c represents the current carrying ability of a superconductor and is a very important property for industrial applications. If the current density $J > J_c$, the specimen will not stay in a superconductive state and will show resistance. In fact, J_c is the property that limits the most power applications for high T_c superconductors. J_c is a function of temperature and magnetic field. It is a microstructure-sensitive property of materials, and is strongly influenced by many materials related factors.

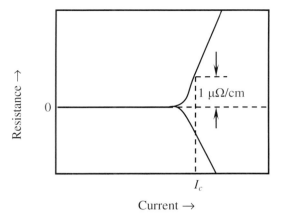

Fig. 10.6 J_c measurement: the criterion of I_c.

J_c measurement is similar as T_c, but the specimen is kept at a constant temperature (e.g. at 77 K in liquid nitrogen). The current that passes through the conductor is increased and a *I-V* curve is plotted. The voltage will follow $V = 0$ line as the specimen stays in the superconductive state. The transition of the voltage is used to measure I_c, and a criterion of 1 μV per cm specimen is commonly used as the transition point, as shown in Fig. 10.6.

J_c is strongly affected by the materials processing and microstructure, while T_c is not so sensitive to the microstructure. For industrial applications, especially for power transmission or storage, $J_c = 10^4 - 10^5$ A/cm^2 is typically required. Large amounts of work have been concentrated on J_c improvement.

10.2.8 *The relations between T_c, J_c and H_c*

T_c, J_c and H_c are related each other as shown in Fig. 10.7. A material is in the superconductive state only when the T_c, J_c and H_c are all inside the shaded surface. When any of T_c, J_c or H_c is outside the surface, the specimen will be in the normal state (i.e. non-superconducting state).

For any practical applications, all the three properties need to be considered. Therefore, the operating temperature for a superconductor has to be considerably lower than T_c to allow certain currents and fields.

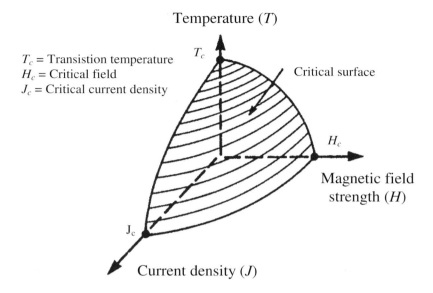

Fig. 10.7 The relationship of T_c, J_c, and H_c for a superconductor.

The relation between H_c and T_c is:

$$H_c = H_0 \left[1 - (T/T_c)^2\right], \tag{10.10}$$

where H_0 is the critical magnetic field of the superconductor at 0 K.

Bulk superconductors are often used in magnetic fields. For example, superconductors are used for making magnets to produce high fields. When superconductors are used for electricity transmission, the high current produces its own field. Therefore, the field dependence of J_c is an important consideration.

10.3 Theories of Superconductivity

The early stage theories, such as London theory and Ginzburg – Landau theory, are phenomenological theories which take some experimental results, make a simple assumption and calculate the consequences. If they agree with experiment, the phenomenological theory is considered a success. These theories explain the phenomena observed in experiments, but not the mechanisms of superconductivity in a microscopic level. For instance, Fritz London considered that superconducting electrons were in a special state called "closed shells" in an atom.

In diamagnetic materials, the energy gap between the valance band and conduction band is large. People believed that this also happens in superconductors. This energy gap was experimentally observed late. Because the "Energy Band Theory" was so useful to explain the electron conduction mechanisms in conductors, semiconductors and insulators, it was natural that people tried to use it to explain superconductivity.

It was believed that when a superconducting electron moving through a lattice of positive ions, it would create a "wake" of positive charge, which would attract another electron along its path. This attractive force overcomes the mutual repulsion of electrons. This concept was further developed when people discovered that moving electrons in a superconductor have a coherent manner over a long distance in the direction of movement.

Quantum mechanics showed that electrons in superconductors bind together to form "bosons". The "boson" is a fundamental particle which can be viewed as the combination of two electrons of mutually opposed spin. The key property of a boson is that any number of these particles could share the same energy state and they are released simultaneously at T_c.

All these ideas are not complete; some of them are not entirely true. But they helped to understand superconductivity. There were three important phenomena that need to be explained in the theory of superconductivity:

1. The presence of the energy gap and all the superconducting electrons were in a similar energy state.
2. The electron-lattice interaction and large distance coherence.
3. The interaction based on boson.

J. Bardeen, L.N. Cooper and J.R. Schrieffer at University of Illinois developed the BCS superconductivity theory in 1957, essentially a microscopic theory of superconductivity. They received Nobel Price for Physics in 1972 for this discovery. BCS theory was based on Cooper's theory which stated that a pair of electrons close to the Fermi level will couple into a Cooper pairs through interaction with the crystal lattice. This pairing results from a slight attraction between the electrons related to lattice vibrations. These electron pairs are called Cooper Pairs.

The BCS theory considers a simultaneous paring of all the conduction electrons at T_c. The first electron of a Cooper pair caused a delayed contraction of the lattice. This resulted in an increase in positive charge density, which attracted the second electron of the pair, as shown schematically in Fig. 10.8. This occurs at many points in the lattice and leads to a large scale movement of Cooper pairs, resulting in zero-resistance electrical conduction.

The mathematical derivation of BCS theory is complex, but an American professor expressed it in an easy understanding, analogous way.

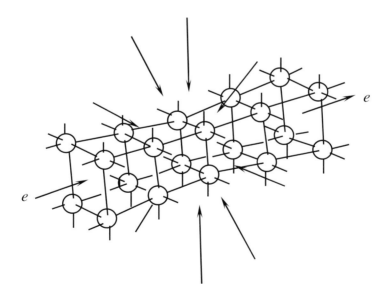

Fig. 10.8 A schematic drawing to show the BCS theory of superconductivity.

Consider a large ballroom is packed with dancers, dancing cheek to cheek. Suppose each dancer is doing his or her dance, the dancers will collide with each other and with any objects scattered on the floor. If there is a pressure on the whole group, the dancers will move toward one side of the room. With dancing, the motion will be random and chaotic, and a lot of energy will be lost in collision. This represents the electrons in a normal metal. Resistance comes from the collision of electrons with each other and with imperfections in the crystal structure.

Now suppose the dancers are paired in couples, each pair dancing together. The partners comprising each pair are not dancing face to face, but are separated by a hundred other dancers (long distance coherence). Consequently, if every couple is going to dance together, everybody in the hall must dance together. The long distance coherence results in a simple coherent motion, with order extending all the way across the hall. There will be no collision and energy loss. The super-conductive state is something like this.

The BCS theory has been very successful in explaining a wide range of experimental results, and even predicted some new development (e.g. the development of Nb superconductor alloys).

10.4 Conventional (Low-T_c) Superconductors

Conventional superconductors may include pure elements, alloys, intermetallics and some chemical compounds. Their T_c is relatively low, with the highest T_c of 23.2 K for Nb_3Ge. Most of them were discovered before 1986.

10.4.1 *Pure elements*

Fig. 10.9 shows the distribution of superconductive elements in the periodic table. We can see the following features from the figure:

1. A large number of elements were found to be superconducting, therefore, superconductivity is not a rare phenomenon for pure elements.

2. T_c varies over a wide range although they are all quite low (lower than 10 K). Some elements are superconducting only when the temperature drops to very close to 0 K (e.g. T_c of W = 0.0154 K).

3. Purity is an important factor that can affect superconductivity. Some elements are not superconducting at normal purity (~99.9%) but show superconducting at a higher purity.

4. High pressure helps some elements become superconducting (e.g. Si, P, Ge, As, Se, Sb, Te, Bi, Y, Cs, and Ba).

5. Some elements were found to be superconductive only in thin film state (Li and Cr).

6. Those elements with incomplete d or f shells (Fe, Co, Ni, and some rare earth elements) have atomic magnetic order. They do not show superconductive state. This implies that the superconductivity and atomic magnetism are mutually exclusive.

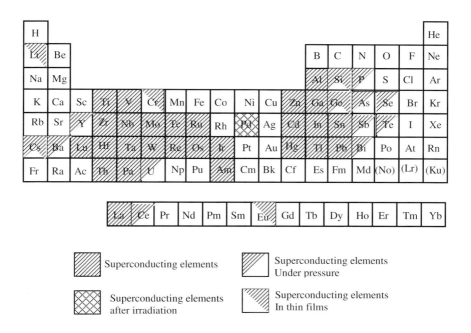

Fig. 10.9 Superconducting elements in the Periodic Table (after Vonsovsky, Izyumov and Kurmaev, "Superconductivity of Transition Metals", Translation, Springer-Verlag, 1982, p2).

10.4.2 *Alloys, intermetallics and chemical compounds*

Conventional superconductors can be alloys, intermetallic compounds and chemical compounds.

10.4.2.1 *Solid solution alloys*

Nb-25Ti (T_c = 10 K) is the most extensively used conventional super-conductor in industry. It has been made into wires, bars for producing high magnetic fields. Nb-25Zr (11 K), Mo-Tc (11 K), and Mo-Re (14 K) are all in this group. They are solid solution alloys.

These alloys are type II superconductors. They have much higher T_c, J_c and H_c than elemental superconductors. They also have relatively good mechanical and manufacturing properties.

10.4.2.2 *Intermetallics*

Examples from this group of superconductors include Nb_3Ge (23.2 K, the highest T_c before 1986), Nb_3Sn (18.1 K, the most important material in this group), and V_3Si (17 K). They have a wide range of crystal structures and relatively high T_c than alloys type superconductors. Mechanical properties are generally poor because of the bonding nature.

Table 10.2 The structures, properties and processing of Nb-Ti and Nb_3Sn.

Nb + 45-52 wt.%Ti	Nb_3Sn
Solid solution alloy	A-15 type intermetallic
T_c ~ 10 K	~ 18 K
H_{c2} ~ 11 T	~ 22 T
Ductile, easy processing (simple co-processing with Cu)	Brittle, difficult processing
Conventional insulating	Complicated insulating
Application limited to 8 T at 4.2 K	Restricted mechanical handling & winding

10.4.2.3 *Chemical compounds*

Examples from this group include NbN (17.3 K), $PbMo_6S_7$ (15.2 K) and MoC (14.3 K). Their T_c values are also quite high compared with alloy type of superconductors. Their structures and compositions are highly different. Two Nb based superconductors are widely used in industry: Nb + 45–52 wt.%Ti, and Nb_3Sn. Table 10.2 compares their structures, properties and processing.

10.5 High Temperature Superconductors

The term "high temperature superconductors" (HTS) generally refers to a group of materials with critical temperatures (T_c) higher than 23 K. Almost all of these materials were perovskite-like ceramic oxides containing layers of Cu-O planes. The first HTS materials were discovered in 1986 by researchers at the IBM Research Laboratory in Rüschlikon, Switzerland, who created a brittle $(La,Ba)_2CuO_4$ ceramic compound that exhibited superconductivity at approximately 30 K. Until recently, the main high T_c superconducting materials include the following systems as shown in Table 10.3.

Table 10.3 Major high-temperature superconducting systems.

Materials	Year discovered	Critical temperature (K)
La-Ba-Cu-O	1986	~30 K
Y-Ba-Cu-O (YBCO)	1987	~90 K
Bi-Sr-Ca-Cu-O (BSCCO)	1988	~115 K
Tl-Ba-Ca-Cu-O (TBCCO)	1988	~125 K
Hg-Ba-Cu-O and Hg-Ba-Ca-Cu-O (HBCCO)	1993	130-155 K

These are all superconducting cuprites (Cu oxide), indicating that superconductivity is clearly related to the Cu-O layered structures. The superconducting transition temperatures T_c are related to the number of CuO layers in the crystal structure:

1 layer Cu-O	$T_c \sim 30$ K
2 layer Cu-O	$T_c \sim 90$ K
3 layer Cu-O	$T_c \sim 110$ K
4 layer Cu-O	$T_c \sim 120$ K

The more CuO layers a structure has, the higher its T_c will be. Unfortunately, the more CuO layers the structure has, the less stable it will be. Even a three layer structure is not stable and tends to decompose. This is always a difficult problem in synthesis of high T_c superconductors.

10.5.1 *Y-Ba-Cu-O (YBCO) superconductor system*

YBCO system is the most extensively studied high temperature superconducting system. Its basic form is $YBa_2Cu_3O_{7-\delta}$, or abbreviated YBCO-123 or Y123, which has a T_c of 93 K. This compound was discovered by Chu and Wu in 1986-1987 and is the first compound for which T_c exceeded the boiling point of liquid nitrogen (77 K at 1 atm).

The YBCO family has three important compounds:

$YBa_2Cu_3O_7$	Y123	$T_c \sim 92$ K
$YBa_2Cu_4O_8$	Y124	$T_c \sim 85$ K
$Y_2Ba_4Cu_7O_{15}$	Y247	$T_c \sim 92$ K

Y247 can decompose into Y123 + Y124:

$$Y_2Ba_4Cu_7O_{15} = YBa_2Cu_3O_7 + YBa_2Cu_4O_8 \qquad (10.11)$$

Y124 can decompose into Y123 + CuO:

$$YBa_2Cu_4O_8 = YBa_2Cu_3O_7 + CuO \qquad (10.12)$$

Then, the T_c of Y124 and Y247 will be similar as Y123.

The YBCO superconducting systems have layered perovskite-like and highly anisotropic crystal structure.

10.5.1.1 *Structure of Y123*

$YBa_2Cu_3O_{7-\delta}$ (Y123) has two crystal structures. These structures are an oxygen deficient perovskite with ordered vacancies, consisting of three unit cells, with four different layers stacked sequentially as BaO-CuO-BaO-CuO$_2$-Y-CuO$_2$-BaO-CuO-BaO. In the CuO layer, each Cu atom is coordinated by four oxygen atoms, which is different from the five oxygen atoms surrounding the Cu atom in the CuO$_2$ layer (see Fig. 10.10).

1. When $\delta < 0.5$, YBa$_2$Cu$_3$O$_{7-\delta}$ has an orthorhombic structure with $a =$ 3.827, $b =$ 3.882 and $c =$ 11.682 Å. This phase is superconductive; and

2. When $\delta \geq 0.5$, YBa$_2$Cu$_3$O$_{7-\delta}$ has a tetragonal structure with $a = b =$ 3.9018, and $c =$ 11.9403 Å. This phase is non-superconductive.

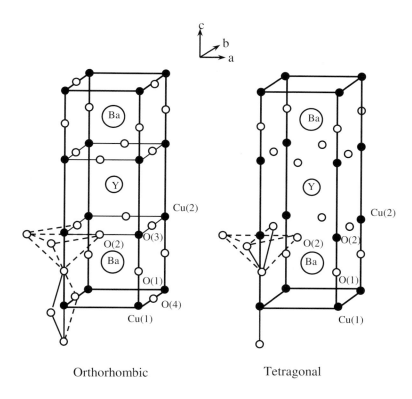

Fig. 10.10 Crystal structure of YBa$_2$Cu$_3$O$_{7-\delta}$.

The main difference in the two structures is in O(4) plane that more oxygen atoms make *b* direction longer than *a*. In order to obtain good superconducting properties, it is necessary to maintain the high oxygen contents in the Y123 phase. There is an equilibrium relation between the temperature and oxygen contents, as shown in Fig. 10.11. A special annealing at ~450°C in O_2 is employed to increase the oxygen content, which is called "oxygen loading" or "oxygenation" process.

$YBa_2Cu_4O_8$ (YBCO124) was originally discovered as a lattice defect in YBa2Cu3O7. Its crystal structure is closely related to Y123, but with one additional CuO chain layer in the unit cell. Y247 was first observed by J. Karpinski *et al* as an impurity phase in the system of $YBa_2Cu_3O_{6+x}O_2$. Its structure in the *c*-direction consists of alternating blocks with CuO single-chains (123-units) and with CuO double-chains (124-units).

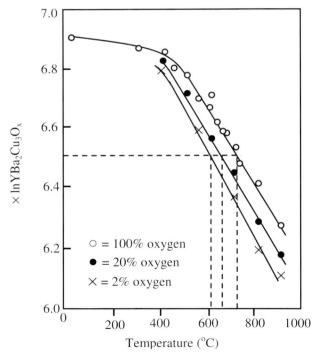

Fig. 10.11 Oxygen heat treatment of $YBa_2Cu_3O_x$ (after D. Robins, "Introduction to Superconductivity", IBC Tech. Serv. Ltd., 1989, p63).

10.5.1.2 *Property of Y123*

Y123 and Y247 have higher T_c than Y124. The T_c is a function of oxygen contents as discussed above. An oxygen annealing process is commonly needed to ensure high oxygen contents in order to achieve the highest possible T_c.

High T_c superconducting oxides including Y123 are Type II superconductors. The lower critical field H_{c1} is low, but H_{c2} is high. H_{c2} is 200 – 300 T at 0 K, 20–40 T for the field perpendicular to the *c*-axis at 77 K, and 34–80 T for polycrystalline Y123 at 77 K. Therefore, YBCO system suits the applications that require high magnetic fields.

As for all high T_c superconductors, the J_c of YBCO is strongly influenced by the processing methods and microstructures. Thin film and bulk YBCO specimens have very different J_c. The best Y123 films have a J_c above 5×10^7 A/cm^2 at 4 K in a field of 0–1 T, and ~5×10^6 A/cm^2 at 77 K and 0 T. Single Y123 crystals also have reasonably good J_c, 1.5 $\times 10^6$ A/cm^2 at 4 K (~1/30 of the best film), 1×10^4 A/cm^2 at 77 K (~1/500 of the best film). However, polycrystalline bulk Y123 compounds have a J_c only 100–1000 A/cm^2 at 77 K and 0 T, and 10–100 A/cm^2 at 77 K and 0.1 T, 10^5 times smaller than thin films.

The low J_c in bulk oxide superconductors is mainly caused by the grain boundary weakness (so called "weak link" problem), and lacking of vortex pinning centres.

10.5.1.3 *Processing of Y123*

Y123 powders are normally prepared through a solid-state reaction route. This process involves repeated grinding and sintering of the parent oxides (Y_2O_3 and CuO) and carbonates ($BaCO_3$), followed by calcination at high temperature (around 950°C). Further annealing under oxygen atmosphere from 950 to 400°C with a 24 hr plateau thermal treatment at 400°C is needed to produce the desired orthorhombic superconducting $YBa_2Cu_3O_7$ phase.

A variety of melt processing techniques, based on a peritectic reaction occurring between 1000 and 1080°C, has been developed for the fabrication of large grain YBCO superconducting systems. Melt texture growth (MTG) process and melt powder melt growth (MPMG) process have been developed to optimize the property, the microstructure and the material shape required for applications. MTG uses the stoichiometric Y123 sintered precursor while MPMG involves powder oxide precursors with 40 mol% molar excess of Y_2BaCuO_5 (Y211) phase and 0.3–0.5 wt% of Pt. The melt processing techniques are successful in overcoming the J_c limitations associated with grain boundaries' weak links and increasing the flux pinning force through the introduction of pinning centres in the microstructure.

Single crystals of high T_c superconductors possess an extremely high degree of crystalline perfection and play an important role in research studies of intrinsic key properties (H_{c1}, H_{c2} and J_c) and crystallographic structure as well as in technological applications. Currently, single crystals of Y123 can be grown by a variety of crystal growth techniques, including flux growth, top-seeded solution growth (TSSG) and travelling solvent floating zone (TSFZ).

10.5.2 *Bi-Sr-Ca-Cu-O (BSCCO) superconductor system*

BSCCO is another important high T_c system, discovered by H. Meada *et al* (Japan) in 1988. This system has been attracting great attention due to peculiar relationships between the structure and the superconductivity characteristics.

There are three compounds in the BSCCO family:

$Bi_2Sr_2CuO_{6+y}$ (Bi2201)   CuO layer n = 1 $T_c \leq 30$ K

$Bi_2Sr_2CaCu_2O_{8+y}$ (Bi2212)   n = 2 $T_c \sim 85$ K

$Bi_2Sr_2Ca_2Cu_3O_{10+y}$ (Bi2223)   n = 3 $T_c \sim 110$ K

Bi2201 phase has a pseudotetragonal symmetry with lattice parameters $a \approx b \approx 5.4$ Å and $c \approx 24.4$ Å. The unit cell of Bi2201 contains four formula units and is a stack of atomic planes in the sequence of $(BiO)_2$ / SrO / CuO_2 / SrO / $(BiO)_2$ / SrO / CuO_2 / SrO / $(BiO)_2$.

Bi2212 and Bi2223 have a similar crystal structure of perovskite with tetragonal unit cells (pseudo-tetragonal symmetry). The difference is that Bi2212 has 2 layers of Cu-O and 2223 has 3 layers of Cu-O. Fig. 10.12 shows their crystal structures. The c-direction lattice parameters of Bi2212 and Bi2223 are 30.9 Å and 37.9 Å, respectively, much greater than a and b.

The critical temperature of Bi2201 strongly varies from 10 to 30 K due to the difficulty in obtaining pure single phase composition. Substitutions of Sr for La, Pr, Nd and Sm in Bi2201 increase the critical temperature. The critical temperature of Bi2212 is near 85 K and some crystals can show T_c up to 96 K. This is a result of different oxygen content and cation intersubstitution. The critical temperature of Bi2223 is close to 110 K.

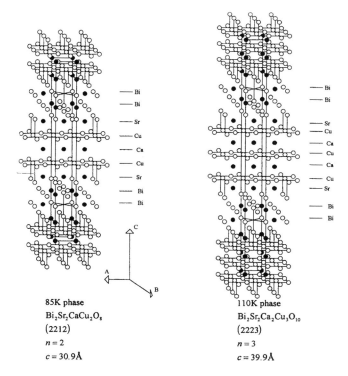

85K phase
$Bi_2Sr_2CaCu_2O_8$
(2212)
$n = 2$
$c = 30.9 Å$

110K phase
$Bi_2Sr_2Ca_2Cu_3O_{10}$
(2223)
$n = 3$
$c = 39.9 Å$

Fig. 10.12 Crystal structure of BSCCO superconductors (a) $Bi_2Sr_2Ca_1Cu_2O_8$ and (b) $Bi_2Sr_2Ca_2Cu_3O_{10}$ (after Sheahen, "Introduction to High Temperature Superconductivity", Plenum Press, 1994, p147).

The critical current density in Bi based superconductors is different in different direction. The critical current density along the c-axis is much lower than the critical current density along a- & b- planes. For example, single crystalline Bi2212 and Bi2223 have their critical current density above 10^6 A/cm^2 along a- & b-planes and 10^2–10^3 A/cm^2 along the c-axis. The low J_c along the c-axis is the result of a weak Josephson interaction between every pair of CuO$_2$ double or triple layers.

Bulk materials of Bi2212 and Bi2223 phases can be produced by sintering of high quality calcined powders which were fabricated through solid state reaction and wet chemical routes. Single crystals, melt-solidified tapes and wires of Bi2212 have been prepared by melt-solidification process. Synthesis of single-phased Bi2223 is very difficult. Recently, large Bi2223 single crystals with similar size to Bi2212 were grown by the floating zone (FZ) method using a very slow growth rate (< 0.05 mm/h).

Bi compounds are of special importance for high temperature superconducting applications. In comparison with YBCO, BSCCO superconductors have the following advantages:

1. BSCCO (Bi2223) has a higher T_c (~110 K) than YBCO (~90 K), which is very important for the applications at 77 K;
2. Due to the weak bonding between the two adjacent Bi-O planes, BSCCO crystals cleave easily along the (001) planes, making it easy to be textured by mechanical deformation (rolling or pressing);
3. Bi2212 and Bi2223 very often co-exist. Pb doping (to partially replace Bi) helps the formation and improves the stability of Bi2223 phase;
4. No "oxygenation" process is needed because the oxygen content in BSCCO is relatively stable; and
5. BSCCO compounds have better environmental stability in water containing atmospheres.

However, BSCCO has weaker behaviour than YBCO in magnetic field. A magnetic field often reduces J_c to a higher extent than YBCO.

10.5.3 *Tl-Ba-Ca-Cu-O (TlBCCO) superconductor system*

Tl-Ba-Ca-Cu-O (TlBCCO) superconductor system has the largest family of high temperature superconductors. Up to the discovery of mercury based cuprates, Tl2223 held the record of the highest critical temperature. This system forms two distinct structure series with the general formulas $Tl(Ba$ or $Sr)_2Ca_{n-1}Cu_nO_{2n+3}$ and $Tl_2Ba_2Ca_{n-1}Cu_nO_{2n+4}$. The two families can be both described by the general formula $Tl_mBa_2Ca_{n-1}Cu_nO_{2n+m+2}$ where *m* can be 1 or 2 and *n* = 1, 2, 3 or 4 for bulk samples prepared by solid state chemistry and arrive up to 5 and 6 for samples synthesized under very high pressure or for thin films.

The stoichiometric $TlBa_2CuO_5$ (Tl1201) and $TlBa_2CaCu_2O_7$ (Tl1212) phases have a tetragonal crystal structure. The T_c value of Tl1212 is very variable and strongly depends on the conditions utilized for the synthesis. $TlBa_2Ca_2Cu_3O_9$ (Tl1223) has a structure in the group P4/mmm a = 3.8429, c = 15.871 Å. Superconductivity at 110 K was first identified in this phase. Tl2201 crystallizes with two structural modifications, one orthorhombic, and the other tetragonal. The latter ($Tl_2Ba_2CuO_6$) is superconducting with T_c up to 90 K for optimal oxygen content. Tl2212 phase, $Tl_2Ba_2CaCu_2O_{8+\delta}$, shows an almost commensurate structural modulation and superconductivity at 86 K. The Tl2223 phase possesses the highest T_c of the family, 128 K, which is rather less dependent on the oxygen content and on the thermal history.

Thallium compounds have a higher T_c than YBCO and BSCCO superconductors. Their J_c and H_c are also high. But much less research has been conducted with this system. The main reason for this is probably that Tl compounds are highly poisonous. Tl sulphate is water soluble, colourless, odourless and nearly tasteless, but highly poisonous. It is also easy to be vaporised, poisonous if inhaled, and can even be absorbed through skin. Good ventilation and isolating operations therefore are needed to reduce exposure.

10.5.4 *Hg-Ba-Ca-Cu-O (HgBCCO) superconductor system*

$HgBa_2CuO_{4+\delta}$ (Hg1201) was the first HgBCCO system reported to show superconductivity in 1993. Subsequently, $HgBa_2CaCu_2O_{6+\delta}$ (Hg1212) and $HgBa_2Ca_2Cu_3O_{8+\delta}$ (Hg1223) were reported to have critical temperatures of 125 and 135 K, respectively. All superconducting phases in the $Hg_1Ba_2Ca_{n-1}Cu_nO_{2n+2}$ system crystallize with a tetragonal cell.

The $Hg_1Ba_2Ca_{n-1}Cu_nO_{2n+2}$ system can be prepared through a two-step procedure. In the first step, Ba-Ca-Cu-O precursor powders are prepared by reacting stoichiometric mixtures of simple oxides, nitrates, and/or carbonates of Ba, Ca, and Cu. In the second step, Ba-Ca-Cu-O precursor powders are reacted with HgO in an evacuated quartz tube. The synthesis requires independent control of mercury and oxygen partial pressures. It is also sensitive to trace quantities of moisture and carbon dioxide and demands freshly prepared oxide precursors. Partial substitution of Hg by several dopants could significantly improve the ease of formation and enhances the stability of the superconducting phases without affecting the T_c. Due to the high volatility of mercury, sample heating is normally carried out in sealed reactor.

10.6 Magnesium Diboride

Magnesium diboride, MgB_2, was known to the scientific community since the early 1950s. Its superconducting behaviour was however discovered in January 2001 to have a critical temperature of ~ 40 K. Although the T_c is not very high, this discovery certainly revived interest in the field of superconductivity, especially in the study of non-oxides.

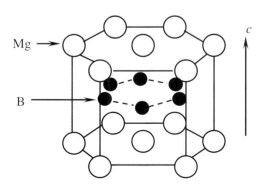

Fig. 10.13 Crystal structure of MgB_2.

MgB_2 has a simple hexagonal AlB_2-type structure (see Fig. 10.13). It contains graphite-type boron layers that are separated by hexagonal close-packed layers of Mg. Mg atoms are located at the centre of hexagons formed by borons and donate their electrons to the boron planes.

MgB_2 powders could be prepared through high temperature reaction between B and Mg powders, and are available commercially though the T_c and transition in the superconducting state are not as good as the materials prepared in lab. MgB_2 films can be achieved using common physical vapour deposition techniques, such as pulsed laser deposition (PLD), co-evaporation, magnetron sputtering and Mg diffusion method, on a variety of substrates including SiC, Si, $LaAlO_3$, $SrTiO_3$, MgO, Al_2O_3 and stainless steel. The nature of substrate and the type of processing method have great influence on the superconducting properties of MgB_2 thin films. For example, it had been reported that films on sapphire show the highest T_c and sharpest transitions by Mg diffusion method. For the same substrate, Al_2O_3, the thin films prepared by PLD have lower T_c and wider transitions than the films prepared using Mg diffusion method.

A simple route, reaction of B (with different geometries) with Mg vapour at a temperature of ~900°C, can also be used to prepare powders, sintered pellets, wires, tapes and films. For instance, Mg diffusion into B wires is a relatively easy method to rapidly convert already commercially existing B wires into superconductive MgB_2 wires. Another wire fabrication technique is called powder-in-tube (PIT). The PIT approach can be used to fabricate metal-clad MgB_2 wires/ribbons using various metals, such as: stainless steels (SS), Cu, Ag, Ag/SS, Ni, Cu-Ni, Nb, Ta/Cu/SS and Fe. This method usually consists of the following procedure: (1) packing of MgB_2 powder of stoichiometric composition in metal tubes or sheaths; (2) transforming of tubes into wires and ribbons; and (3) heat treatment at 900 – 1000°C.

MgB_2 has an unusual high critical temperature of about 40 K among binary compounds. Table 10.3 lists the superconducting properties of this material. MgB_2 is a simple and cheap material. It is promising for use in superconducting low to medium field magnets, electric motors and generators, fault current limiters and current leads.

Table 10.3 Superconducting property of MgB_2.

Property	Values & Descriptions
Critical temperature	$T_c = 39\text{-}40$ K T_c follows a quadratic or linear dependence on applied pressure, decreasing monotonically Substitution by Al, Mn, C, Co, Fe and Ni reduces T_c
Upper critical field	$H_{c2}//ab(0) = 14\text{-}39$ T $H_{c2}//c(0) = 2\text{-}24$ T
Lower critical field	$H_{c1}(0) = 25\text{-}48$ mT
Critical current densities	
$J_c(H)$ in bulk	10^6 A/cm^2 (0 T) 10^4 A/cm^2 (6 T) 10^2 A/cm^2 (10 T)
$J_c(H)$ in powders	3×10^6 A/cm^2
$J_c(H)$ in wires and tapes	7×10^5 A/cm^2 (0 T) 10^5 A/cm^2 (5 T)
$J_c(H)$ in thin films	1×10^7 A/cm^2 (0 T) 10^4 A/cm^2 (14 T)
Highest $J_c(H)$ at different temperatures	$J_c(4.2$ K, 0 T$) > 10^7$ A/cm^2 $J_c(4.2$ K, 4 T$) = 10^6$ A/cm^2 $J_c(4.2$ K, 10 T$) > 10^5$ A/cm^2 $J_c(25$ K, 0 T$) > 5 \times 10^6$ A/cm^2 $J_c(25$ K, 2 T$) > 10^5$ A/cm^2

10.7 High Temperature Superconducting Thin Films

Primarily, the motivation for the fabrication of high temperature superconducting thin films is to meet the requirements of superconducting electronic applications since thin films have superior properties compared to bulk materials. For instance, superconducting films are indispensable for applications in microwave and radio frequency components and for integrated junction circuits (infrared sensors, Josephson junctions, low loss microwave cavities and filters, transition edge bolometers, flux flow transistors, flux transformers, and dc and rf superconducting quantum interference devices (SQID). In addition, superconducting films have been used extensively in basic studies of physical properties of these materials.

10.7.1 *Techniques for thin film deposition*

Growth of high temperature superconducting thin films has normally been achieved using chemical and physical methods. The most commonly used chemical techniques include metalorganic chemical vapor deposition (MOCVD), spray pyrolysis and spin coating. These chemical deposition methods are complex in process and it is difficult to control the composition, structure and property of films.

It is apparent that physical methods are capable of producing films with better properties. Typical physical methods for superconducting film deposition include atomic layer epitaxy (ALE), electron beam evaporation, molecular beam epitaxy (MBE), pulsed laser deposition (PLD) and magnetron sputtering. In the operation of physical methods, several processing parameters are critical to the deposition of high quality thin films. They are: substrate temperature, gas composition, deposition rate, distance between the anode and the cathode, orientation of the substrate, substrate preparation conditions, and quality of the sputtering targets.

10.7.2 *Substrate materials*

Substrates for epitaxial growth of high temperature superconducting thin films need to meet some requirements so that the film-substrate system is suitable for device applications.

1. Lattice match: A lattice mismatch of < 15% is normally required for epitaxy. The in-plane lattice spacing of the superconducting film should closely match that of the substrate. In general, substrate materials having perovskite crystal structures, usually oxides, are used to support the growth of high quality superconducting thin films. The substrates are also expected to have a similar thermal expansion coefficient with superconductors to avoid film cracking.
2. Chemical inertia: The substrate should be inert during deposition and application so the possible chemical interactions between

substrate and superconductor will be minimized and unwanted interfacial layers will then not form. and

3. Low microwave losses: the dielectric constant and the loss tangent should be as small as possible.

Perovskite materials ($LaAlO_3$, $LaGaO_3$, $NdGaO_3$, $SrTiO_3$), non-perovskite oxides (Al_2O_3, MgO, yttria stabilized zirconia (Y_2O_3-ZrO_2)), metals (Ag and Hasteloy (Ni-Cr-Mo)) and semiconductors (Si and GaAs) are suitable substrates for the growth of superconducting films.

10.7.3 *Typical high temperature superconducting films*

10.7.3.1 *YBCO films*

Experimental research is focused on YBCO since this system can be synthesized in the form of high quality thin films using various deposition techniques. It is also desirable for many electronic applications. YBCO can be grown in the form of thin film with two most possible orientations: the *c*-axis of the film lies along the substrate surface plane or lies perpendicular to the substrate surface plane. $LaAlO_3$, $SrTiO_3$ and MgO are the most commonly used substrates for growth of YBCO films with smooth surfaces and interfaces. YBCO films with $T_c > 90$ K and $J_c > 5 \times 10^6$ A/cm^2 at 77 K had been routinely prepared by a number of physical and chemical deposition techniques.

10.7.3.2 *BSCCO films*

It is more difficult to obtain epitaxial BSCCO films with good superconducting properties than for YBCO films. Physical methods, such as PLD and magnetron sputtering (DC & RF), are preferred experimentally for deposition of BSCCO deposition than chemical approaches. The formation of a specific phase is highly dependent on the growth conditions, typically including substrate temperature, oxygen partial pressure, target quality and growth rate. In most cases, post-annealing at elevated temperatures is necessary for phase formation and crystallinity improvement. Even though, pure-phase materials with higher T_c and J_c are difficult to realize.

Epitaxial $Bi_2Sr_2CaCu_2O_8$ (Bi2212) films on (100) MgO had been prepared using PLD. They have a T_c of ~71 K and J_c (4.2 K) of 5×10^6 A/cm^2. $Bi_2Sr_2Ca_2Cu_3O_{10}$ (Bi2223) films prepared using MOCVD have been reported to have a T_c of 97 K and J_c (77 K) of 3.8×10^5 A/cm^2. Some films with higher T_c values (~100 K) were obtained by spray pyrolysis. However, these films are not completely epitaxial and phase pure.

10.7.3.3 *TBCCO films*

The growth of epitaxial TBCCO thin films is challenging because of the complex phase relationship and the high volatility/toxicity of Tl oxides. Hybrid techniques therefore were employed. In the first stage, precursors containing Ba, Ca and Cu oxides are deposited onto substrate by a number of chemical and/or physical methods, such as evaporation, laser ablation, sputtering and chemical vapor deposition. Tl is then added during a thermal activated vapor phase diffusion process. This is achieved by placing the precursor films in a sealed gold or platinum crucible along with powders or pellets that have a composition similar to that of the desired phase and serve as the Tl source.

$Tl_2Ba_2CaCu_2O_8$ (Tl2212) films have been obtained with BaCaCuO precursors with $T_c = 105$ K and J_c (77 K) $= 1.2 \times 10^5$ A/cm^2. $Tl_2Ba_2Ca_2Cu_3O_{10}$ (Ti2223) films with nearly pure phase structure have been synthesized on $LaAlO_3$ substrate by annealing sputter deposited BaCaCuO precursors. They films were reported to have a T_c of 120 K. $TlBa_2CaCu_2O_7$ (Tl1212) thin films have been obtained with $T_c = 90$ K and J_c (77 K) $= 2 \times 10^5$ A/cm^2.

10.7.3.4 *HgBCCO films*

HgBCCO systems are similar to TBCCO system, in composition and structure. Their preparation techniques are similar as well.

HgBCCO epitaxial films can be obtained by annealing of precursor films in Hg atmospheres. The precursors are commonly deposited using two typical approaches. In one approach, a layer-by-layer mixing of HgO and $Ba_2CaCu_2O_x$ layers is realized by alternative laser ablation of two separate targets. The precursor film is then annealed in an evacuated quartz tube with bulk Hg-cuprate and $Ba_2CaCu_2O_x$. $HgBa_2CaCu_2O_6$ films have been grown on $SrTiO_3$ with $T_c = 124$ K and J_c (5 K) $= 1 \times 10^7$ A/cm^2. In the 2nd approach, Hg-free precursor films are post-annealed in the presence of Hg vapor. Epitaxial $HgBa_2Ca_2Cu_3O_8$ films have been grown on $LaAlO_3$ with $T_c = 128$ K and J_c (77 K) $= 1.4 \times 10^7$ A/cm^2.

10.8 Typical Application of Superconducting Materials

The development of new technologies or materials is for applications. Superconductor research has been related to applications from the beginning. Many researchers in this field have been spending most of their time to improve various properties of superconductors in an effort to put them in industrial applications.

Applications of superconductors, both at the present time and in the future, can be classified into the following main groups:

1. Superconducting magnets for generating high magnetic fields,
2. Electric power utilities - electricity transmission and storage etc,
3. Transportation - levitation trains and ships,
4. Electronics and instrumentation, and
5. Super computers.

10.8.1 *Superconductor magnets*

Superconductors, both high and low temperature, are promising for the production of magnetic fields *per se* for research and applications in medical, pharmaceutical, petrochemical, mining and other industries. Superconductor magnets are being used to generate high magnetic fields. Their major advantages over conventional magnets are (1) a small magnet can provide a huge magnetic field, and (2) the energy loss is minimal. In a conventional magnet, a high current passes through copper conductors generates enormous heat. A powerful cooling system has to be used to dissipate the heat. The current and potential industrial applications of superconductor magnets include:

1. Nuclear magnetic resonance (NMR) and magnetic resonance imaging (MRI): they are currently the dominant industrial application of superconductors and, in particular, superconducting magnet can provide a highly homogeneous and very stable magnetic field. MRI is mainly used for medical diagnostics and research. It shows the image of soft tissues while X-ray machine can only tell materials with very different densities like bone and soft tissue. Low temperature superconductors, such as Cu sheathed multi-filamentary Nb-Ti and Nb-Sn wires, are in use for many high field high resolution magnets. Melt-processed YBCO may also be applied as a permanent magnet material though many practical problems need to be tackled.

2. Magnetic separation: magnetic separation is a technology that allows materials with different magnetic properties to be separated by using inhomogeneous magnetic fields. For many years magnetic separation has been extensively used by the mining industry. A typical Marston-type magnetic separator consumes a large amount of power. A great potential for cost reduction would be possible when superconductor technology reaches a level sufficient to allow superconducting solenoids to become a viable alternative to Cu coils. Commercial superconducting systems in operation today use low temperature superconductors in their magnet coils, such as Nb-Ti alloy.

3. Radio-frequency devices (gyrotrons) for navigation etc.

4. Magnetic shielding.

Superconductor magnets are also used in instrumentations for scientific research. For example, superconducting super colliders (SSC) need extremely powerful magnets. SSC, as a particle accelerator, is designed for studying the fundamental questions about the universe.

Currently, high T_c superconducting materials have not yet found their way into large scale applications as magnets. This is mainly due to the fact that their critical current density in a magnetic field is too low. There are also difficulties in working with high T_c materials, which is often related to the ceramic nature of these materials.

10.8.2 *Electric power utilities*

The applications in this area include:

1. Electric power generation,
2. Electricity transmission,
3. Electricity application, and
4. Electricity storage.

Power applications have tremendous economical importance to our energy industry. However, it often requires high current density, high magnetic field, long distance transmission, mechanical strength and operational safety. A number of difficulties including materials, engineering and economy problems need to be overcome.

10.8.2.1 *Superconducting generators and motors*

Superconducting wiring can reduce the energy loss of a large generator or motor to ~50%. This means that the energy efficiency will go up from 98% to 99%. Various types of AC generators of 20–300 MVA have been constructed and tested by US and Japan companies using low-temperature superconductors. Some proton-types of superconducting motors have also been built for demonstration purposes. The Naval Surface Warfare Center in the United States developed a 300 kW Nb-Ti low T_c superconductor drum type homopolar machine installed in a boat to demonstrate the feasibility of marine propulsion by superconducting machines. The Nb-Ti coils were then replaced by BSCCO high temperature superconductor windings. The output power was 122 hp at 28 K and 320 hp at 4.2 K.

10.8.2.2 *Power transmission*

Copper transmission lines consume ~5% of the all power generated. Thus, one of the most attractive future applications for superconductors is low loss transmission of electric power using superconducting cables. Superconducting transmission cables can provide two to five times the current transmission capability of a conventional cable with the same cable diameter.

The early work on superconducting transmission cables using low T_c superconducting Nb_3Sn tape. Due to the extremely low operation temperature (liquid helium), this system cannot attract enough attention from the electric power sector and be used for practical application though the testing results were successful. The discovery of high temperature superconductors with the promise of liquid nitrogen operation revitalized the interest for superconducting transmission within the electric power industry.

A typical superconducting cable system has four main components:

1. Electrical core: which contains the current carrying conductor, the dielectric insulation and the grounded shield. The superconducting materials used here can be both low T_c and high T_c superconductor tapes;
2. Cryogenic envelope: which isolates the 'ambient' external surrounding from the low temperature superconducting electrical core;
3. Terminations: which provide a thermal, current, and voltage transition from the cold core to the outside power system at the cable end; and
4. Refrigeration system: which conditions the cryogenic fluid used to cool the cable.

The first complete prototype cable system was developed by Pirelli Cavi & Sistemi and EPRI, under the US Department of Energy Superconductivity Partnership Initiative (SPI). This activity culminated in the successful demonstration of a 50 m 400 MVA 115 kV pipe-type high T_c superconducting cable system capable of carrying 2 kA$_{rms}$ AC current.

At the present time, critical current density (J_c) of high T_c superconductors for short specimens (1–10 cm) is just about enough to meet the requirement of the power transmission (10^5–10^6 A/cm^2 at 77 K), but long wires or cables (> 50 m) can barely reach ~10^4 A/cm^2 at 77 K, and are also not stable enough for long distance transmission. The cost of running and maintaining the liquid nitrogen cooling systems is also an important consideration for long distance power transmission.

Long superconducting wires and cables have manufacturing problems. It is difficult to obtain uniform microstructures and properties over a long length. Other problems include mechanical strength and the effects of bending on J_c, environmental and thermal stabilities, and joining technologies etc.

If a part of a superconducting cable becomes non-superconducting for any reason, heat will be generated and concentrated there very quickly, and damage the nearby part of the cable. Conventional conductors have to be provided to stand a short time (a few second at least) before the system shuts down. Then a backing-up system is needed while repairs are conducted. There is still a long way to go before the Cu and Al wires can be replaced by superconductors.

10.8.2.3 *Energy storage*

Energy storage is an attractive method to maximize the generation and transmission capacity of an electric power system. It can store electric power during off-peak periods and then provide support during high demand conditions to offset the need for larger generation or transmission capacity. In addition, energy storage can provide system benefits to improve voltage stability, frequency control, power compensation and provide rapid response power during momentary faults or complete power interruptions. Energy storage can be achieved through hydroelectric, pumped gas storage, batteries, etc.

Superconducting magnetic energy storage (SMES) is using large superconductor magnets for energy storage, i.e., the energy is stored in the magnetic field of an inductor. The maximum energy for an inductively stored power device is dependent on the ability to carry very high currents. A superconducting coil capable of carrying high currents without power dissipation is therefore an attractive energy storage candidate. A SMES system includes the superconducting magnet, a solid-state AC-DC power conditioning system and a closed-cycle refrigerator.

Major programmes on SMES were started in 1970s in Japan and Europe with the participation of most of the major electric power companies, i.e. Hitachi, Mitsubishi, Toshiba, Siemens, ABB, etc. In the late 1980s, the interest in the SMES technology shifted to the very large 20 MWh (72 GJ) magnet. The Japanese International Superconductivity Technology Center (ISTEC) started the design and evaluation of a small-scale SMES pilot plant at a 100 kWh/20 MW rating in 1991. The small-scale SMES, using current LTS technology, is more applicable to power system stabilization.

Flywheel energy storage (FES) is another storage technology that can be beneficial from superconductivity. The energy stored in flywheel is the kinetic energy stored in the rotor after spin up to a speed determined by the structural limits of the rotor. The utilisation of superconductors in FES is actually to provide a levitation force and a non-contact frictionless bearing in a flywheel. This concept was suggested in the early 1990s after the discovery of high T_c superconductors. The levitation force for the bearing can be improved by increasing J_c. The levitation pressures between a high quality YBCO disc and Nd-Fe-B permanent magnet at 77 K can be as high as ~280 kPa. An array of superconducting discs and permanent magnets can thus provide sufficient force to support a fairly substantial rotor mass. A flywheel system using superconductor bearings would have projected turn-around efficiency in excess of 90%.

10.8.3 *Transportation tools*

In 1960s, a MIT scientist Dr. H. Kolm suggested a train that would fly along a frictionless track supported by a powerful magnetic field. It is called the "magnetically levitated" train, or for short "Maglev". These levitation trains are based on the Meissner effect. As shown in Fig. 10.14, a simple superconducting levitational system comprises a permanent magnet positioned over a superconductor. When the superconductor is approached by the magnetic field generated by the permanent magnet, screen currents are set up on its surface, which create an equal but opposite force to support itself in air. Currently, the material of choice for superconducting levitation is high T_c YBCO system and rare earth based magnets, because they exhibit high magnetic irreversibility fields at liquid nitrogen temperatures and have the capacity to grow large grains.

Electrodynamic levitation of high-speed trains depends on repulsive forces between moving magnets and the eddy currents induced in a conducting guideway or loops to produce passively stable levitation (Fig. 10.15). The repulsive levitation force is inherently stable with distance, and comparatively large levitation heights (20 – 30 cm) are attainable by using superconducting magnets. A drag force is associated with the eddy currents.

In 1987, the Japanese demonstrated a prototype maglev train called MLU-002. The train travelled 520 km/hour, flies about 10 cm above the track, rubber wheels touch the track only during lift-off and landing periods. This programme is supported by the Japanese National Railway System, in cooperation with some big companies including Fuji, Hitachi, Mitsubishi and Toshiba, as an example of collaboration efforts between the government and private sectors. A three-car test train achieved the design speed of 550 km/h on a 42 km long test track in Yamanashi in December 1997, and 581 km/h in December 2003.

These Maglev trains use high temperature superconductors in its onboard magnets, cooled with inexpensive liquid nitrogen. This allows for a larger gap and repulsive-type electrodynamic suspension (EDS). They are currently the fastest trains in the world. Many Maglev train systems have been proposed in several countries of North America, Asia, and Europe. For example In April 2007, Central Japan Railway Company announced the plan to begin commercial maglev service between Tokyo and Osaka using the Superconductive Magnetically Levitated Train in the year 2025. However wider acceptance of Maglev trains will still depend on research that can improve the efficiency of the electromagnets and the cost of maintaining the field through further development of high temperature superconductors.

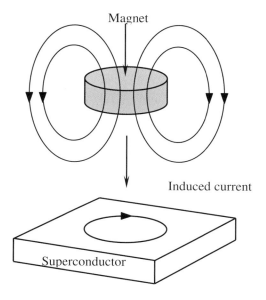

Fig. 10.14 Superconducting levitation.

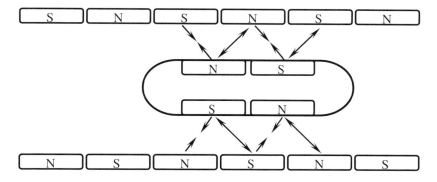

Fig. 10.15 Schematic of electrodynamic levitation train.

10.8.4 *Electronics and thin film applications*

Bulk high T_c superconductor materials have two difficult problems that need to be solved: low J_c and poor mechanical properties (brittleness). Thin films are better in these two aspects. They have a much higher J_c, and are supported by strong substrates. It is possible that the thin films may find industrial applications before the bulk superconductor materials.

10.8.4.1 *Josephson effect*

Brian Josephson (Cambridge) discovered this effect theoretically in 1962 based on the coherence of the electron pair. He predicated that a current will flow through a superconductor-insulator-superconductor junction even when the voltage across the junction is zero. If a small voltage is applied across the junction, the current will oscillate across the junction at a frequency of 2 eV/h. Anderson and Rowell confirmed the Josephson effect experimentally in 1963. Josephson junctions can be developed using both low and high temperature superconductors, including Nb, NbN and YBCO thin films. This effect is one of the two essential discoveries that led to today's extensive capability in superconductor electronic (SCE) devices and components.

Josephson effect has been used to develop electronic measurement devices, including:

1. Electric measurement: voltmeters with a sensitivity of 10^{-15} V,
2. Sensitive magnetometers for biological, medical, and geophysical applications, and
3. Switches and memories in supercomputers.

10.8.4.2 *SQUID*

SQUID is an acronym for superconducting quantum interference device. It is based on two physical phenomena, namely Josephson tunnelling and flux quantization, and is fundamentally a superconducting ring with a Josephson junction on it. The ring acts as a storage device for magnetic flux. When an outside magnetic field is applied, the superconducting ring will respond by generating a superconducting current. There are two main versions of SQUID, i.e, DC SQUID and RF SQUID. The DC SQUID consists of a superconducting loop interrupted by two resistively shunted junctions; and the voltage across the parallel junctions represents the output signal. The RF SQUID has only one junction in the loop; and the read-out is realized by a resonant LC circuit coupled to the loop.

SQUIDs have been made with both low and high T_c superconducting materials. In general, materials for low T_c DC SQUIDs are multi-layers of thin films of Nb-Al. In the case of high T_c SQUIDs, YBaCuO is the most frequently used material. However, the reproducibility of the high T_c junctions is not as good as for the low T_c junctions.

These devices are the most sensitive detectors for magnetic flux currently known. All physical quantities that can be converted to a magnetic flux can be measured with extreme sensitivity. SQUIDs can be used, for instance, for measuring magnetization, magnetic susceptibility, magnetic fields, current, voltage and small displacements. Practically SQUIDs can detect the very small magnetic fields such as those produced by the human heart (10^{-10} T) and brain (10^{-13} T).

Magnetometers for geophysical applications can detect disturbance to the earth's magnetic field caused by a submarine deep in the ocean. A widespread application of SQUIDs is in the field of non-destructive testing and evaluation. Physical failures such as cracks, corroded areas or small areas with (magnetic) impurities in materials can be detected with SQUIDs. SQUID using low T_c materials may also open a broad field of particle detection research.

10.8.4.3 *High performance computers*

Since computer speed is dependent in part on the time required to transmit signals within and between different circuit elements, switches are often the limiting factor. Josephson junctions are able to perform switching functions with incredible speed, around 10 times faster than ordinary semiconductor circuits, and with very little power consumption. In 1970s, IBM researchers developed a reproducible process for thin film Josephson junctions as part of a program to develop high performance computers based on the voltage-latching properties of Josephson tunnel junctions. Superconductor devices (Josephson junction and SQUID devices) can be integrated with thin film resistors and thin film transmission-line interconnections into complex, monolithic, integrated circuits (ICs) and packed tightly together without generating much heat. This advantage also makes them ideal candidates for use in the high speed logic components of a super-fast and much smaller computer.

Summary:

Superconducting materials are at the "leading edge" of materials science and engineering. The applications of superconductors are making revolutionary changes in our modern technological society. These materials enable us to improve the efficiency of electricity transmission and usage, and to build smaller and better electrical equipment, and faster and more powerful computers. The most important properties for a superconductor are critical transition temperature (T_c), critical current density (J_c) and critical magnetic field (H_c). Some other properties such as mechanical strength, environmental stability and joining properties are also important for applications. Material chemistry, microstructure and processing techniques are the primary factors to influence these properties. Research in this area is concentrated on both developing new superconductors with higher T_c and improving the properties of existing superconductors in an effort to use them in industries.

Important Concept:

Superconductivity: The phenomenon that some materials exhibit zero electrical resistance below certain temperatures.

Critical transition temperature, T_c: The temperature above which a superconductor regains its normal electrical resistance.

Critical current density, J_c: The current density above which superconductivity disappears.

Critical magnetic field, H_c: The magnetic field above which superconductivity disappears.

Meissner effect: The expulsion of magnetic flux in a superconductor, or the phenomenon that a superconductor will not allow a magnetic field to penetrate its interior.

Type I superconductor: A group of superconductors that exhibit a simple diamagnetic behaviour in *B-H* relation. There is a vertical drop at the critical field H_c, above which the superconductor becomes normal abruptly.

Type II superconductors: The superconductors have more complicated magnetic behaviour that B begins to increase gradually at H_{c1}, and reaches the value as it in the normal state at H_{c2}.

High temperature superconductors: Superconducting oxides with T_c generally above 77 K discovered after 1986.

BCS theory: A theory of superconductivity that is based on the simultaneous paring of all the conduction electrons at T_c. The first electron of the pair caused a delayed contraction of the lattice, which attracted the second electron of the pair to follow. When all the conduction electrons paired up at many points in the lattice, the coherent movement of the pairs does not produce any energy loss.

Magnetic resonance imager (MRI): A medical diagnostics and research equipment that can show the image of soft tissues. Superconductors are used to create very high magnetic field for MRI.

Superconducting super colliders (SSC): A large scale particle accelerator which is designed to study the fundamental questions about the universe.

Maglev: A magnetically levitated vehicle that can fly along a frictionless track supported by a powerful magnetic field.

Josephson effect: A current can flow across a superconductor-insulator-superconductor junction without an applied voltage.

Questions:

1. Discuss the fundamental electric and magnetic properties of superconductors.
2. What are the major differences in the structure and property between low and high temperature superconductors?
3. What are the major structural characteristics of high temperature superconductor?

4. List the major techniques used for the preparation of thin films of high temperature oxide superconductors ad discuss the influences of processing on property.

5. Understand the conventional BCS theory of superconductivity.

6. (1) Draw a schematic *B-H* plot to show the difference between Type I and Type II superconductors. (2) Which type of superconductors has wider applications? Why?

7. A bulk Bi-Sr-Ca-Cu-O superconductor has a critical current density (J_c) of 1×10^5 A/cm^2 at 77 K. (a) What is the maximum supercurrent that could be carried in a wire of 1 mm diameter at 77K? (b) What will happen if the wire is used at 30 K? (c) There is a 0.1 mm diameter wire and a 1 mm diameter wire of identical superconductor, which will have the higher J_c? Why?

8. A Nb wire is to be used to produce a superconducting coil to create a strong field. Estimate the maximum field could be produced at the liquid helium temperature. Nb has a T_c of 9.25 K and a critical magnetic field of 1.57×10^5 A/m.

Chapter 11

Ionic and Mixed Ionic/Electronic Conductivity

11.1 Point Defects

A point defect can be defined as a lattice site which contains an atom, ion or molecule which would not be present on that site in a perfectly stoichiometric material. Generally, point defects can be defined as being either vacant lattice sites or interstitial atoms. However, this term can also be used to include foreign atoms substituting for lattice sites. Point defects are generally equilibrium constituents of all solids, and thus they obey the usual constraints of thermodynamics. Figure 11.1 summarizes most of the common types of point defects present in a number of materials; vacancy, divacancy, interstitial (self interstitial and impurity interstitial), and substitutional impurity atom, which will be described later.

We will initially consider the most important defect mechanisms for intrinsic point defects present in the pure material. Throughout this text we will use the most commonly used notation, the Kroger-Vink notation, to describe the defect chemistry involved in the different mechanisms. In this notation, the defect in question is represented by a major symbol signifying the species. For example, V signifies a vacancy, while A is a cation and B is an anion in the system AB. A subscript signifies the site occupied by the species, while a superscript signifies the virtual or effective charge. These are the charges on the defect in respect to the ideal unperturbed crystal of the defect ($^\bullet$ = positive charge, $'$ = negative charge, and ^{x} = neutral). Electron and electron defects (holes) are given by e' and $h^\bullet$ respectively. An interstitial is given by the letter "i". Typically, most materials produce these types of defects via one of two mechanisms; Schottky defect or Frenkel (or anti-Frenkel) defect, which we will now describe.

Thus, in a system AB, for example,

A^x_A = (Neutral) A on an A site

B^x_B = (Neutral) B on a B site

A^x_B = (Neutral) A on a B site

V'_A = vacancy on an A site with an effective charge of –1.

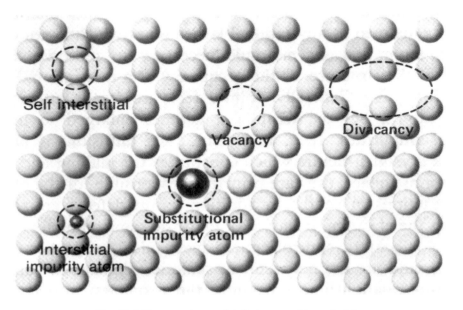

Fig. 11.1 Common types of defects present in materials.

11.1.1 *The Schottky defect*

The Schottky defect can be described as a pair of vacant lattice sites, an anion site and a cation site. The formation of a Schottky defect involves the formation of two vacant lattice sites, and therefore the formation of two extra atoms on the surface of the crystal to compensate for the vacancies. The Schottky defect is the principle defect found in alkali halides, for example, NaCl. The constituents of the Schottky defect in NaCl are V'_{Na} and $V^{\bullet}_{Cl}$. The Schottky defect mechanism, for NaCl, is summarised in Fig. 11.2a. As can be seen in the figure, there are equal numbers of anion and cation defects to preserve the local electroneutrality as far as possible (drawn as shaded squares). This takes place in the bulk of the crystal and on the surface. In simple binary compounds, such as MgO and NaCl, the Schottky disorder consists of pairs of defects. However, for more complicated systems, the number of vacancies formed in one Schottky defect will be greater than two.

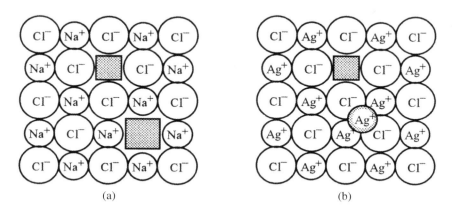

(a)   (b)

Fig. 11.2 Schottky (a) and Frenkel (b) defect mechanisms.

11.1.2 *The Frenkel and anti-Frenkel defect mechanism*

The Frenkel defect is formed when an atom is displaced from its original site into an interstitial site which is usually empty. The Frenkel defect mechanism is summarised in Fig. 11.2b, for the AgCl crystal. The material containing a majority of this type of defect is AgCl (which has the NaCl crystal structure). In AgCl, the interstitial Ag^+ ion is tetrahedrally surrounded by four Cl^- ions, and by four Ag^+ ions The defect consists of V'_{Ag} and $Ag^{\bullet}_i$ species. CaF_2 shows predominantly anti-Frenkel defects, where the F^- ions occupy the interstitial site, thus $V^{\bullet}_F$ and F'_i species are formed. Many fluorite-type materials also have predominantly Frenkel defects, including ZrO_2, as will be discussed later.

Although both types of defect are known to occur in simple ionic crystals, it is usual for the concentration of one type to exceed that of the other. Thus, Schottky defects predominate in the rock-salt alkali halides, cation Frenkel in AgCl and AgBr, and anion or anti-Frenkel defects in CaF_2 and in the fluorites generally.

11.1.3 *Concentration of point defects*

The defect density of these intrinsic defects can be estimated using simple statistical thermodynamics. The concentration is temperature dependant and follows the simple thermodynamic relationship:

$$n/N = \exp(-G/kT) \tag{11.1}$$

where n is the number of either Schottky or Frenkel defects, and N is the total number of sites available, and G is the free energy of formation of the defect. If we consider a Schottky defect in a material MX, then the quasi-chemical reaction can be written as:

$$MX = V_M + V_x + MX_{surface} \tag{11.2}$$

Now, the total number of ways (Ω) that we can pick these n defects on the N sites is given by:

$$\Omega = \frac{N!}{(N-n)!n!} \tag{11.3}$$

where n! in the denominator occurs, because the order in which the vacancies are formed is immaterial.

Hence, the total change in the free energy of formation (ΔG) for the anion vacancy (AV) and cation vacancy (CV) is given by:

$$\Delta G = n_{CV}\Delta G_C + n_{AV}\Delta G_A - kT \ln\left[\frac{N_C!}{(N_C - n_{CV})!n_{CV}!}\right]\left[\frac{N_A!}{(N_A - n_{AV})!n_{AV}!}\right] \tag{11.4}$$

where n_{cv} and n_{av} are the numbers of cation and anion defects respectively, and N_C and N_A are the total number of anion and cation sites available.

Now, using Stirling's approximation to expand $\ln x! = x \ln x - x$ which is good for large x, and minimising the free energy change with respect to n_{CV}, we have,

$$\Omega = N \ln N - (N-n) \ln (N-n) - n \ln n \tag{11.5}$$

and thus,

$$\frac{\partial \Delta G}{\partial n_{CV}} = 0 = \Delta G_C + \Delta G_A + kT \ln \left(\frac{n_{CV}}{N_C - n_{CV}} \right) \tag{11.6}$$

or,

$$\frac{n_{CV}}{(N_C - n_{CV})} = \exp \left(-\frac{\Delta G_C + \Delta G_A}{kT} \right) \tag{11.7}$$

This can also be repeated for the anion vacancies. Thus, the number of Schottky defects is calculated on thermodynamic grounds and the number has an activation energy. It should however be remembered that the number of anionic and cationic vacancies does not have to be equal when we are dealing with a more complex system. Thus, we can include a term in the calculations for systems such as $M_a X_b$, where the system would form a + b vacancies.

The same type of calculations can be undertaken for the formation of Frenkel defects and the result, as expected, is very similar. If a system has N normal sites, N′ interstitial sites, and n Frenkel defects, then we can show, using the same assumptions as above, that the defect concentration can be given by equation 11.8, where n << N, N′.

$$n = (NN')^{\frac{1}{2}} \exp \left(-\frac{\Delta G_F}{2kT} \right) \tag{11.8}$$

where ΔG_F is the change in free energy of formation of the Frenkel defect. It should be noted that the Frenkel defect always produces an interstitial site and a vacancy together.

In much the same way as the concentration of Schottky defects, the Frenkel defect concentration is thermally controlled.

11.2 Line Defects

When we describe line defects, we usually consider the dislocation. Dislocations are stoichiometric defects present in materials in very large numbers; typically $10^7/cm^2$ in annealed materials, and much higher in worked materials. Their presence has a profound influence on the mechanical strength of the material, but they may also influence its electrical properties.

To describe a dislocation, let us consider the two main types that occur in a crystal:

- Edge dislocation
- Screw dislocation

11.2.1 *Edge dislocation*

This is described in Fig. 11.3. Here, displacement of the lower part of the crystal has occurred in the direction of the arrow (b) on the slip plane ABCD. The line FE occurs because there is disregister between the atoms in the planes above and below the slip plane. In effect, an extra half plane of atoms is present in the upper part of the crystal (drawn as ⊥ on Fig. 11.3). This is described in Fig. 11.4 and is known as a positive edge dislocation. In both Figs. 11.3 and 11.4, the Burgers vector (b) is drawn. This is described as the amount of slip which has taken place, and the direction it has occurred in, or the width of the edge dislocation (see 11.4). If the extra half plane of atoms was present below the slip plane, then the system would be known as a negative edge dislocation (and drawn as ⊤). In edge dislocations, the dislocation line is perpendicular to the direction of slip.

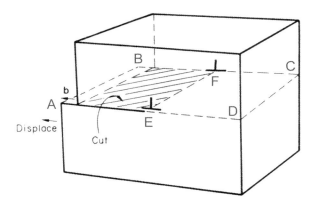

Fig. 11.3 Diagrammatic representation of an edge dislocation present in a material.

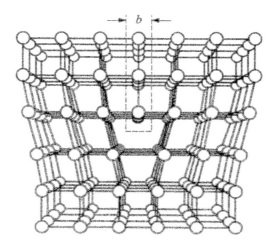

Fig. 11.4 Diagrammatic representation of a positive edge dislocation.

11.2.2 *Screw dislocation*

When the dislocation line is parallel to the direction of slip, then a screw dislocation occurs, as described in Fig. 11.5.

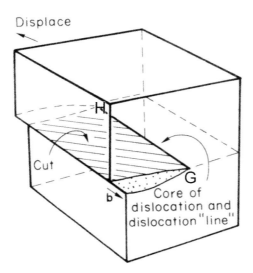

Fig. 11.5 Diagrammatic representation of a screw dislocation present in a material.

The dislocation line, GH, marks the boundary between slipped and unslipped material. Again the Burgers vector is described as showing the direction and amount of slip having taken place. The atoms around the dislocation line are arranged in a spiral ramp which advances, in this case, in a clock-wise direction (known as right-handed). The spiral, when it intersects with the crystal surface, will produce a step.

11.2.3 *Mixed dislocation*

In general, real crystals are neither pure slip nor pure screw, but contain both, as shown in Fig. 11.6. At J, the dislocation line is parallel to the slip direction, and thus it is screw in nature, while at K it is perpendicular and therefore edge in nature. The dislocation line between J and K is known as a mixed dislocation.

In general, more dislocations are introduced into a crystal when a stress is applied, which subsequently contributes to its strength. This is known as work hardening and can be relaxed by heat treatments.

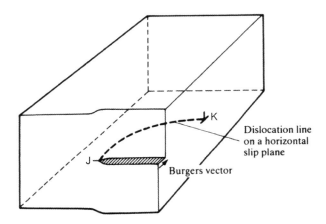

Fig. 11.6 Diagrammatic representation of a mixed dislocation present in a material.

11.3 **Planar Defects**

The external surface of the crystal, or material in general, can be classified as a planar defect. The properties of these surfaces depend upon the fact that they are not perfect but exhibit a variety of point and linear imperfections. The usefulness of the material as a catalyst, for example, may depend upon the imperfections present. Other planar defects include stacking faults, crystallographic shear planes, and grain boundaries.

In this chapter, we will only consider point defects in any detail. However, it should be pointed out that the mobility of these point defects is affected by the concentration and type of other linear and planar defects present.

11.4 Diffusion Mechanisms

To describe the process in which atoms move through the crystal we usually consider the concept of diffusion. The easiest way of describing diffusion through a crystal lattice is to consider atoms (possibly foreign) actually jumping from one site to another.

Let us consider that a crystal which contains particles of type i, has C_i impurity atoms per cm^3, and that the impurity atoms move from one side of the crystal to the other, in a +z direction. Thus, C_i will increase in this direction, allowing for $\partial C_i/\partial z$ to be a positive number. If the sample is held at constant temperature, we would expect that the atoms would diffuse from the high concentration side to the low concentration side, due to diffusion. Now, if J_i is the flux of atoms (that is the number of atoms crossing a unit area in a unit time) we can show that

$$J_i \propto \partial C_i/\partial z,$$

or

$$J_i = -D_i{'} \partial C_i/\partial z \qquad (11.9)$$

where $D_i{'}$ is known as the individual partial or intrinsic diffusion constant. The units of $D_i{'}$ are cm^2/s, and the minus sign is present because the flux is in the opposite direction to the concentration gradient. This is known as Fick's First Law, which basically shows that the driving force for diffusion is simply the concentration gradient.

In general, particles in solids react chemically and do not move independently. Thus, $D_i{'}$ is one of many types of chemical diffusion coefficients.

Fick's first law only describes mass transport by a spatial variation in the concentration of the mobile species i. This is a special case of mass transport in response to a wide range of driving forces. The driving forces can arise from electric fields, temperature gradients, chemical reactions, etc. Generally, the gradient in any given thermodynamic quantity will give rise to a driving force which will lower the gradient free energy of the system.

However, this becomes much more complicated, as the system not only contains mass transport, but also electrical and heat transport. Thus, a more general phenomenological approach is needed.

Therefore, we can write a more generalised expression for the transport of the ith species in a system, with n components.

$$J_i = \sum_n L_{in} X_n + L_{io} X_o \qquad (11.10)$$

Where the transport coefficients are given by L, and the driving forces are given by X., while X_o and L_{io} are specifically associated with the transport of heat.

Now the fundamental gradient involved is not the concentration gradient, as described by Fick's law, but the chemical gradient.

Thus,

$$J_i = -L_i \frac{d\mu_i}{dz} \qquad (11.11)$$

Where μ_i is the chemical potential of the species, i, and L_i is the transport coefficient, given by

$$L_i = D_i' \left(\frac{dC_i}{d\mu_i} \right) \qquad (11.12)$$

Now the chemical potential, μ_i, for an ideal system, can be given by

$$\mu_i = \mu^o_i + kT \ln a_i \qquad (11.13)$$

Where a_i is the activity of the system.
Thus,

$$L_i = \frac{D_i' C_i d \ln C_i}{kT d \ln a_i} \qquad (11.14)$$

This now allows, D_i' is defined as the individual partial diffusion coefficient. It is usual to define a particle diffusion coefficient, D_i

$$D_i = D_i' \frac{d \ln C_i}{d \ln a_i} \qquad (11.15)$$

If we say that $a_i = \gamma_i C_i$, where γ_i is the activity coefficient of species i,

$$D_i = D_i' \left[1 + \left(\frac{d \ln \gamma_i}{d \ln C_i} \right) \right]^{-1} \qquad (11.16)$$

Hence, from equations, 11.11, 11.14 and 11.15

$$J_i = -\frac{D_i C_i}{kT} \frac{d \mu_i}{dz} \qquad (11.17)$$

In the dilute limit, γ_i tends to 1, and D_i tends to D_i'; thus, equation 11.17 tends towards Fick's law (11.9).

For mass transport in solid state materials we should consider an electrochemical potential, η_i, which is defined as

$$\eta_i = \mu_i + Z_i |e| \phi \qquad (11.18)$$

Where Z_i is the particle charge, and ϕ is the electrical potential. Thus, in ionic solids the transport equation should be written as

$$J_i = \frac{D_i C_i}{kT} \frac{d \eta_i}{dz} \qquad (11.19)$$

If we define the mobility υ_i as the drift velocity per unit field strength, then it is related to the particle velocity v_i by $v_i = \upsilon_i Z_i |e|$, and the particle diffusion coefficient ($D_i = kT v_i$) is related to the mobility by

$$D_i = \left[\frac{kT}{Z_i |e|} \right] \upsilon_i \qquad (11.20)$$

This equation shows that the diffusion coefficient is proportional to the mobility of the species through the material. It is a form of the Nernst-Einstein equation, which we shall be using later to describe the effect of the diffusion coefficient on the ionic conductivity of the material.

11.5 Diffusion as a Random Walk

Mass transport in solids occurs by making transitions between the well-defined positions the atoms take up in a crystal. They do this in such a way that the time of transit between sites is much less than the residence time at any particular position.

This diffusion can be considered to be via hopping in a random way on a lattice of sites distributed in space. If an individual atom is labelled, its motion can be followed and related to the phenomenological tracer diffusion coefficient. For a cubic lattice, the tracer diffusion coefficient (D^*) is given by

$$D^* = 1/6 f \Gamma r^2 \qquad (11.21)$$

Γ is the jump frequency to a nearest neighbour site
r is the jump distance
f is a correlation factor that takes into account non-random walk.

The hopping involves particles crossing an energy barrier, and the diffusion coefficient, thus thermally activated, takes on the form of an Arrhenius equation.

$$D = D_o \exp\left[\frac{-Q}{kT}\right] \qquad (11.22)$$

where, Q is the activation enthalpy of the ion movement, D_o is a pre-exponential factor, k is the Boltzmann constant, and T is the absolute temperature.

D_o can be expressed by $D_o = (1/3)(Ze)^2 n_d^2 \, \omega_o$, where Ze is the charge of the conducting ion, n is the defect density (interstitials or vacancies), d is the unit jump distance of the ion (usually the closest ionic pair) and ω_o is the attempt frequency.

So far, we have considered a number of phenomenological equations. It is important now to consider how atoms actually move from one site to another.

Atoms (either intrinsic (host) or extrinsic (foreign)) can diffuse through the material via one or more of the following mechanisms:

11.6 Vacancy Mechanism

All types of substitutional diffusion have been shown to operate through a vacancy mechanism. In this type of mechanism, a diffusive jump takes place via an exchange of position with an adjacent vacancy. Thus, a flux of diffusing ions or atoms in any particular direction requires an equal number of vacancies diffusing in the opposite direction. Figure 11.7a summarises the vacancy mechanism.

11.7 Interstitial Mechanism

When atoms which are either dissolved in the host material or make up the host material occupy interstitial sites, then diffusion of the atom between the interstitial sites can occur. It is quite apparent that the atom is usually a lot smaller than the host. Typical examples are carbon in steels, and dissolved gases in metals. The interstitial mechanism is summarised in Fig. 11.7b.

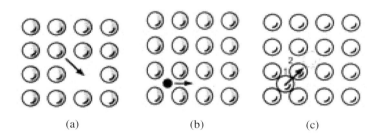

(a)   (b)   (c)

Fig. 11.7 Diffusion mechanisms (a) vacancy; (b) interstitial; (c) interstitialcy.

11.8 Interstitialcy Mechanism

This mechanism is considered to be a combination of a substitutional and interstitial mechanism, and is summarised in Fig. 11.7c. The principle of the interstitialcy mechanism is that an interstitial atom (labelled 1 in the figure) displaces another atom (labelled 2) from its original substitutional site into an adjacent interstitial position. Due to the type of diffusion mechanism, any measurement of the ionic conductivity of a material showing the interstitialcy mechanism will lead to twice the value of D.

11.9 Other Mechanisms

11.9.1 *The ring mechanism*

The ring mechanism operates when three or four atoms rotate as a group, and the distortion required for a simple exchange between two-nearest neighbours is greatly reduced over that which could be expected. The anomalies observed in, for example, the diffusion coefficients of some bcc metals, could be explained by assuming a ring mechanism.

11.9.2 *The Crowdian mechanism*

In this mechanism an extra atom has been placed in a close-packed plane, and the distortion is taken up by displacing several atoms rather than just the nearest neighbours. The energy required to move the configuration is very small. The configuration resembles an edge dislocation, described above, which has a very low energy for motion.

There are a number of other mechanisms for diffusion, but the ones listed above are those required for the arguments used throughout this text.

11.10 Conductivity

In solid state materials we usually refer to the term *conductivity* (σ), which, traditionally, has the units S/cm (1 Siemens equals $1\Omega^{-1}$). Note, that although S/cm is not a standard SI unit, we usually use it in solid state electrochemistry, simply due to the size of the units involved (in SI we would state the value in S/m)

The partial electrical conductivity (σ_i) of a species i in a material is related to the mobility by

$$\sigma_i = Z_i |\,e|\, C_i\, \upsilon_i \qquad (11.23)$$

where, C_i is the concentration of the *i*th charge carrying species, Z_i is its valency, and υ_i is its mobility; e is the electronic charge.

Now, equation 11.23 can be put into equation 11.20, showing that the diffusion coefficient described earlier is related to the electrical conductivity of the material

A correlation factor (f) may need to be put into equation 11.24 if it is to be related to the isotopic diffusion.

$$\sigma_i = \frac{C_i(Ze)^2}{kT} D_i \qquad (11.24)$$

11.11 Introduction to Solid Electrolytes

These are a class of materials sometimes called superionic conductors or fast ion conductors, containing either ionic (anionic or cationic) conductivity or mixed ionic/electronic (p or n-type) conductivity. We will initially discuss ionic conductors, and then mixed conductivity later on in the chapter.

Most crystalline materials have low ionic conductivities because the atoms cannot escape from their lattice sites, even though they are able to vibrate around that site. However, fast ion conductors have either anion or cation sites which, due to their high degree of disorder, are free to move throughout the structure. They can, in effect, be considered as intermediate between a liquid (with an irregular array, but with mobile ions) and a crystalline solid (which has a 3-dimensional array but with essentially immobile ions).

The effect of temperature is very marked in these materials. Often they are only ionic conductors at high temperature, and may go through a phase change at lower temperatures. For example at approximately $730^\circ C$ pure Bi_2O_3 is stable in a face centred cubic form (fcc) known as δ-Bi_2O_3, and shows an oxygen conductivity in excess of 1S/cm. Below 730°C however, it reverts to monoclinic α-Bi_2O_3, which is a p-type electronic conductor (see Fig. 11.8). as will be described in Section 11.12.7. It is also possible to stabilise γ and α-Bi_2O_3 (also shown in Fig. 11.8), although these phases are metastable. δ-Bi_2O_3 has a very disordered structure, and is sometimes referred to as a semi-liquid.

Other solid electrolytes may show a gradual increase in defect concentration with increasing temperature (for example CeO_2), which shows no phase transition with increasing temperature, but at approximately 500°C has a high enough defect concentration for the material to be considered a good fast ion conductor. In the following sections, we will consider a number of the more important ionic conductors, with particular reference to their properties and uses.

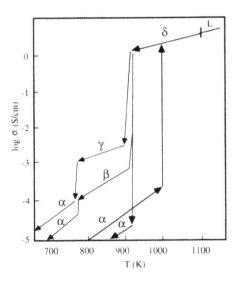

Fig. 11.8 Phase changes observed in Bi_2O_3.

11.12 Oxygen Ion Conductors

The history of anionic conductors in the solid state can be considered to have started in the early 1800's from the work of Faraday (1936), who noted the remarkable ionic properties of PbF_2. The work of Nernst (1900) however, at the turn of the century, was really the turning point in the use of solid state materials when he produced the Nernst Glower in 1900. Nernst proposed that an ionic conductor such as ZrO_2, when lightly doped with a small amount of CaO or Y_2O_3, could be used as the light source of an electric lamp. The material finally chosen by Nernst was 85wt% ZrO_2/15wt% Y_2O_3, and was later known as the Nernst mass.

It appeared that the Nernst glower could operate for a long period of time on DC current; although electrolysis was found to occur. This was explained by the fact that the Nernst mass was an oxide ion conductor at high temperature and any oxygen lost at the anode was balanced by oxygen taken into the Nernst glower at the cathode. A diagrammatic representation of the Nernst Glower is given in Fig. 11.9.

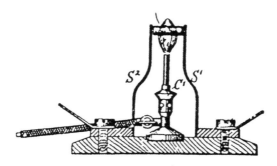

Fig. 11.9 Diagrammatic representation of the Nernst Glower (US patent 653349).

In Fig. 11.9, the ionic conductor was heated by a Bunsen burner. This type of oxygen ion conductor is known as stabilised zirconia. Investigations using stabilising oxides such as CaO, MgO and Y_2O_3 have been studied for many years and have been exploited for use as practical devices to be run at 1000°C. Kiukkola and Wagner (1957) carried out the initial studies of electromotive force measurements to determine the free enthalpy's of the reactions at the high temperatures employed. Tubandt (1932) carried out the transference number measurements on barium and lead halides, and showed that their values were almost unity at high temperature.

To explain the defect structure of oxygen ion conductors, let us consider the case of MO_2 (where M is a metal cation (such as Zr^{4+}, or Ce^{4+})) containing intrinsic defects. In oxygen ion conductors, the predominant defect reactions are those which preserve stoichiometry, ie Frenkel and Schottky disorder. Now, in systems with a fluorite-type structure, it is generally accepted that the predominant intrinsic defects are the *anion* or anti-Frenkel defects; Vö, and $O_I^{''}$. Thus, the following defect reactions are observed for these systems.

The (anti-) Frenkel defects can be destroyed or created at other intrinsic defects, such as dislocations, grain boundaries or surfaces. The effect of partial pressure on the defect equilibrium is very marked, and determines the use of the material as an oxygen ion conductor. This will be discussed later, when we consider electronic conductivity in materials.

$$M_M^x = V_M'' + M_i^{\bullet\bullet} \tag{11.25}$$

$$2O_o^x = 2V_o^{\bullet\bullet} + 2O_I'' \tag{11.26}$$

11.12.1 *Thorium oxide (Thoria)*

Thoria (ThO_2) exists in the cubic fluorite-type structure, and is usually used as the model material when describing defect chemistry. In the pure form it exhibits p-type conductivity. It shows p-type conductivity (predominant defects are the electron holes) at high oxygen partial pressure, although interstitial anions are also present.

The defect equilibrium at high oxygen partial pressures can be given by:

$$O_{2(g)} \leftrightarrow 2O_i'' + 4h^{\bullet} \tag{11.27}$$

Thus, the equilibrium constant can be given by:

$$K_1 = [O_i^{2-}]p^2/PO_2^{1/2} \tag{11.28}$$

Therefore, as the oxygen partial pressure is increased, the electron hole concentration increases to preserve electroneutrality. If $[O_i'']$ is effectively large and constant, then at high oxygen partial pressures, $p \propto PO_2^{1/4}$.

Figure 11.10 shows a plot of total conductivity of ThO_2 as a function of oxygen partial pressure. At oxygen partial pressures below approximately 10^{-8} atm., the conductivity of ThO_2 is effectively independent of oxygen partial pressure, and thus the majority charge carrier is the oxygen ion. However, the value is still far too low to be of practical use. No n-type conductivity has been observed even at partial pressures as low as 10^{-20} atm.

carrier is the oxygen ion. However, the value is still far too low to be of practical use. No n-type conductivity has been observed even at partial pressures as low as 10^{-20} atm.

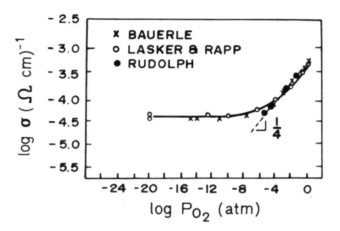

Fig. 11.10 A plot of total conductivity of ThO_2 as a function of oxygen partial pressure (after S. Geller, Ed. Solid Electrolytes, W.L. Worrell, Oxide Solid Electrolytes, Chapter 6, p143, Springer-Verleg).[1]

Dopants such as CaO, MgO, Y_2O_3, Yb_2O_3 and Gd_2O_3 are added to introduce anion defects into the ThO_2 lattice, and thus increase the ionic conductivity. Figure 11.11 shows a typical plot of ionic conductivity versus impurity or doping cation concentration, for ThO_2 doped with Y_2O_3. The dissolution of Y_2O_3 into the fluorite-type lattice is given by:

$$Y_2O_3 \rightarrow 2Y_{Th}' + 3O_o^x + V_o^{\bullet\bullet} \qquad (11.29)$$

Each addition of two yttrium ions introduces one oxygen vacancy. The solubility of the added dopant depends upon temperature and the dopant species. As the temperature is increased, the solubility increases. At approximately 1700°C, for example, between 10 and 20 mole% of CaO and Y_2O_3 can be dissolved in ThO_2. Thus, as the dopant concentration is increased, the conductivity is increased. This is valid at relatively dilute

[1] Work of, J.E. Baurle, J. Chem. Phys., 45(1966) 4162, M.F. Lasker, and R.A. Rapp, Z. Phys. Chem. NF. 49 (1966) 198, J. Rudolph, Z. Naturforsch, A14 (1959) 727.

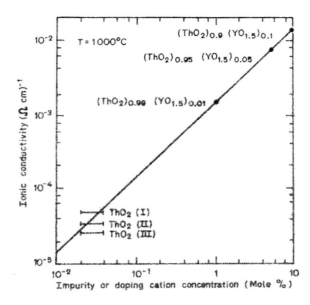

Fig. 11.11 Typical plot of ionic impurity or doping cation concentration versus ionic conductivity for ThO_2 doped with different concentrations of Y_2O_3.

11.12.2 *Zirconium Oxide (Zirconia)*

Stabilized zirconia is the system which has been studied more than any other for ionic devices, and thus the literature on the material is vast.

Pure ZrO_2 exhibits 3 polymorphs, cubic zirconia, tetragonal zirconia, and monoclinic zirconia.

Cubic zirconia is stable from approximately 2370°C to the melting point of the material (2170°C). It has the Fm $\bar{3}$m space group, and the crystal structure is shown in Fig. 11.12a.

Tetragonal zirconia is stable from approximately 1170°C to 2370°C, where it transforms into cubic zirconia. Tetragonal zirconia has a $P4_2$/nmc space group, and the crystal structure is shown in Fig. 11.12b.

The tetragonal to monoclinic transformation, on cooling, is accompanied by a 3 to 5% change in volume, which can cause the ceramic to crack during sintering. This is known as a martensitic reaction and, although the word martensite is usually used for quenched steels, it is just as applicable for this reaction.

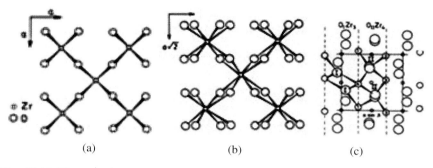

(a)   (b)   (c)

Fig. 11.12 The crystal structure of the 3-phases of zirconium dioxide (zirconia); (a) Projection of a layer of ZrO_8 groups in the cubic ZrO_2 on the (100) plane; (b) Projection of a layer of ZrO_8 groups on the (110) plane of tetragonal zirconia; (c) Projection of the crystal structure of monoclinic zirconia along the C_m-axis showing layers of O_IZr_3, and $O_{II}Zr_4$ polyhedra. (after E.C. Subbarao, in Science and Technology of Zirconia 1, Ed. A.H. Heuer, and L.W. Hobbs, Amer. Ceram. Soc., (1981), Ohio, USA, p1–24).

The monoclinic phase is stable from room temperature up to approximately 1170°C (although this temperature varies depending upon what literature is read) at which point it transforms, reversibly, into tetragonal zirconia. Monoclinic zirconia has a $P2_1/c$ space group, and the crystal structure of the system is shown in Fig. 11.12c. The main features of the structure are:

- Sevenfold co-ordination of Zr with a range of bond lengths and angles.
- Layers of triangularly co-ordinated O_I-Zr_3 and tetrahedrally co-ordinated O_{II}-Zr_4.
- Zr atoms are located in layers parallel to the (100) planes, separated by O_I and O_{II} atoms.
- The layer thickness is wider when the Zr atoms are separated by O_I atoms than by O_{II}.

Monoclinic zirconia shows predominantly electronic conductivity. The addition of metal oxide stabilisers (MO and M_2O_3) to zirconia can stabilise the tetragonal and/or cubic phases to lower temperatures, and thus avoid the destructive tetragonal to monoclinic transformation.

Many zirconia stabilisers are known, including CaO, Y_2O_3, Yb_2O_3, Sm_2O_3, Sc_2O_3, Gd_2O_3 and Nd_2O_3. The stabilisers form solid solutions with zirconia, and substitute for the Zr^{4+} in the lattice sites. As a result of their lower valency (2+ or 3+), they create vacancies in the oxygen sub-lattice, as described in equation 11.30, for example.

$$Y_2O_3 \rightarrow 2Y_{Zr}{}' + 3O_o{}^x + V_o{}^{\bullet\bullet} \qquad (11.30)$$

The vacancies are responsible for the observed high ionic conductivity. If sufficient dopant is added to the zirconia, then the cubic phase can be fully stabilised. If insufficient dopant stabiliser is added to the zirconia, then the material will consist of 2 or more phases. These phases exist in a metastable form. The amount of dopant required to typically stabilise zirconia depends on the stabiliser itself. However, the amounts are typically: 12 – 13 mole% for CaO, 8–9 mole% for Y_2O_3 and Sc_2O_3, and 8 to 12 mole% for the other rare earth oxides (Dy_2O_3, Yb_2O_3, Gd_2O_3, Sm_2O_3, etc).

The ZrO_2-Y_2O_3 system is the most well known, and really the only system which has been fabricated into devices such as solid oxide fuel cells and oxygen sensors.

Figure 11.13 shows a typical phase equilibrium diagram for the ZrO_2-Y_2O_3 system. It should however be noted that there is still a lot of controversy relating the phase equilibria of the system, possibly due to the various measuring techniques. Thus, this phase equilibrium diagram is one of many which have been constructed for this system.

Figure 11.13 shows a number of features, including the presence of the stable ordered compound $Zr_3Y_4O_{12}$. There is also a eutechtoidal decomposition reaction where cubic zirconia (F_{SS}) converts to monoclinic zirconia (M_{SS}) and $Zr_3Y_4O_{12}$.

The effect of dopant concentration on the ionic conductivity of the system is quite marked. Figure 11.14 shows the conductivity as a function of dopant concentration for a number of different dopants. There also appears to be a correlation between the minimum amount of dopant required to stabilise the cubic phase, and the maximum conductivity observed. This observed drop in conductivity has been examined for many years, and a number of models have been proposed to explain it. However, in general, the models discuss varying degrees of interaction between defect pairs, ordering of vacancies, varying defect configurations, and for concentrated solutions, complex defect pair associations.

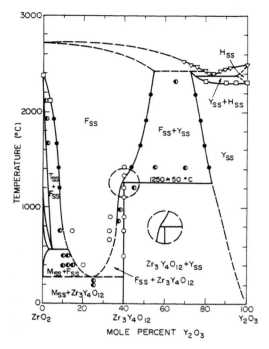

Fig. 11.13 Phase equilibrium diagram for the system ZrO_2-Y_2O_3 (after V.S. Stubican, R.C. Vink, and S.P. Ray, J. Amer. Ceram. Soc., 61 (1978) 18–21).

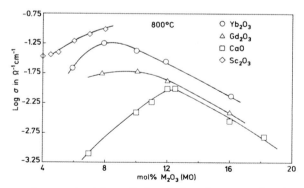

Fig. 11.14 shows the conductivity of doped-zirconia as a function of dopant concentration for a number of different dopants, at 800°C (after SPS Badwal, Chapter 11 in Materials Science and Technology, Volume 11, Structure and Properties of Ceramics, Ed M. Swain, VCH Publishers, (1994), Germany, p567–633 (1994)).

Table 11.1. Ionic conductivity of some zirconia-based electrolytes.

Composition	Conductivity (S/cm) at 1000°C
$(ZrO_2)_{0.88}(Y_2O_3)_{0.12}$	0.11
$(ZrO_2)_{0.90}(Y_2O_3)_{0.10}$	0.12
$(ZrO_2)_{0.92}(Y_2O_3)_{0.08}$	0.15
$(ZrO_2)_{0.92}(Sc_2O_3)_{0.08}$	0.18–0.31
$(ZrO_2)_{0.87}(CaO)_{0.13}$	0.06
$(ZrO_2)_{0.90}(Gd_2O_3)_{0.10}$	0.02 (800°C)
$(ZrO_2)_{0.92}(Yb_2O_3)_{0.08}$	0.2
$(ZrO_2)_{0.81}(In_2O_3)_{0.19}$	0.04 (800°C)

The ionic conductivity for a number of different doped-ZrO_2 systems is shown in Table 11.1, at 1000°C. This is the usual temperature at which a device made from the material is typically run. With the relatively high oxygen ion conductivity at 1000°C and the good stability, together with reasonable mechanical properties and the relative ease and cheapness of manufacture, doped-ZrO_2 materials are quite difficult to beat when alternative fast-ion conductors for an electrochemical device are examined.

11.12.3 *Hafnium oxide (Hafnia)*

Pure HfO_2 undergoes a number of phase transformations with increasing temperature and cannot be stabilised in the pure form as a cubic-fluorite structure. The cubic-fluorite structure can only exist when a dopant, such as Y_2O_3, Gd_2O_3, La_2O_3 or CaO is added. HfO_2-Y_2O_3 is the most extensively examined material with 7 – 8 mole% Y_2O_3 required to stabilise the cubic-fluorite structure. In this system, the defect structure consists of a filled cation sub-lattice and anion vacancies, with a maximum in conductivity observed at approximately 8mole% Y_2O_3. As can be seen in Fig. 11.15, the ionic conductivity is low when compared to the ZrO_2–based systems, and the electrolytic domain is very narrow; 1 to 10^{-16} atm. has been reported, at 1000°C. Thus HfO_2-based materials have little or no use as electrolytes when compared to other systems such as doped-ZrO_2 and doped-ThO_2.

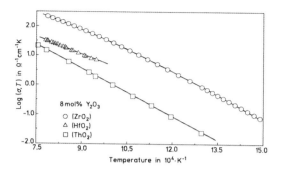

Fig. 11.15 A plot of total conductivity of 8 mol% Y_2O_3-stabilise HfO_2, ZrO_2, and ThO_2 as a function of temperature, in the form of an Arrhenius plot (after Badwal 1994).

11.12.4 *Cerium oxide (Ceria)*

In the pure state, CeO_2 exists in the cubic fluorite structure, as shown in Fig. 11.16. The structure is relatively open and has a large tolerance for high levels of atomic disorder, introduced either by doping, or by reduction-oxidation (redox) reactions. Moreover, CeO_2 has an ionic radius which allows for a variety of dopants to be added with either smaller or larger ionic radii.

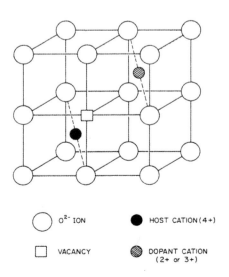

Fig. 11.16 The cubic fluorite structure of CeO_2.

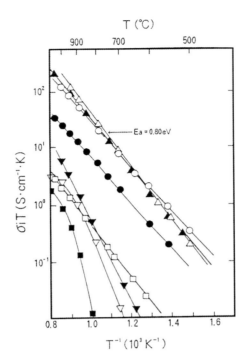

Fig. 11.17 Conductivity of doped-CeO_2 as a function of temperature: ($\square$) $(CeO_2)_{0.9}(BaO)_{0.1}$; (O) $(CeO_2)_{0.9}(SrO)_{0.1}$; (■) CeO_2; (●) $(CeO_2)_{0.7}(SrO)_{0.3}$; (▲) $(CeO_2)_{0.7}(CaO)_{0.3}$; (▼) $(ZrO_2)_{0.85}(CaO)_{0.15}$; ($\triangle$) $(CeO_2)_{0.9}(CaO)_{0.1}$; ∇ $(CeO_2)_{0.9}(MgO)_{0.1}$. (after H. Yahiro, T. Ohuchi, K. Eguchi, and H. Arai, J. Mater. Sci., 23 (1988) 1036).

Pure CeO_2 is a mixed conductor, and has been shown to have equal proportions of hole, electron and oxygen ion conductivities; the ionic conductivity is significantly increased if impurities (such as CaO or SrO) are found.

The addition of dopants such as CaO, SrO, MgO, and BaO have been extensively studied and the ionic conductivities, as a function of temperature, are shown in Fig. 11.17, with CaO-doped ZrO_2 for comparison.

The greatest dopant effect on the ionic conductivity is seen for CaO and SrO, which also lowers the activation energy, the affect of rare-earth dopants on the ionic conductivity is even more marked, and the highest conductivity reported was for $Ce_{0.8}Sm_{0.2}O_{1.9}$, which appears to have the highest conductivity of all the ceria based solid solutions.

The effect of the dopant on the parent material can be explained by looking at the formation of defect pairs.

We can consider an interaction between the anion vacancies ($V_o^{\bullet\bullet}$) and the aliovalent metal cation (M_{Ce}'), introduced via the following reaction:

$$M_2O_3 + CeO_2 = 2M_{Ce}' + V_o^{\bullet\bullet} + 3O_o^x \qquad (11.31)$$

Thus, the total activation energy for ionic conduction (ΔH_t) can be expressed by the sum of the enthalpy of migration of an oxygen ion (ΔH_m) with the association energy of the complex defect $[M_{Ce}' - V_o^{\bullet\bullet}]$. It has been shown that the association energy is dependent upon the dopantion in the system CeO_2-M_2O_3. It has been suggested that the association energy incorporated a strain energy term due to the mismatch between the host and the dopant materials.

The defect binding energies have also been examined and Fig. 11.18 shows that the association enthalpy varies with the trivalent ion doped in CeO_2, for both experimental data and calculations. It is quite clear that there is good agreement between the two sets of data. It has also been suggested that because the defect binding energy is dependant upon the relative ionic radii of the host and the dopant, then a plot of relative lattice parameters of the dopant and host should yield some interesting information; in fact there does appear to be a requirement to allow the radii of the parent and host materials to be closely matched. This explanation describes the affect of dopant type on the parent material, in relation to its ionic conductivity.

Table 11.2 below shows the conductivity and activation energy of a number of doped ceria based electrolyte systems. A large amount of ambiguity exists in relation to these results and has been reported, although it is clear that the conductivity is approximately 1 order of magnitude greater than that of doped ZrO_2-systems.

The limiting factor in the use of CeO_2-based systems is that the ionic conductivity regime is very narrow, and that at relatively high oxygen partial pressures the material is susceptible to reduction, and can easily develop n-type electronic conduction. For example, $(CeO_2)_{0.95}(Y_2O_3)_{0.05}$ has an ionic domain ($t_i > 0.99$) at 10^{-13} atm. oxygen partial pressure at $600°C$, and only 10^{-6} atm. oxygen partial pressure at $1000°C$. Fig. 11.19 shows a series of typical electrolytic domains for a number of fluorite oxides.

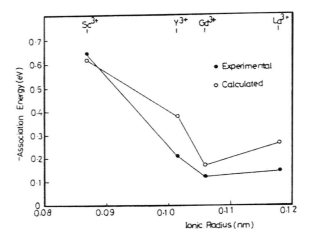

Fig. 11.18 Association enthalpy of CeO$_2$ as a function of trivalent dopant ion (after V. Butler, C.R.A. Catlow, B.E.F. Fender, and J.H. Harding , Solid State Ionics, 8 (1983) 109).

Table 11.2 Conductivity of a number of doped-ceria electrolyte materials.

Composition	Conductivity (S/cm) at 1000°C
$(CeO_2)_{0.92}(Y_2O_3)_{0.08}$	0.0091 (600°C)
$(CeO_2)_{0.95}(Y_2O_3)_{0.05}$	0.145
$(CeO_2)_{0.90}(CaO)_{0.10}$	0.10
$(CeO_2)_{0.89}(CaO)_{0.11}$	0.0064 (600°C)
$(CeO_2)_{0.90}(Gd_2O_3)_{0.10}$	0.25
$(CeO_2)_{0.82}(Gd_2O_3)_{0.18}$	0.235
$(CeO_2)_{0.82}(Nd_2O_3)_{0.18}$	0.23
$(CeO_2)_{0.82}(La_2O_3)_{0.18}$	0.154

The effect of dopants on the critical oxygen partial pressure can also play a role. Figure 11.20 shows a plot of radius of dopant ion versus the critical oxygen partial pressure; the value becomes a minimum at approximately 0.11nm for the dopant ion. Thus, in maximising the oxygen ion conductivity by looking at the dopant size, it is also important to look at the dopant size when considering the size of the ionic domain. Double doping has also been investigated to try and increase the size of the electrolytic domain.

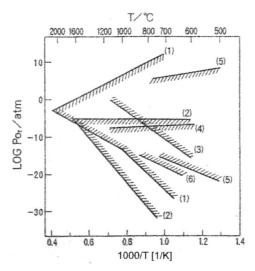

Fig. 11.19 Typical electrolytic domain boundaries for a number of fluorite materials: (1) calcia-stabilised zirconia; (2) yttria-doped thoria; (3) $(CeO_2)_{0.95}(Y_2O_3)_{0.05}$; (4) $(CaO)_{0.15}(La_2O_3)_{0.85}$; (5) $(Bi_2O_3)_{0.73}(Y_2O_3)_{0.27}$; (6) $Ce_{0.8}Gd_{0.2}O_{0.19}$ (after H. Inaba, and H. Tagawa, Solid State Ionics, 83 (1996) 1).

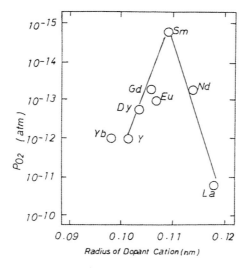

Fig. 11.20 A plot of radius of dopant ion versus the critical oxygen partial pressure for metal (Yb, Dy, Gd, Y, Eu, Sm, Nd, and La) oxide doped CeO_2 systems (after H. Yahiro, K. Eguchi, and H. Arai, Solid State Ionics, 36 (1989) 71).

11.12.5 *Perovskites*

In 1971, it was found that by substituting the A or B sites of the perovskite system, ABO_3, with aliovalent metal oxides, a range of ionic and electronic conductivities could be realised. It was discovered that a reasonably high ionic conductivity could be realised for $La(Ca)AlO_3$, for example, of 5 x 10^{-3} S/cm at 800°C. Though not a necessarily high oxygen ion conductor, these results did pave the way for a lot of new work in examining the possible use of perovskite-based materials as oxygen ion conductors.

In 1994, a paper was published which showed that the oxygen ion conductivity of doped-$LaGaO_3$, when suitably doped, had an oxygen-ion conductivity approximately 3 times that of yttria-stabilised zirconia at 800°C; approximately 0.14 S/cm. It was also found to have an ionic transport number close to unity, and was stable to very low oxygen partial pressures. The best system, to date, has been found to be $La_{0.8}Sr_{0.2}Ga_{0.85}Mg_{0.15}O_{3\pm d}$. Figure 11.21 shows a typical plot of oxygen ion conductivity of some lanthanum-gallate materials, compared to a number of other systems. Other systems which have been considered include doped-$NdGaO_3$, doped $PrGaO_3$.

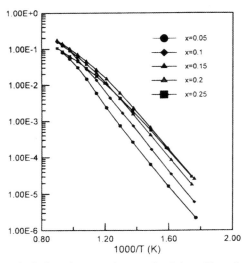

Fig. 11.21 shows a typical plot of oxygen ion conductivity of $La_{1-x}Sr_xGa_{1-y}Mg_yO_{3-d}$, with a fixed value of x + y = 0.35 (after P. Huang, and A. Petric, J. Electrochem. Soc., 143 [5] (1996) 1644).

11.12.6 *Pyrochlores*

Pyrochlores have the general formula $A_2B_2O_7$ and have a defect fluorite superstructure. They have a relatively high intrinsic oxygen ion conductivity, without the necessity to use a dopant ion. Most of the studies have examined the materials where A = Ln, where Ln is a lanthanoid element such as Gd, Nd, Sm, Tb, or Er, and B = Zr or Ti. Examples of these systems include $Gd_2Ti_2O_7$, which can be acceptor doped with, for example, Ca on the A-site, producing a material with high oxygen ion conductivity ($>10^{-2}$S/cm at 1000°C) and negligible electronic conductivity over a wide range of oxygen partial pressures and temperatures. In general, higher ionic conductivities are obtained by high dopant concentrations, and close host/dopant radii size matching. For example, $Gd_2(Zr_{0.4}Ti_{0.6})_2O_7$, has been shown to have an oxygen ion conductivity of $10^{-3.5}$S/cm at 800°C, and an ionic domain stable down to an oxygen partial pressure of 10^{-21} atms.

11.12.7 *Bismuth oxide*

At temperatures greater than 730°C, Bi_2O_3 is stable in a face centred cubic form (fcc), known as δ-Bi_2O_3, and shows an oxygen conductivity in excess of 1S/cm. Below 730°C, however, it reverts to monoclinic α-Bi_2O_3, which is a p-type electronic conductor. It should also be noted that on cooling δ-Bi_2O_3, there is a large hysteresis and one of two other metastable phases is possible: tetragonal β-Bi_2O_3 (formed at 650°C) or body centred cubic γ-Bi_2O_3 (formed at 639°C). These phases may revert to α-Bi_2O_3 on further cooling, but γ-Bi_2O_3 has been observed at room temperature. Table 11.3 shows the temperature regions of the stable and metastable forms of Bi_2O_3.

The formation of δ-Bi_2O_3 is accompanied by a very large transition enthalpy (approximately 29.6 kJ/mole), which is due to a very large change in the degree of disorder in the material. This large increase in disorder is responsible for the high ionic conductivity of δ-Bi_2O_3. δ-Bi_2O_3 has been shown to have a fluorite-type structure, but with a large number of defects in the oxygen sublattice. There is a certain amount of controversy surrounding the actual structure of the disordered phase, but the three most common models quoted are those by Sillen (who considered the fluorite structure contained ordered defects in the oxygen sublattice), Gattow and Schroeder (who considered an oxygen sublattice with a 75% statistical occupation of the sites), and the Willis model (which replaced each anion site in the fluorite lattice by four equivalent sites displaced in the, <111> direction). The 3 structures given in Fig. 11.22 summarize the three models described above.

Table 11.3 Temperature regions of the stable and metastable forms of Bi_2O_3.

Phase	α	δ	β	γ
Phase stability range (K)	<1002	1002–1097	603–923	773–912
Temperature (K)	298	1047	916	298
Structure	monoclinic	fcc	tetragonal	bcc
a (nm)	0.58496	0.56595	0.7738	1.026
b (nm)	0.81648			
c (nm)	0.75101		0.5731	
β (°)	112.977			

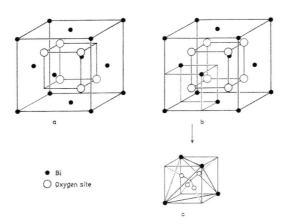

● Bi
○ Oxygen site

Fig. 11.22 summarises the 3 models for the defect structure of Bi_2O_3 ; (a) Sillen model; (b) Gattow and Schroeder model; (c) Willis model.

In order to stabilize the fcc form to lower temperatures (as quenching the phase in is not possible), the addition of aliovalent metal oxides have successfully been introduced. Here, highly conductive oxide ion conductors of the fcc phase, or an alternative rhombohedral phase were formed. In general, the rhombohedral phase is formed with dopants with relatively large rare-earth (Ln_2O_3) ionic radii and relatively low x in $(Bi_2O_3)_{1-x}(Ln_2O_3)_x$, where Ln = Nd, Sm or Gd. The fcc phases are usually stabilized by cations which are smaller than the Bi^{3+} ion. The ionic conductivity of the doped-Bi_2O_3 has been studied in great detail, and is summarised in Table 11.4, which shows the oxide ion conductivity of a number of doped systems, all of them being fcc in nature. In general, the minimum amount of dopant required to stabilise the fcc phase (as shown in Fig. 11.23a), is required for the maximum conductivity, and thus the maximum conductivity is reached with the minimum amount of dopant. Fig. 11.23b shows the minimum value of x (D_{min}) required to stabilise the fcc-phase in $(Bi_2O_3)_{1-x}(Ln_2O_3)_x$ versus the ionic radius of Ln^{3+}. What is apparent is that the value for D_{min} for Y^{3+} is at approximately 0.2, which is also observed in Fig. 11.23a to be the value of maximum conductivity at 500°C and 600°C.

Figure 11.23b, however, also shows that at temperatures of 700°C and above, the conductivity shows no such maximum; this has been described as being due to the stabilisation of the δ-phase at these temperatures, whereas at temperatures below 700°C, the δ-phase is proposed as being metastable and reverts to a rhombohedral phase with time. This phenomena was still under investigation at the time this book was written.

Table 11.4 The ionic conductivity of a number of doped-Bi_2O_3 based electrolytes.

Composition	Temperature (°C)	Structure	Conductivity (S/cm)
Bi_2O_3	800	fcc	2.3
$Bi_{0.75}Y_{0.25}O_{1.5}$	600	fcc	4.38×10^{-2}
$Bi_{0.65}Y_{0.20}O_{1.5}$	600	fcc	2.5×10^{-2}
$Bi_{0.65}Gd_{0.35}O_{1.5}$	650	fcc	5.6×10^{-2}
$Bi_{0.8}Tb_{0.2}O_{1.5}$	650	fcc	0.28
$Bi_{0.715}Dy_{0.285}O_{1.5}$	700	fcc	0.14
$Bi_{0.75}Ho_{0.25}O_{1.5}$	650	fcc	0.17
$Bi_{0.80}Er_{0.20}O_{1.5}$	600	fcc	0.23
$Bi_{0.80}Er_{0.20}O_{1.5}$	700	fcc	0.37
$Bi_{0.75}Tm_{0.25}O_{1.5}$	650	fcc	8.0×10^{-2}
$Bi_{0.65}Yb_{0.35}O_{1.5}$	700	fcc	6.3×10^{-2}
$Bi_{0.7}Gd_{0.3}O_{1.5}$	700	fcc	1.0×10^{-2}
$Bi_{0.75}Lu_{0.25}O_{1.5}$	650	fcc	3.7×10^{-2}

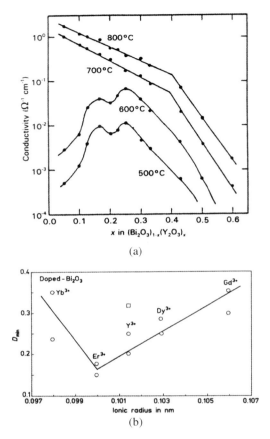

(a)

(b)

Fig. 11.23 (a) Effect of conductivity on doped-Bi_2O_3 as a function of dopant concentration (after T. Takahashi, H. Iwahara and T. Arao, J. Appl. Electrochem. Soc., 10 (1980) 677); (b) Minimum value of x required to stabilise the fcc phase in $(Bi_2O_3)_{1-x}(Ln_2O_3)_x$ versus the ionic radius of Ln^{3+} (after M.J. Verkerk and A.J. Burggraaf, J. Electrochem. Soc., 128 (1981) 75).

The oxygen ion conductivity of doped-Bi_2O_3 is very high and is among the highest known solid state ion conductors. However, the effect of oxygen partial pressure is quite marked. At high temperatures, Bi_2O_3 – based systems are prone to reduction at low oxygen partial pressures, and may even form metallic Bi. Figure 11.24 shows the typical oxygen partial pressure dependence of $(Bi_2O_3)_{0.73}(Y_2O_3)_{0.27}$ on the electron, electron hole, and ion conductivity. The electron and hole conductivities correspond to a $PO_2^{-1/4}$ and $PO_2^{1/4}$ relationship respectively. This would indicate that the following defect equilibria were present:

$$O_2 + 2V_o^{\bullet\bullet} \leftrightarrow 2O_o^x + 4h^{\bullet} \tag{11.32}$$

$$2O_o^x \leftrightarrow 2V_o^{\bullet\bullet} + 4e^{'} + O_2 \tag{11.33}$$

At high oxygen partial pressures, the equilibrium concentrations of the electrons and electron holes are much smaller than that of the oxide ion conductivity. However, at relatively high oxygen partial pressures the majority charge carrier becomes the electron, due to the reduction of the material. In fact, Bi_2O_3-based systems are only stable down to approximately 10^{-6} to 10^{-7} atm. oxygen partial pressure at 650°C. It has been shown that the decomposition of $(Bi_2O_3)_{1-x}(M_2O_3)_x$ can be described as:

$$(Bi_2O_3)_{1-x}(M_2O_3)_x \rightarrow (Bi_2O_3)_{1-x-\alpha}(M_2O_3)_x + 2\alpha Bi + 2\alpha/2\ O_2 \tag{11.34}$$

Thus, the potential use of the material is limited, particularly in a solid oxide fuel cell or related system, where low oxygen partial pressures are found at the anode. However, work has been done to reduce the instability phenomena by fabricating double layered ceramics with doped zirconia, and looking at double or treble dopant systems.

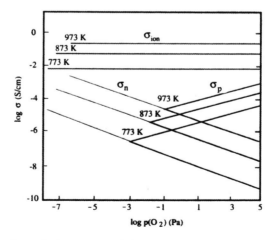

Fig. 11.24 shows the typical oxygen partial pressure dependence of $(Bi_2O_3)_{0.73}(Y_2O_3)_{0.27}$ on the electron, electron hole, and ion conductivity (after T. Takahashi, T. Esaka, and H. Iwahara, J. Appl. Electrochem., 7 (1977) 303).

The latest generation of bismuth oxide based systems should be mentioned. These are known as BIMEVOX (**Bi**smuth **Me**tal (dopant) **V**anadium **Ox**ide) solid electrolytes. They denote a family of oxides based on the $Bi_2VO_{5.5}$ system. The phase is a solid solution of Bi_2O_3-$VO_{2.5}$, which has a composition range from approximately 66.7% to 70.4%.The conductivity of some of the systems studied is shown in Fig. 11.25, with metal ions being Cu, Ni, Co and Zn, although other metals such as W, Mo, Ti, Sn and Pb have also been studied.

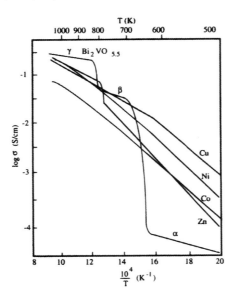

Fig. 11.25 The ionic conductivity of some BIMEVOX ($Bi_2V_{0.9}M_{0.1}O_{5.5-x}$ for M = Cu, Co, Ni, Zn) systems, as a function of temperature ($Bi_2VO_{5.5}$ is shown for comparison) (after G. Mairesse in Fast Ion Transport in Solids, ed. B. Scrosati, Kluwer, Academic Publishers Amsterdam, 1993, p271).

11.13 Alkali Ion Conductors

In alkali halides, such as NaCl, the cationic species is considerably more mobile than the anionic species. Thus, Na^+ ions, for example, are considered to be the main current carriers in NaCl. The magnitude of the ionic conductivity in NaCl is very dependent upon the number of cation vacancies present, which in turn depends upon the impurity concentration (and thus the stoichiometry of the system), and its thermal history. Vacancies can be created either by heating the sample, which increases the number of intrinsic vacancies, or by doping the sample with aliovalent dopants, which increases the number of extrinsic vacancies by preserving the charge balance of the system as a whole.

For example, when a small percentage of $CaCl_2$ (Ca^{2+} ion) is introduced in the lattice of NaCl, the following equilibrium reaction occurs

$$CaCl_2 \rightarrow Ca_{Na}^{\bullet} + 2Cl_{Cl}^{x} + V_{Na}^{'} \qquad (11.35)$$

The vacancies produced are extrinsic, as they would not be present in the pure material. At low temperatures, the number of intrinsic vacancies is very small, and unless the material is extremely pure, which is very unlikely, they will be much less than the number of extrinsic vacancies. Thus, when a plot of conductivity (or diffusion coefficient for that matter) is plotted against temperature, two regions are observed, as seen in Fig. 11.26. At low temperatures, an extrinsic region occurs, which follows the Arrhenius relationship examined above, and the temperature dependence of the conductivity depends only upon the mobility of the cations. At higher temperatures a change in slope is observed, where the extrinsic region occurs. In this region, the conductivity depends upon both the vacancy concentration and the mobility of the cations. The simple behaviour described in Fig. 11.26 is somewhat idealised, and there are usually a number of other factors which need to be taken into account. Figure 11.27 shows a plot of NaCl doped with $CdCl_2$. It is quite apparent that at lower temperatures, the conductivity deviates downwards from the ideal extrinsic line. This has been attributed to the formation of defect complexes such as anion vacancy/cation vacancy pairs, or cation vacancy/aliovalent cation impurity pairs. As a result, the activation energy is increased, with a subsequent effect on the Arrhenius behaviour. There is still much controversy over the activation energy values for Na^+ ion migration; the values range from 0.65eV to 0.85eV. The variance is probably due to the effect of different concentrations of other defects, such as dislocations, which affect the diffusion of the cations.

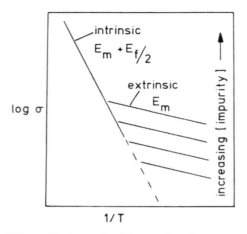

Fig. 11.26 Plot of ionic conductivity as a function of temperature.

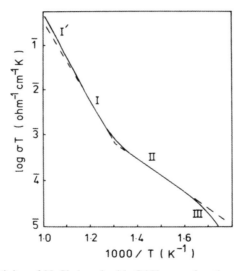

Fig. 11.27 Conductivity of NaCl doped with CdCl$_2$ as a function of temperature (after A.R. West, Solid State Chemistry and its Applications, Wiley, 1984).

Not all impurities increase the ionic conductivity. K$^+$ and Br$^-$ lead to little, if any, effect unless they are present in large concentrations. If impurities were added producing more sluggish vacancies, then the ionic conductivity would be reduced, as would be observed when a divalent anion was dissolved in NaCl. However, this effect is unlikely due to the insoluble nature of divalent anion species in NaCl, but it is very important when discussing systems such as AgCl.

11.14 Halide Ion Conductors

Several types of halide systems which have the CaF_2–structure, can be classed as fast ion conductors at high temperatures because they have high halide ion conductivity. One of the best examples is PbF_2 in which the conductivity is found to be approximately 5 S/cm above 500°C.

11.15 Ag-ion conductors

The predominant defect in AgCl is the Frenkel defect, i.e. interstitial Ag^+-ions and Ag^+ ion vacancies. It has been shown that the interstitial Ag^+-ions are more mobile than the vacancies, and are thus considered to be the majority charge carrier in AgCl. Two diffusion mechanisms are possible, which include either the interstitial mechanism, where Ag^+-ions jump into an adjacent interstitial vacancy, or the interstitcialcy mechanism, which, as was described in Section 11.7 involves a knock-on effect whereby an Ag^+ ion pushes an adjacent Ag^+-ion into the next interstitial vacancy.

It is interesting to note that the diffusion mechanism in NaCl is via a vacancy mechanism, where the Na^+-ions move from one corner of the crystal cube to another, via a transient interstitial. However, in AgCl, the Ag^+-ion moves from the interstitial site of one cube into the adjacent interstitial site of another.

The affect of aliovalent dopants is also different in AgCl, as compared to NaCl. If, for example, Cd^{2+} is added to AgCl, the number of cation vacancies is increased, but because the diffusion mechanism is via interstitials, the affect of the dopant ion is to reduce the concentration of the more mobile species, and thus reduce the ionic conductivity.

AgI is also known to be a fast ion conductor, as shown in Fig. 11.28. As can be seen from the figure, it undergoes a phase transition at 146°C where it transforms from β-AgI to α-AgI. α-AgI has an Ag^+-ion conductivity in excess of 1 S/cm, and an activation energy of 0.05eV. The bcc structure of α-AgI allows relatively easy diffusion of the Ag^+-ion, and in fact the conductivity decreases slightly upon melting. The bcc crystal structure contains I⁻ ions at the corners and body centred position of the cube, and the Ag^+-ions appear to be distributed randomly over a total of 36 sites of tetrahedral and trigonal co-ordination. The tetrahedral sites are effectively linked together by shared faces, and the trigonal sites lie at the centre of the faces of the AgI_4 tetrahedra. The Ag^+-ions move from one site to another in what appears to be a liquid like manner. AgI has a far superior ionic conductivity to AgCl, and AgBr, which is possibly due to the different structure of α-AgI. Figure 11.28 also shows the conductivity of $RbAg_4I_5$, for comparison (as discussed later). β-AgI, which is stable below 146°C, has the wurtzite structure, with hexagonal close packed I^- ions and Ag^+ ions in the tetrahedral sites.

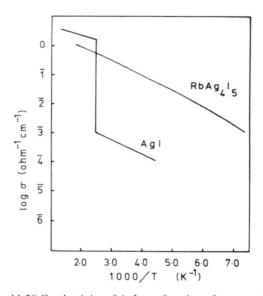

Fig. 11.28 Conductivity of AgI as a function of temperature.

11.15.1 *Stabilized α-AgI*

Many attempts have been made to try and stabilize the high temperature modification of AgI to lower temperatures, using dopants. The most successful is that of the addition of RbI, where the Rb partially replaces the Ag ions. The phase diagram is shown in Fig. 11.29, and the phase $RbAg_4I_5$ was shown to have an ionic conductivity at room temperature of 0.25 S/cm, with an activation energy of only 0.07eV. From the phase equilibrium diagram, however, it can be seen that $RbAg_4I_5$ decomposes into AgI and Rb_2AgI_3 below 27°C, particularly in moist air or in I-vapour.

The structure of $RbAg_4I_5$ is quite different to α-AgI, but it does contain a random distribution of silver atoms over a network of tetrahedral sites.

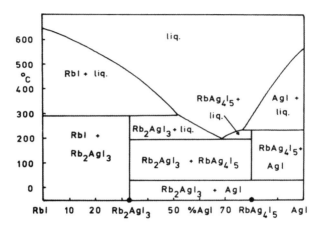

Fig. 11.29 Phase equilibrium of AgI and RbI (after T. Takahashi, J. Appl. Electrochem. 3 (1973) 79).

11.16 β-Al$_2$O$_3$

β-Al$_2$O$_3$ is the name for a family of compounds of the general formula M$_2$O.nX$_2$O$_3$ (n = approximately 5 – 11), where M is a monovalent cation such as Cu, Ag, Ga, In, and X is a trivalent cation such as Al, Ga or Fe. The two most important systems in the series are Na$^+\beta$-Al$_2$O$_3$, and Na$^+\beta''$-Al$_2$O$_3$, which have nominal compositions of Na$_2$O·11Al$_2$O$_3$ and Na$_2$O·5Al$_2$O$_3$ respectively, although neither have true-stoichiometry. The Na$^+$-β-Al$_2$O$_3$ has much higher amounts of Na than the formula suggests, while Na$^+$-β''-Al$_2$O$_3$ is sodium deficient.

Interest in the material started in the 1960's when the Ford Motor Company found that the Na$^+$ ion was mobile at room temperature, and that the Na$^+$ ion could be exchanged for other mobile cations.

The crystal structure of beta-Al_2O_3 consists of Al^{3+} and O^{2-} ions packed in spinel blocks, as shown in Fig. 11.30(a). Two dimensional loosely packed layers of Na^+ ions and oxygen ions separate the spinel blocks. The reason for the high mobility is that the Na^+ ion is smaller than the O^{2-} ion, and there are more available sites than Na^+ ions. The sodium containing layers also act as mirror planes for the close packed oxygen layers on either side (11–30b).

In the mirror plane there are seven alternative positions for the Na^+ ions. The lower energy position is called the Bevers-Ross (BR) position (seen in Fig. 11.30c). However, mid-oxygen positions (mO) or anti-BR positions can be occupied. The occupancy of the various sites and the total number of M^+ ions depends on the size of the cation.

The two modifications of beta-Al_2O_3 are known as β and β''. The major difference between the two modifications is the sequence in which the spinel blocks are stacked. In β-Al_2O_3, there are two spinel blocks in the unit cell, while in β''-Al_2O_3 there are three blocks in the unit cell.

β''-Al_2O_3 is rhombohedral, with the space group R_{3m}, and a c-axis of 3.395nm, while β-Al_2O_3 is hexagonal. Just as in β-Al_2O_3, the blocks in β''-Al_2O_3 are linked by Al-O-Al bonds, with all the Na^+ ions situated in these loose-packed layers, thus giving rise to the high ionic conductivity of these materials.

In β-Al_2O_3, the sodium ions occupy the centre of triangular prisms of six oxygen atoms. However, when moving through the conduction layer, the sodium ions must pass between the two oxygen atoms which are 0.238nm apart. Na^+ (diameter 0.190 nm) is small enough to pass through, but K^+ (diameter 0.266nm) and the other larger alkali ions will not pass through very easily. The geometry of β''-Al_2O_3, on the other hand, does not have this problem, due to the staggering of the layers, as described in Fig. 11.30. It must be mentioned that the conduction of the Na^+ ion is limited to the loosely packed layers, and thus conduction is virtually limited to these two-dimensional planes. However, there is much controversy concerning the fact that a polycrystaline material, with randomly oriented crystal grains, can only conduct in 2-dimensions.

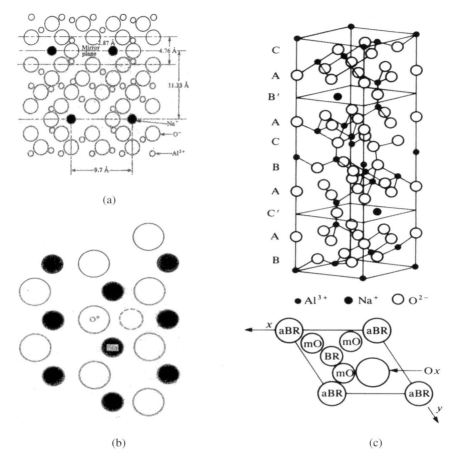

Fig. 11.30 The crystal structure of β-Al₂O₃; (a) Crystal structure, showing a plane parallel to the c-axis; (b) Structure of β-Al₂O₃ showing the site occupation in the conducting plane; (c) The arrangement of atoms in the mirror plane.

β'''-Al₂O₃, and β''''-Al₂O₃ have also been investigated, but will not be described here.

The ability of the Na⁺ ion to move through the structure of β-Al₂O₃, under the application of an electric field, means that the material has been classed as a Na⁺-ion conductor. The principle reason for the high conductivity is the large number of available sites for conduction, and the low activation energy for ion migration.

The conductivity mechanism in this material has been considered, and a number of theories have been proposed.

- A disorder-order phenomenon involving the atoms in the conduction plane has been observed, which was later found to be temperature dependant; the disordered state having the high ionic conductivity.
- The interstitial mechanism for the ion diffusion in β-Al_2O_3 was later discarded, and instead it was suggested that an interstitialcy mechanism took place.
- A similar mechanism involving mO-mO interstitialcy configurations combined with regular lattice positions was later suggested.
- The above mechanisms were then suggested to be incorrect and instead the ions were considered to move in pairs, with all three cation sites being involved.

Whatever the correct mechanism, there still appears to be much controversy, and this material is still very much under the spot-light in terms of understanding its behaviour. However, what is clear is that the conductivity of beta-Al_2O_3 is sensitive to the sodium content, the relative proportion of β to β'', and, thus, the non-stoichiometry.

It has been shown that a number of cations can be used to replace the Na^+ ion, either totally or in part. In β-Al_2O_3, the Na^+ ion can be exchanged with a number of different monovalent and divalent cations, while in β''-Al_2O_3 monovalent, divalent or trivalent ions (including Li^+, K^+, Ag^+, Rb^+, Tl^+, Cu^+, Ca^{2+}, Sr^{2+}, Pb^{2+}, Gd^{3+}, Nd^{3+} etc.) can be exchanged with the Na^+ ion, with a subsequent change in the cell dimensions.

The ionic conductivity for a number of these systems is described in Table 11.5. It is quite apparent that substitution of the Na^+ ion has a very distinct effect on the ionic conductivity. However, the highest conductivity is observed for the Na^+-ion, due to its ideal cationic size, while the divalent and trivalent cations are considerably more sluggish than the monovalent cations. It has been suggested that the ionic transport number of these materials is approximately unity, and little electronic conductivity is observed.

Table 11.5 Ionic conductivity and activation energy for a number of β and β''-Al_2O_3 single crystals.

Sample	Conductivity (S/cm) at 25°C	Conductivity (S/cm) at 300°C
$Na^+ \beta–Al_2O_3$	0.014	0.152
$Ag^+ \beta–Al_2O_3$	0.0067	0.087
$K^+ \beta–Al_2O_3$	6.5×10^{-5}	
$Tl^+ \beta–Al_2O_3$	2.2×10^{-6}	
$NH_4^+ \beta–Al_2O_3$	1.0×10^{-6}	0.001
$Li^+ \beta–Al_2O_3$	1.3×10^{-4}	
$Na^+ \beta''–Al_2O_3$	0.16	1.7
$K^+ \beta''–Al_2O_3$	0.13	0.54
$Ag^+ \beta''–Al_2O_3$	3.8×10^{-3}	0.073
$Li^+ \beta''–Al_2O_3$	5×10^{-3}	

The ionic conductivity of β-Al_2O_3 single crystals is found to be of the order of four to five times lower than that of β''-Al_2O_3, at similar temperatures, as shown in Fig. 11.31. This is suggested to be due to the higher concentration of defects present in the latter material. The conductivity of $Na^+\beta''$-Al_2O_3 is significantly higher than $Na^+\beta$-Al_2O_3, possibly due to their crystal structures, higher concentration of Na^+ in the conduction planes, and higher mobility of Na^+ in $Na^+\beta''$-Al_2O_3.

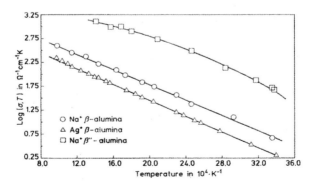

Fig. 11.31 The ionic conductivity of $Na^+\beta$-Al_2O_3, $Ag^+\beta$-Al_2O_3 and $Na^+\beta''$-Al_2O_3 single crystals (after SPS Badwal, Chapter 11 in Materials Science and Technology, Volume 11, Structure and Properties of Ceramics, Ed M. Swain, VCH Publishers, (1994), Germany, p567–63 (Badwall 1994)).

In polycrystalline β-Al_2O_3 and β''-Al_2O_3 materials, there are other factors needing to be taken into account, including the proportion of β to β'', the microstructure, the percentage theoretical density, the effect of dopants and/or impurities, and the presence of basal plane defects.

The proportion of β to β'' is, however, dependant upon the firing temperature employed in the fabrication of the ceramic material. In fact, β''-Al_2O_3 decomposes to β-Al_2O_3 and $NaAlO_2$ at approximately 1450°C. However the addition of MgO or Li_2O raises the decomposition temperature, which is important if a highly dense ceramic needs to be prepared. Figure 11.32 shows a typical plot of resistivity of β-Al_2O_3 as a function of the ratio of β to β''.

Due to the two-dimensional nature of the conduction, as explained above, the effect of grain size and orientation is very marked. For example, a material which has a totally random crystal orientation has been shown to have a conductivity 2/3 lower than a sample which has a specific orientation. However, it should be explained that the main factor affecting the conductivity, as is found in all these ionically conducting ceramics, is the fabrication route.

It has been found that additives strongly affect the final beta-Al_2O_3 ceramic material. For example, additions of TiO_2, ZrO_2, CaO, SiO_2, B_2O_3 and SrO have been found to enhance the sintering, allowing for dense materials to be formed at temperature lower than 1750°C. CaO, SrO and SiO_2, however, cause a decrease in the ionic conductivity, possibly due to the formation of an inactive glassy phase, or a blocking of the conduction planes which segregates at the triple points and significantly alters the grain boundary resistivity.

β-Al_2O_3 and β''-Al_2O_3 are mainly used as probes to measure the metal activity, and in sodium sulphur batteries. In sodium sulphur batteries, the Na^+-β-alumina is usually used as the electrolyte. The system is a secondary battery used for electric vehicles and load levelling applications, for example. It consists of a molten sodium anode and a molten sulphur impregnated graphite felt as the cathode. The anode and cathode are separated by the $^+$-β-alumina electrolyte, usually in the form of a close-ended tube. The cell discharge reaction is

$$2Na + xS \rightarrow Na_2S_x$$

where x depends on the level of the charge (5 during the early stages of discharge).

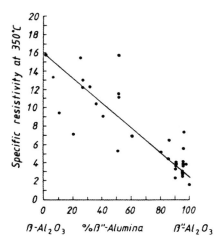

Fig. 11.32 A typical plot of resistivity of β-Al_2O_3 as a function of the ratio of β to β'' (after R. Stevens and J.P.G. Binner, J. Mat. Sci., 19 (1984) 695).

11.17 Proton Conductors

The proton, or hydrogen ion (H^+), is the only ion which has no electron shell of its own. Consequently, it strongly interacts with the electron density of its environment, which then takes on some of its character. These ions are not subject to the usual ion-repulsion effects that make other ions spatially separate individuals.

As a result of the small radius of the proton, the isolated H^+ ion is not present in solids under equilibrium conditions. Due to its strong polarising power, H^+ is covalently bonded to one or two electronegative ions or atoms in the surrounding system.

There are at least three different bonding mechanisms for the proton:

- the acceptor site for the proton may be an ion of the immobile lattice; for example the formation of a hydroxogroup
- the proton may be attached to a mobile ion; for example an oxygen ion (forming a hydroxyl ion, OH^-)

- the proton may be attached to a mobile molecule, such as water or ammonia.

There are a number of mechanisms which describe the motion of the proton; the two most important are described as:

11.17.1 *Vehicle mechanism*

Here the proton diffuses with a so-called "vehicle", such as H_3O^+. The counter diffusion of the unprotonated species, in this case H_2O, allows for the net flow of protons in the opposite direction. Such a mechanism is not possible where the vehicle and vehicle+proton have charges of the same sign, as is the case of OH^-/O^{2-} system.

11.17.2 *Grotthuss mechanism*

This mechanism has often been used to describe the reasons protons in ice have a higher mobility than in water. In this mechanism, the proton can be considered to be attached to a polyatomic ion (the vehicle) in the lattice, such as PO_4^{3-}. Here the vehicles show pronounced local dynamics, but reside on their sites; the protons are transferred from one vehicle to another over very long distances.

The rates describing the two limiting proton-conducting mechanisms above are the rates of diffusion (Γ_{dif}) and reorganisation (Γ_{reorg}) of the proton, and the proton-transfer rate (Γ_{trans}).

In the next section we will examine the main types of proton conductors which have gained favour over the last few years.

11.17.3 *Water containing systems*

This work was instigated by the chlor-alkai industries, such as Dupont and ICI. In the 1960's, the development of chemically resistant cation exchange membranes occurred, the most attractive of these being NAFION, which was commercialised by Dupont.

This polymer has a perfluorinated backbone and side chains terminated by strongly acidic $-SO_3H$ groups.

$$\text{—(CF}_2\text{CF}_2)_n\text{—C—O(CF}_2\text{—CFO)}_m\text{CF}_2\text{CF}_2\text{SO}_3\text{H}$$
$$\underset{\displaystyle \underset{\displaystyle |}{\text{CF}_2}}{|} \qquad\qquad \underset{\displaystyle \text{CF}_3}{|}$$

Figure 11.33 shows an Arrhenius plot of proton conductivity of some water containing compounds. The high proton conductivity in $H_3OUO_2AsO_4\cdot 3H_2O$, for example, is observed above a phase transition within the water layer of the structure, which also contains the excess protons. This material acts as a model compound for proton conductivity in acidic hydrates.

Other layered acidic phosphates and phosphonates of zirconium have been studied in both dry and wet form, and they do show some residual conductivity even after dehydration.

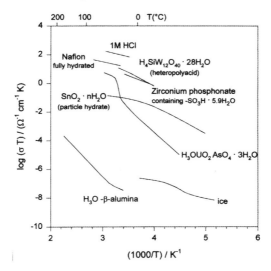

Fig. 11.33 Arrhenius plot of proton conductivity of some water containing compounds (after K. Kreuer, Chem. Mater., 8 (1996) 610).

Very high proton conductivity is observed in heteropolyacid hydrates built from keggin anions, which are held together by acidified hydrogen-bonded water structures. Highly dispersed systems, such as structures built out of hydrated acidic particles or xerogels, may also show high proton conductivities.

β-Al$_2$O$_3$ can also be prepared in its H$_3$O$^+$-form, although the proton conductivity is quite low.

There are many other types of proton conducting hydrates, although it should be remembered that the conductivity is always dependent upon the presence of water, and that the host acts as a Bronsted acid towards the water of hydration generally loosely bound in the structure. This retains the water of hydration only up to temperatures not significantly higher than 100°C.

11.17.4 *Oxo acids and their salts*

Oxo acids, such as H$_3$PO$_4$, H$_2$SO$_4$, or HClO$_4$ dissociate in aqueous solutions according to their PK$_A$, generating protons and thus proton conductivity. They can, however, also show proton conductivity in the absence of water due to self-dissociation. The proton conductivity of these systems is, however, very small. In the 1980's, the discovery of CsHSO$_4$ occurred, which showed very high proton conductivity above a first-order improper ferroelastic phase transition. The high proton conductivity has also been observed in RbHSO$_4$ and the systems containing large cations (see Fig. 11.34).

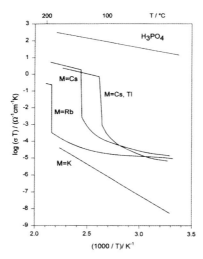

Fig. 11.34 Proton conductivity of a number of Oxo-salts, of composition MHSO$_4$ (M = Cs, Tl, Rb and K), H$_3$PO$_4$ is shown for comparison (after K. Kreuer, Chem. Mater., 8 (1996) 610).

11.17.5 *High temperature proton conductivity*

The oxide first shown to have predominantly proton conductivity was acceptor doped ThO_2 at low oxygen partial pressure (high hydrogen partial pressure) and temperatures above 1200°C.

In the 1980's, many perovskite (ABO_3) systems were studied. When the A-site is substituted with an appropriate lower valency cation, oxygen vacancies are formed to maintain electroneutrality

$$\tfrac{1}{2}O_2 + V_o^{\cdot\cdot} = O_o^x + 2h^{\cdot} \qquad (11.36)$$

In a hydrogen atmosphere, the electron hole may react with the hydrogen to produce a proton

$$\tfrac{1}{2}H_2 + h^{\cdot} = H^+ \qquad (11.37)$$

or they may react with water vapour

$$H_2O + 2h^{\cdot} = \tfrac{1}{2}O_2 + 2H^+ \qquad (11.38)$$

Other reactions that may occur in the lattice include

$$H_2O + V_o^{\cdot\cdot} = 2H^+ + O_o^x, \qquad (11.39)$$

And

$$H_2O + O_o^x + V_o^{\cdot\cdot} = 2(OH)_o^{\cdot} \qquad (11.40)$$

The protons produced in the cases described above are considered to be interstitial in nature, although the hydroxide ions are considered to migrate via a vacancy mechanism or between sites adjacent to the oxygen ions. Recently, however, it was shown that because of the small size of the proton, it does not occupy a true interstitial site but attaches itself to an oxide ion, thus forming a hydroxyl ion $OH_o^{\cdot}$, as described in equation 11.40.

The two most widely studied high temperature proton conductors are the solid solutions of $SrCeO_3$ and $BaCeO_3$.

11.17.6 *Doped BaCeO₃ and SrCeO₃*

$BaCe_{0.85}Ca_{0.15}O_{3-x}$, for example, exhibits a proton conductivity of approximately 2×10^{-3} S/cm at 900°C, with an activation energy of 0.54eV. The structure of $BaCe_{0.9}Gd_{0.1}O_{2.95}$ is described in Fig. 11.35, and was shown to be orthorhombic with the lattice constants a = 0.8773nm, b = 0.6244nm, and c = 0.622nm. The reason for the high protonic conductivity was explained by studying the material using thermogravimetry in moist N_2. At 900°C, the system evolved O_2, followed by a reaction with the water vapour according to reaction 11.39. Thus, the material contains a concentration of mobile protons.

From the equilibrium equations for reactions 11.41 and 11.42 given below, it is apparent that protonic and p-type conductivity depend upon the water vapour and the oxygen partial pressure:

$$p = K_{yy}[V_o^{\bullet\bullet}]^{1/2}(P_{O2})^{1/4} \qquad (11.41)$$

$$[H^+] = K_{xx}[V_o^{\bullet\bullet}]^{1/2} (P_{H2O})^{1/2} \qquad (11.42)$$

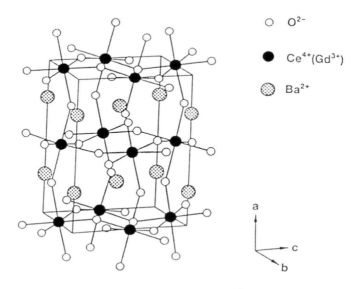

$\bigcirc$ O^{2-}

$\bullet$ $Ce^{4+}(Gd^{3+})$

$\textcircled{\tiny ::}$ Ba^{2+}

Fig. 11.35 The structure of $BaCe_{0.9}Gd_{0.1}O_{2.95}$ (after N. Bonanos, Solid State Ionics, 53–56 (1992) 967).

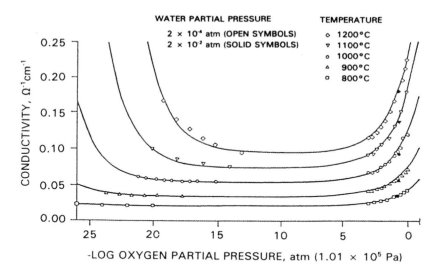

Fig. 11.36 Conductivity of $BaCe_{0.9}Gd_{0.1}O_{2.95}$ as a function of oxygen partial pressure (after N. Bonanos, Solid State Ionics, 53–56 (1992) 967).

However these relationships do not always follow, as shown in, for example, $BaCe_{0.9}Gd_{0.1}O_{2.95}$ as a function of oxygen partial pressure, as described in Fig. 11.36. Because the water vapour pressure has increased by a factor of 100, the proton conductivity should have increased by a factor of 10, according to equation 11.43. This is obviously not the case, and thus we must consider non-classical proton hopping as a possible mechanism.

The temperature dependence of the conductivity of $BaCe_{1-x}M_xO_{3-\delta}$ (for various dopants, M) is described in Fig. 11.37(a). In hydrogen atmospheres, doped $SrCeO_3$ exhibits high proton conductivity. Examples of Yb_2O_3, Y_2O_3, and Sc_2O_3-doped materials are shown in Fig. 11.37 (b). $SrCe_{0.95}Yb_{0.05}O_{3-x}$, for example, was shown to have a protonic conductivity of 2×10^{-3} S/cm in $N_2/5\%H_2$ at 600°C, and an activation energy of 0.59eV.

Other oxides showing high protonic conductivity include doped zirconates ($CaZrO_3$, $BaZrO_3$, and $SrZrO_3$), $BaTh_{0.9}Gd_{0.1}O_{3-x}$, and $Ba_2GdIn_{1x}Ga_xO_3$.

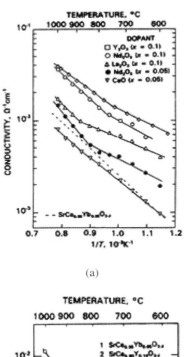

(a)

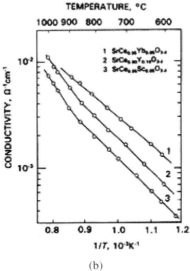

(b)

Fig. 11.37 (a) The temperature dependence of the conductivity of $BaCe_{1-x}M_xO_{3-\delta}$ (After H. Iwahara, H. Uchida, K. Ono, and K. Ogaiki, J. Electrochem. Soc., 135 (1988) 529); (b) Proton conductivity of Yb_2O_3, Y_2O_3 and Sc_2O_3-doped $SrCeO_3$ as a function of temperature (after H. Iwahara, T. Esaka, H. Uchida, and N. Maeda, Solid State Ionics, ¾ (1981) 359).

11.17.7 *Organic/inorganic systems*

H_2SO_4 and H_3PO_4, for example, form narrow compounds in narrow composition ranges with organic molecules exhibiting basic groups (eg $C_6H_{12}N_2$ or $C_6H_{12}N_4$). These compounds can have moderate proton conductivity. However, most of the work has concentrated on oxo-salts blended with a variety of polymers. For example polyacrylamide and sulphuric acid have been shown to have reasonable proton conductivity, although not as high as pure-H_3PO_4. Other systems include polyethylene oxide and ammonium salts, and organic molecules (such as heterocyclic bases, pyrazoles, imidazoles, and alkyls) into the lamellar structures of acidic phosphates and phosphonates.

11.18 Ionic Glasses

Glasses are defined in this sense as bulk solids which have been quenched from the melt, and exhibit a glass transition temperature (T_g). The most frequently studied systems are based on:

- elemental glasses (eg Ge, Si, As, Te, C, B, Sb, Se)
- chalcogenide glasses (compounds containing S, Se, and/or Te)
- transition metal-oxide glasses (eg V, Ti, and Fe oxides)

The electrolytic properties of glass were demonstrated over 100 years ago by Warburg, who electrolysed Na^+ and other alkali cations through the walls of thin glass tubes, and showed that the Faraday's laws were obeyed. In fact most of the work to-date has concentrated on using these electrical properties of glass in glass electrodes in the field of analytical chemistry. The main challenge to the development of the glass electrolyte is the instability of the electrode/electrolyte interface, and the possibility of getting chemical and electrochemical side reactions.

In the majority of systems studied, cation mobility is the main ion-mobility of interest. The cation mobility can be described using a so-called Warren-Biscoe model, shown in Fig. 11.38. Cations are placed in the holes of the glass structure, which shape is determined by a partially broken silicate structure. This form of monopolar conduction is prevalent in a wide range of borate, phosphate and silicate glasses of varying stoichiometries. O^{2-} has also been observed under special circumstances.

Anionic conduction is really only found in fluoride glasses. For example, $62ZrF_4$-$30BaF_2$-$8LaF_3$ has an anionic conductivity of 3×10^{-6} S/cm at 300°C, which is very similar to the cationic conductivity in $20Na_2O$-$80SiO_2$.

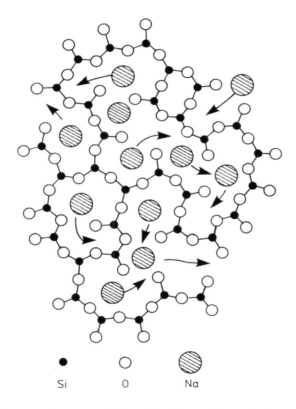

Fig. 11.38 A diagrammatic representation of the Warren-Biscoe structure of alkali silicate glasses, showing the possibility of localised and extended cationic motion within an anionic framework (after M.D. Ingram, Chapter 14 in Materials Science and Technology, Volume 9, Glasses and Amorphous Materials, Ed J. Zarzycki, VCH Publishers, (1991), Germany, p716–750 (1991)).

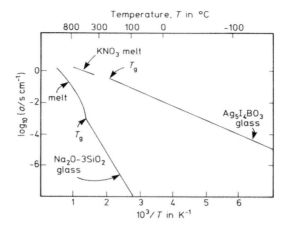

Fig. 11.39 An Arrhenius plot of log conductivity versus temperature for a normal alkali silicate system, a superionic silver iodoborate glass, and a highly conducting molten salt (after M. Ingram and A.H.J. Robertson, Solid State Ionics, 94 (1997) 49).

Over the last 20 years or so, there has been increasing interest in glassy ionic materials. The major breakthrough occurred when AgI-Ag_2SeO_4 was found to contain highly conductive glassy phases. This was the start of the development of many silver iodide/silver oxosalt compositions, with conductivities of approximately 10^{-2} S/cm at room temperature. These glasses are purely Ag^+-ion conductors; Fig. 11.39 shows the Arrhenius plot of log conductivity versus temperature for a normal alkali silicate system, a superionic silver iodoborate glass, and a highly conducting molten salt. The sodium silicate glass is a reasonably good conductor (approximately 10^{-3} S/cm at T_g), but falls to below 10^{-10} S/cm at room temperature. Silver iodoborate glass, on the other hand, is an extremely conductive material, with a conductivity of 10^{-1} S/cm at T_g.

A number of features are apparent. Firstly, the $Na_2O{\cdot}3SiO_2$ glass contains a glass transition temperature below which the plot shows a straight line (and obeys the Arrhenius equation), although the material can be said to be an insulator. Above the glass transition point the material deviates from the straight line, and is acting as an electrolyte. In fact, most common glasses, including the silicates, borates and phosphates, have conductivities at the glass transition temperature varying from 10^{-4} to 10^{-2} S/cm. Secondly, the $Ag_5I_4BO_3$ glass is an example of the new type of system described above, containing AgI as the main constituent. The conductivity of this system is very high at the glass transition temperature, with a value of 5 x 10^{-1} S/cm. This plot is virtually an extension of the line for molten KNO_3. This would be indicative of liquid-like ion mobilities being trapped in the rigid glass matrix.

The chalcogen systems contain either a single chalcogen element (eg S, Se and Te), or are alloys or compounds of two such chalcogens. They typically have band gaps in the range of 1-2.5eV, and are, thus, considered semi-conductors. Typical values vary between 10^{-17} and 10^{-11} S/cm at 300K, although higher values have been observed.

11.19 Solid Oxide Fuel Cell

As an example of an electrochemical system that utilized an ionic conductor, we will now study the solid oxide fuel cell (SOFC) in detail. A solid oxide fuel cell (SOFC) is an all solid-state electrochemical device producing both electricity and waste heat directly from the electrochemical conversion of a fuel with an oxidant. Because it produces electricity by electrochemical means, there is no necessity to have any moving parts, and thus the SOFC overcomes the Carnot limitation inherent in all heat engines. Hence, the SOFC has high-energy efficiency, which can be further increased by using waste by-product heat in applications such as co-generation (greater than 50% electrical efficiency is possible, with efficiencies approaching 75% with co-generation).

The SOFC also has a number of advantages over conventional systems including its modularity whereby single units can be stacked, its flexibility, and its ability to use a number of different fuels. It has a number of disadvantages including cost (far greater than conventional systems/kW at the moment), no long term data (although some systems have been running for more than 10,000 hours), raw-material availability (possibly a problem in the long term), and it has no real history, unlike the turbine engine. It should however be noted that the fuel cell was invented by Sir William Grove over 150 years ago, but was never commercialised possibly due to the invention of the turbine and its use and development in the turbine-driven aeroplane.

11.20 Characteristics of the SOFC

The SOFC system consists of a ceramic electrolyte, usually yttria stabilised zirconia, with an air electrode (cathode) and a fuel electrode (anode) coated on either side. The electrolyte must be of a high percentage theoretical density so that the two fuels are physically separated and have no opportunity of mixing. The cell (anode/electrolyte/cathode) is operated at a high temperature primarily to allow the ionic conductivity of the electrolyte to be high enough to produce a reasonable current density, although other factors such as reaction kinetics must also be considered.

Fuel (currently H_2, although other hydrocarbon fuels are being studied) is fed to the anode, where it undergoes an oxidation reaction and releases electrons to an external circuit. Oxidant (either air or pure oxygen) is fed to the cathode where it is reduced and accepts electrons from the external circuit. The flow of electrons around the external circuit produces DC electricity. The oxygen is transported as oxygen ions across the electrolyte via the vacancy mechanism, as described earlier. Fig. 11.40 summarises the SOFC operation.

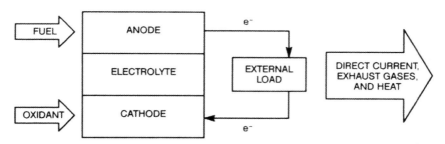

Fig. 11.40 Diagrammatic representation of the elements of the SOFC.

11.21 Cathode, Anode and the SOFC Reactions

As explained above, the electrolyte is usually based on yttria stabilised zirconia, which is an oxygen ion conductor. Other oxygen-ion conductors have been studied, including for example, doped-CeO_2 systems, perovskite-based systems, and doped-Bi_2O_3. Most of these other oxides, although they are superior oxygen ion conductors, are prone to reduction at low oxygen partial pressures (as found at the anode), and thus show n-type electronic conductivity which reduces the efficiency of the overall system. Doped-LaGaO$_3$ is a new material which does show promise as a SOFC electrolyte, because it does not appear to be reduced at the anode. Much work is still necessary, as preliminary studies appear to show it to be mechanically weak.

Now let us look at the reactions occurring in the SOFC cell. At the cathode, the reduction of oxygen occurs, via equation 11.43.

$$O_2 + 4e' = 2O^{2-} \qquad (11.43)$$

The main function of the cathode is to provide reaction sites for reaction 11.43 to take place. The cathode, therefore, must be a good electronic conductor with a high surface area, and catalytically active towards this reaction. The cathode must also have the following requirements; stability at high temperature and under the oxidising gas present; compatibility with the other cell components; similarity of thermal expansion coefficient to that of the other components (otherwise it is prone to peeling off the electrolyte); retention of its porosity (and thus number of reaction sites) during the life-time of the cell; and retention of its catalytic activity during the life time of the cell.

Because of the high temperature and oxidising environment, most metals cannot be used. Only noble metals will withstand the environments found at the cathode, but these materials are too expensive for a commercial system. Many oxide materials have a high electronic conductivity, but they are either incompatible with the electrolyte or have a thermal expansion coefficient very different to those of the other cell components. Doped-LaMnO$_3$ meets almost all the requirements of the cathode, and is the most commonly used material. Traditionally Sr-doped (on the A-site of the perovskite) is used, due to its good electronic conductivity. Sr enhances the electronic conductivity by increasing the Mn^{4+} content by substitution of La^{3+} by Sr^{2+}, as described in equation 11.44, and increases with increasing Sr content. At Sr contents greater than 20–30mol%, metallic conduction is observed.

$$LaMnO_3 + xSr^{2+} \rightarrow La^{3+}_{1-x}Sr^{2+}_xMn^{3+}_{1-x}Mn^{4+}_xO_3 \qquad (11.44)$$

The electronic conductivity occurs via the small polaron conduction mechanism. Ca doping has a similar effect to that of Sr doping, causing a significant increase in the electronic conductivity, with increasing Ca content.

At the anode, reaction 11.45 takes place, whereby the oxide ion is oxidised, releasing an electron.

$$2O^{2-} = O_2 + 4e'$$ (11.45)

Thus, the SOFC can be regarded as an oxygen concentration cell, and the electromotive force (EMF) is given by equation 11.45.

The EMF is, therefore, dependent upon the oxygen partial pressure at the anode and the cathode. The oxygen partial pressure at the anode is based not upon oxygen being present, but upon the oxidation of the fuel. In this example we will consider H_2. The oxidation of H_2 is given by equation 11.46.

$$H_2 + \tfrac{1}{2}O_2 = H_2O$$ (11.46)

The oxygen partial pressure at the anode is, therefore, given by equation 11.47.

$$P_{O_2} = \left[\frac{P_{H_2O}}{P_{H_2} K_{(11.46)}} \right]^2$$ (11.47)

where $K_{11.46}$ is the equilibrium constant for reaction 11.46. Substitution into equation 11.45 yields equation 11.48.

$$E = E^0 + \frac{RT}{4F} \ln P_{O_2} + \frac{RT}{2F} \ln \frac{P_{H_2}}{P_{H_2O}}$$ (11.48)

where E^0 is the reversible voltage at the standard state, given by equation 11.49.

$$E^0 = \frac{RT}{2F} \ln K_{11.46}$$ (11.49)

Under standard state conditions, E equals the reversible standard voltage at the standard state, E^0, and thus we can say that equation 11.50 is applicable.

$$E^0 = -\frac{\Delta G^0}{4F} \qquad (11.50)$$

Where ΔG^0 is the Gibbs free energy of the reaction given in equation 11.46 (at 1250K this is −178.2 kJ/mol). This value will differ for different combustion reactions, such as CO, CH_4 and CH_3OH.

Now, this reaction occurs at the anode, which, like the cathode, must perform under quite specific conditions. The anode must provide the reaction sites for the electrochemical oxidation of the fuel gas to occur. Hence, the anode must be stable under the very reducing conditions of the fuel atmosphere; have sufficient electronic conductivity; have excellent catalytic activity for the reaction (and remain catalytically active during the life time of the cell); be chemically compatible with the other cell components; have a high surface area to allow a large reaction zone to occur (the anode must also remain with a high surface area, and not degrade with time); have a similar thermal expansion coefficient to that of the other cell components, and be relatively low cost and easily fabricated.

Because of the reducing environments metals can be used, although the high temperatures limit these to Co, Ni and noble metals. Electronically conducting ceramic materials can also be used (provided they do not reduce under the anode atmosphere).

Currently, Ni is the anode of choice, primarily due to its good catalytic activity for hydrogen oxidation, its low cost, and its good stability. YSZ powder is added to the Ni metal in the form of a metal/ceramic composite for two reasons. Firstly, the YSZ inhibits the Ni coarsening during the operation, and secondly the YSZ acts to lower the thermal expansion coefficient of the metal composite (which is much higher than the electrolyte) closer to that of the other cell components. The optimum amount of YSZ depends upon the percolation theory. Too much YSZ, and the sample does not electronically conduct (due to a minimal/no electron pathways), too little YSZ, and the composite has too high a thermal expansion coefficient. The effect of Ni content on the electronic conductivity is shown in Fig. 11.41.

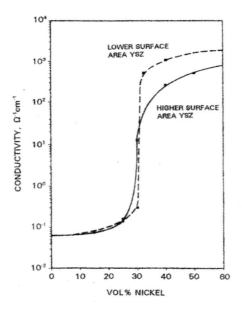

Fig. 11.41 The conductivity of Ni/YSZ as a function of Ni content, at 1000°C (after D.W. Dees, T.D. Claar, T.E. Easler, D.C. Fee, and F.C. M'razek, J. Electrochem. Soc., 134 (1987) 2141).

It is obvious that the percolation threshold is at approximately 30 vol % Ni. Most anodes have approximately 40- 50mol% Ni in the composite, which appears to produce a relatively stable system. However, all anodes are found to degrade with time, due to either slight oxidation of the Ni to NiO, or due to sintering of the Ni particles.

Alternative anode systems are being studied, particularly in respect to internal reformation of hydrocarbons, rather than having to use H_2 as a fuel. Ni, although an extremely good catalyst for the oxidation of H_2, is also very good at cracking natural gas into C. The C is then liable to form whiskers which lift the anode from the electrolyte surface, thus greatly reducing its activity. Other catalytic electrode materials being studied do not cause methane (and other hydrocarbons to crack) and thus may allow for a direct internal reformation reaction to take place. These include, for example, doped-CeO_2-based systems, and perovskites. Currently, however, there are still many problems to be solved when investigating alternative electrode materials and the technology of today either externally reforms the hydrocarbon, or uses a high steam to carbon ratio in the feed stream. Both of these are not totally acceptable, as large efficiency losses are observed.

The reversible EMF produced by the reaction 11.46 is approximately 1.0V (0.924V at 1250K, and 0.997V at 1000K). This EMF is of little practical benefit, and thus single cells are connected in electrical series in what is known as a stack. The height of the stack (or number of cells) varies depending on the design, and the power output required. Many designs have been examined; however, the two most common ones today are the tubular and planar designs, shown in Fig. 11.42.

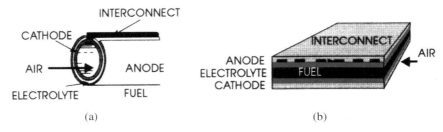

(a)   (b)

Fig. 11.42 The SOFC stack designs (a) tubular; (b) planar (after K. Kordesch, G. Simader, Fuel Cells and Their Applications, VCH, Germany, 1996).

The planar design consists of electrolyte components configured as thin (approximately 50μm) planar plates. On either side of these plates is put the anode and cathode materials. Between the electrolyte plates is what is known as an interconnect. The interconnect has two main purposes. It serves as a bipolar gas separator, contacting the anode and cathode of adjoining cells, and sometimes has ribs on both sides (as shown in Fig. 11.42 (b)) to form gas channels. The properties of the interconnect are quite tight, as it must be electronically conducting, stable in both reducing and oxidising environments and at high temperatures; relatively cheap (as it is usually the thickest component); dense (for the same reasons as the electrolyte, so that the fuel and air gases do not mix); easy to fabricate; compatible with the other cell components; and possess a similar thermal expansion coefficient to those of the other cell components. The interconnect is usually based on the perovskite $LaCrO_3$. To increase the electronic conductivity, A (such as Ca or Sr) and/or B-site (such as Co or Fe) are added, which have the same effect as that of the Mn^{3+} in the cathode, to increase the amount of Cr^{4+} ions.

Once the stack has been fabricated, the system is sealed using a ceramic or glass ceramic seal with a similar thermal expansion coefficient as the cell, and gas manifolds are placed around it, similar to those shown in Fig. 11.43

The tubular design, on the other hand, uses a very different concept. Here, an electrolyte tube (either supported, or unsupported) has electrodes placed on the inside and outside. The most promising design to date is based on the Westinghouse design. In this design (which uses traditional SOFC materials), the electrolyte tube is electro-vapour deposited (EVD) onto a porous cathode support tube. On the outside of the electrolyte/cathode tube is placed the anode, using a dip-coating technique. The interconnect (Mg-doped $LaCrO_3$) is then added using EVD processing, and the tubular cells (which are closed at the bottom) are bundled together, using Ni-felt to connect the anode of one cell to the adjacent anode for parallel connection, and to the Ni-plating on the interconnect for series connection (as shown in Fig. 11.44).

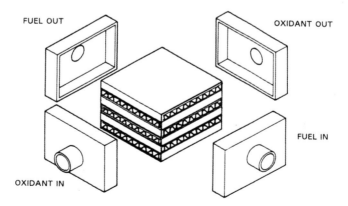

Fig. 11.43 Gas manifolding in the planar SOFC design (after N.Q. Minh and T. Takahashi, Science and Technology of Ceramic Fuel Cells, Elsevier, 1995).

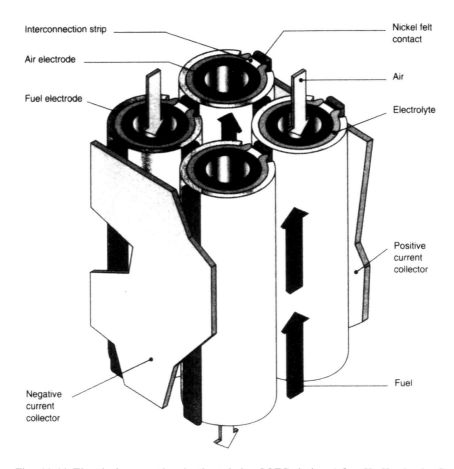

Fig. 11.44 Electrical connection in the tubular SOFC design (after K. Kordesch, G. Simader, Fuel Cells and Their Applications, VCH, Germany, 1996).

As the tubes are themselves already sealed, there is no requirement for a sealant, as required in the planar design. The gas manifolding is relatively straightforward with oxidant (air) being fed down the centre of the tubes via an oxidant plenum. The fuel is fed from a fuel plenum at the bottom and up along the outside of the tubes, where the oxidation reaction takes place. The spent fuel flows through a porous ceramic barrier, enters the combustion chamber and combines with the spent oxidant. This heat is then used to preheat the oxidant entering the cell. The exhaust gases exit the generator at approximately 900°C. The gas-manifolding concept is described in Fig. 11.45.

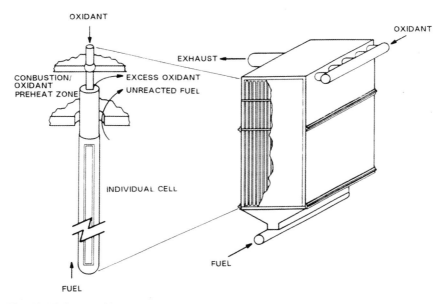

Fig. 11.45 Gas manifolding in the tubular design. (after S.C. Singhal in Proceedings of the 2nd International SOFC Symposium, Editors, F. Grosz, P. Zegers, S.C. Singhal, and O. Yamamoto, Comm. of the European Communities, 1991, p25).

There are a number of other SOFC designs, as the use of ceramic systems allows for a large variety of shapes. The other common types of systems include the segmented cell-in series design; the monolithic design; and the bell-and-spigot configuration. However, the most common two designs are the flat plate planar design and the tubular design, described above. The advantages and disadvantages of both of the designs are summarised in Table 11.6.

Table 11.6 Advantages and disadvantages of the planar and tubular SOFC systems.

Tubular design		Planar Design	
Advantages	Disadvantages	Advantages	Disadvantages
No gas seals	Long current path, therefore high internal cell resistances	Improved performance, and higher power densities	High temperature gas seals required which may react with the cell components
Very robust structure	EVD processes limits the use of suitable dopants	Cross-plane conduction means lower internal resistance losses	Seals may limit the height of the stack
Some freedom in thermal expansion	Cost is some-what inhibitive at the moment	More flexibility in design	Contact resistance can be high in planar designs
Easy cell connections		Easy to fabricate	Stacking thin plates can be difficult, and may limit size
Integration of a high temperature heat exchanger		All components can be fabricated separately	
No gastight seals		Each component can be assessed separately	
No costly interconnect		Can incorporate different materials, such as metallic interconnects	

11.22 Power Output and System Efficiency

SOFC efficiency consists of a number of elements.

- *Thermodynamic efficiency.* This is given by the assumption that the Gibb's free energy of the cell reaction may be totally converted to electrical energy, as is given by equation 11.51

$$\varepsilon_{therm} = \frac{\Delta G}{\Delta H} \tag{11.51}$$

- *EMF efficiency*. This is shown by the fact that the cell voltage, in an operating cell, is always less than the reversible voltage. As the current is drawn from the fuel cell, the cell voltage falls due to a number of losses, as described in Fig. 11.46. The EMF efficiency is given by the ratio of the operating cell voltage under load to the equilibrium cell voltage. This difference is due to a number of polarisation's (or overpotentials), as described in equation 11.51, which shows that the total polarisation losses (η) is the sum of a number of losses.

$$\eta = \eta_a + \eta_d + \eta_r + \eta_{ohm} \tag{11.52}$$

where the subscripts denote; a = charge transfer polarisation; d = diffusion polarisation; r = reaction resistance; ohm = ohmic losses.

Power is obtained from the cell, or stack, by applying a load between the anode and cathode, and hence drawing current. The power output (P) is thus the product of the current drawn and the EMF across the cell, after the losses have occurred.

$$P = I\,(E - R_T I) \tag{11.53}$$

where E is the thermodynamic EMF across the cell, and $R_T I$ are the losses (R_T is the total resistance of the cell, and I the current).

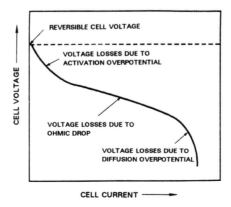

Fig. 11.46 Typical voltage losses in a running SOFC cell.

In summary, SOFC systems offer an attractive option for both large scale and small scale (localised) power generation. They are very efficient systems which can use a large number of oxidisable fuels to produce both electricity and waste heat. SOFC systems are an excellent example of how ionic and electronically conducting materials can be used to fabricate a system which is now becoming a commercial reality. We have only briefly covered the vast topic of fuel cell technology, just to show that there are examples of systems which utilise ionically conducting materials as their main component.

11.23 Introduction to Mixed Ionic/Electronic Conductivity

An important feature of ionic or electronically conducting materials is the way in which their defect concentration is related to external factors such as temperature, partial pressure or activity (such as the oxygen partial pressure in an oxide ion conductor), and other imposed thermodynamic parameters. It has been seen that some ionically conducting materials become electronically conducting (p or n-type) in particular environments. For example, at low oxygen partial pressures δ-Bi_2O_3 becomes an n-type electronic conductor (see section 11.12.7).

11.24 The Defect Equilibrium in MO_2

To describe the effect of the environment on a material, we can consider the case of MO_2 (a metal oxide such as ZrO_2), which is subjected to various oxygen partial pressure regimes (changes in PO_2). We will describe how the defect equilibria is affected by the changes in the oxygen partial pressure surrounding the material[2].

If we consider a system with predominantly Frenkel defects (in fact, in this example we will use anti-Frenkel defects, such as those found in ZrO_2), then the reaction of the oxide with the surrounding oxygen could be written as 11.54 or 11.55.

---

[2] Note that in this example, we are dealing with the partial pressure of a gas. This is not always the case; in fact in many ionic/electronic conductors, the activity of a particular species would have to be considered. However, the concepts developed here are perhaps a little easier to visualise using changes in the partial pressure of oxygen.

$$\tfrac{1}{2}O_2 + 2e' + V_o^{\bullet\bullet} = O_o^x \tag{11.54}$$

$$\tfrac{1}{2}O_2 + V_i = O_i'' + 2h^\bullet \tag{11.55}$$

Now, the equilibrium constant, K, can be obtained for both of the reactions (K_1 (for 11.54) and K_2 (for 11.55) respectively).

$$K_1 = (PO_2^{-1/2})/[V_o^{\bullet\bullet}][e']^2 \tag{11.56}$$

$$K_2 = (PO_2^{-1/2}) [h^\bullet]^2[O_i''] \tag{11.57}$$

The intrinsic Frenkel defect equilibrium in MO_2 is given by the following reaction:

$$nil = 2V_o^{\bullet\bullet} + 2O_i'' \tag{11.58}$$

The intrinsic electronic equilibrium can be given by the following reaction:

$$nil = [h^\bullet] + [e'] \tag{11.59}$$

11.24.1 *Intermediate oxygen partial pressures*

Under regions of intermediate partial pressure, or under near-stoichiometry, it is possible to show that the concentration of oxygen vacancies is approximately equal to the concentration of oxygen interstitials.

Thus, $[O_i''] \approx [V_o^{\bullet\bullet}]$, and is independent of the oxygen partial pressure. Hence, when the number of electrons is greater than the number of electron holes (n-type conductivity) the relationship becomes:

$$[e'] \propto PO_2^{-1/4},$$

And when the number of electron holes is greater than the number of electrons (p-type conductivity), we find that:

$$[h^\bullet] \propto PO_2^{1/4}.$$

In fact we can show that p-type conductivity predominates at high oxygen partial pressure, while n-type conductivity predominates at low oxygen partial pressure.

At non-stoichiometry, the defect equilibrium is quite different.

11.24.2 *Low oxygen partial pressures*

At low oxygen partial pressure, the number of oxygen vacancies ($V\ddot{o}$) predominate. Thus from the electroneutrality equations,

$$2[V_o^{\bullet\bullet}] \approx [e'].$$

Hence, from equation (11.54),

$$K_1 = (PO_2^{-1/2})/4(V_o^{\bullet\bullet})^3 \qquad (11.60)$$

Thus,

$$[V_o^{\bullet\bullet}] \propto PO_2^{-1/6},$$

and by the same assumption,

$$[e'] \propto PO_2^{-1/6}.$$

11.24.3 *High oxygen partial pressure*

At high oxygen partial pressure, the number of oxygen interstitials ($O_i^{''}$) predominates, and using the same argument as shown for low oxygen partial pressure using equation (11.55),

$$[O_i^{''}] \propto PO_2^{1/6},$$

$$\text{and } [h^{\bullet}] \propto PO_2^{1/6}.$$

11.25 Kroger Vink (Brouwer) Diagram

The approximation used above to describe the effect of oxygen partial pressure on the defect equilibria is known as the Brouwer approximation. By combining all three regions, the functional dependence of the defect concentrations over a wide range of oxygen partial pressures can be succinctly graphed in a Kroger Vink, or Brouwer diagram. The example given in 11.24 is summarised in a Kroger Vink diagram shown in Fig. 11.47.

In summary we observe 3 regions:

Region I (low oxygen partial pressure). Here the material is a n-type semi-conductor and shows a $PO_2^{-1/6}$ proportionality.

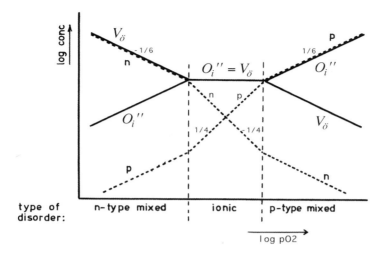

Fig. 11.47 Defect equilibrium diagram of pure (undoped) MO_2 with anti-Frenkel disorder.

Region II (intermediate oxygen partial pressure). Here the majority charge carrier is the oxide ion, the concentration of which remains constant with changing oxygen partial pressure.

Region III (high oxygen partial pressure). In this region the material is a p-type semi-conductor, and shows a $PO_2^{1/6}$ proportionality.

Fig. 11.47 shows the change in defect (electrons, electron holes or ions) concentration with change in oxygen partial pressure. However, the conductivity of the material, as a function of oxygen partial pressure, can be plotted in a similar way. The partial conductivities (conductivities of each of the species involved) are obtained by multiplication of the concentrations (given in Fig. 11.47) with the respective charges and mobilities, as described in equation 11.61.

$$\sigma_\tau = \sigma_i + \sigma_n + \sigma_p = 2e[V_o^{\cdot\cdot}]\mu_{V_o^{\cdot\cdot}} + en\mu_n + ep\mu_p \qquad (11.61)$$

where μ is the mobility of the ion ($V_o^{\cdot\cdot}$), the electron (n) and the electron hole (p), and σ is the conductivity (t = total, i = ionic, n = electronic, p = electron hole).

As the mobilities of the electrons and electron holes are much higher than those of the ionic species, the effect on Fig. 11.47 is to shift the electron and electron hole conductivity upwards in respect to the ionic conductivity (on a logarithmic scale). The effect of oxygen partial pressure on the partial conductivities of MO_2 is described in Fig. 11.48.

The process described in sections 11.24 and 11.25, can be repeated for other defect equilibrium such as Schottky defects (for example, as observed in ThO_2), where a similar diagram is observed. An example of this type of equilibria is seen in Fig. 11.49 which describes the change in defect concentration in an MO oxide with predominantly Schottky disorder, as a function of partial pressure. Note that the ionic defects are oxygen vacancies (Vö) and metal ion vacancies ($V_M^{''}$).

Figures such as these are very useful in determining the externally imposed thermodynamic parameters on the material's defect concentration (or partial conductivity). Thus we can observe what type of conductor the material is (p, n, ionic) by looking at the change in conductivity as a function of the change in the activity of the surrounding atmosphere (change in oxygen partial pressure, for example). If the conductivity does not change as a function of PO_2, in the case of an oxygen ion conductor for example, then the majority charge carrier is an ion, and the material is considered to conduct ionically. Methods of determining the partial conductivity of the material are described below

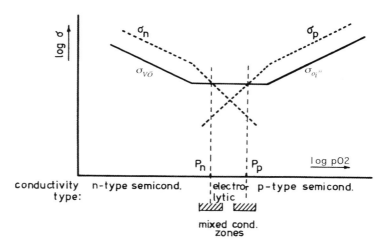

Fig. 11.48. The effect of oxygen partial pressure on the partial conductivity of pure (undoped) MO_2 with Frenkel disorder.

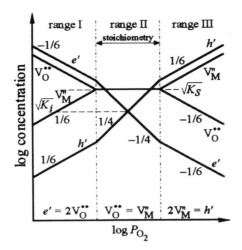

Fig. 11.49 Defect concentration in MO, with predominantly Schottky disorder, as a function of oxygen partial pressure.

11.26 Electrolytic and Ionic Domains

The **ionic domain boundary** is the oxygen partial pressure, in the case of an oxide ion conductor (such as MO_2), at which the ionic conductivity (σ_i) equals the p-type conductivity (known as P_P) or the n-type conductivity (known as P_n). The region between these two boundaries is known as the ionic domain, and can be described as the region in which the material is an ionic conductor.

The **electrolytic domain boundaries** are the oxygen partial pressures at which the ionic conductivity is 100 times the p-type conductivity or 100 times the n-type conductivity. The material is said to be exclusively ionic within the electrolytic domain. The reason to use the number 100, is that the electrons and electron holes are 100 to 1000 times more mobile than ions. Thus, in the case of the ionic domain boundary, the material within the domain could have a high degree of electron or electron hole conductivity (known as a **mixed conductor**), particularly close to the boundaries. However, in the case of the electrolytic domain boundary, the material properties within the domain are almost exclusively ionic.

A typical domain boundary (both ionic and electrolytic) for the system MO_2 is described in Fig. 11.50.

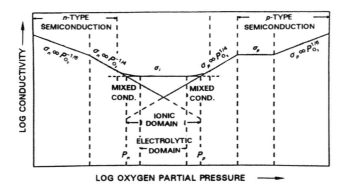

Fig. 11.50 The variation of the conductivity as a function of oxygen partial pressure for MO_2 (with predominantly Frenkel disorder) showing the electrolytic and ionic domain boundaries.

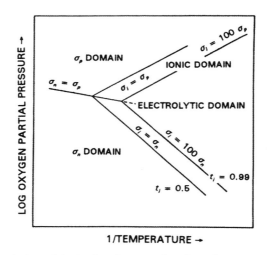

Fig. 11.51 Electrolytic and ionic domains as a function of temperature for the system MO_2 containing predominantly Frenkel defects.

We can also describe the effect of the electrolytic and ionic domain boundaries as a function of temperature. These are described in what are known as Patterson maps, an example of which is given in Fig. 11.51.

The electrolytic domain boundary is a very important factor in establishing whether the material is suitable for use as an electrolyte in a particular device. For example, if the system becomes a p-type semiconductor under an oxygen partial pressure of, say, 10^{-5} atmospheres, then it would not be of use in a solid oxide fuel cell where the partial pressure at the anode is much lower.

11.27 Effect of Dopants on the Defect Equilibria

When a dopant material, such as CaO (with a valency different from that of the parent material) is incorporated into the structure of MO_2, the following equilibrium equation is observed:

$$CaO = Ca_M'' + V_o^{\bullet\bullet} + O_o^x \qquad (11.62)$$

Thus, when CaO is incorporated into the MO_2 lattice, a double positively charged vacancy is produced. If Y_2O_3 were introduced instead (Y^{3+} into a M^{4+} lattice site), then we would observe the following relationship:

$$Y_2O_3 = 2Y_M' + V_o^{\bullet\bullet} + 3O_o^x \qquad (11.63)$$

Thus, when Y_2O_3 is incorporated into the MO_2 lattice, a double positively charged vacancy is also produced, although $2Y^{3+}$ ions are required.

Hence, when aliovalent (different valency) dopants are introduced into the lattice, the concentration of oxygen defects is increased. The effect of the partial conductivities of MO_2, doped with small and large concentrations of defects, as a function of oxygen partial pressure, is shown in Fig. 11.52. It is apparent that the ionic and electrolytic domains increase with increasing defect concentration. The ionic conductivity is also found to increase (which is consistent with equation 11.60). The effect of defect concentration is also described for several materials in earlier.

11.28 Ionic Transport Number

When the ionic conductivity (σ_i) of a material is quoted, the transport (or transference) number (t_i) is also given. This is described in equation 11.64:

$$t_i = \sigma_i/(\sigma_i + \sigma_e + \sigma_h) \qquad (11.64)$$

For an ionic conductor, the value for the transport number is approximately 1.0. The higher the number, the closer the material is to being a "pure" ionic conductor. The electronic (t_n) or electron hole (t_p) transport number can also be examined. In this case, the higher the transport number, the higher the electronic conductivity. Materials with ionic transport numbers much greater than zero, but much less than one, are considered to be mixed conductors.

There are a number of different methods for determining the transport number of a material. However, the most commonly used technique is known as the EMF method, which will be described later, together with a number of other methods.

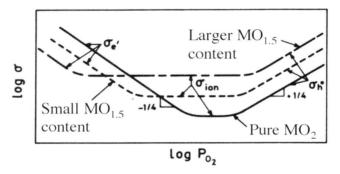

Fig. 11.52 The effect of oxygen partial pressure on the partial conductivity of MO_2 doped with different amounts of dopant (M_2O_3), as a function of oxygen partial pressure

11.29 Three Dimensional Representation

The diagrams given in Figs. 11.47, 11.48 and 11.49 are plotted at constant temperature. However, graphs can be constructed which show the effect of temperature on the defect concentration (or conductivity) and species activity (oxygen partial pressure in the case of the oxide ion). This three dimensional graph, which was first described by Patterson, shows the change of p-type, n-type, and ionic conductivity as a function of temperature and species activity. The domain boundaries are now shown as lines, where two planes intersect (rather than as single points).

An example of this type of graph is shown in Fig. 11.53. If we take a cross section through the graph, at constant temperature, then plots similar to that described in Fig. 11.48 are observed.

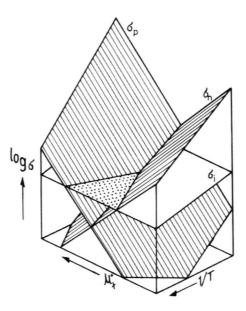

Fig. 11.53 The effect of conductivity as a function of temperature and species activity, as a three-dimensional plot.

In the three dimensional plots, it can be seen that the ionic domain only exists because the line representing the intrinsic electronic conductivity is situated below the ionic conductivity planes. Because the position of this intrinsic line is determined almost exclusively by the electronic band gap, a minimum size for this band gap is required if the materials are to have good ionic conductivity. Fig. 11.54 describes the minimum value of the band gap for a good ionic conductor, as a function of temperature (Curve I). Curve II is plotted with a safety factor of 100. In general a good ionic conductor will have a band gap greater than approximately T/300eV. The colours are displayed on the figure, as they are related to the electronic band gap. In fact, most dark materials are usually good mixed or electronic conductors, while most light materials can be good ionic conductors.

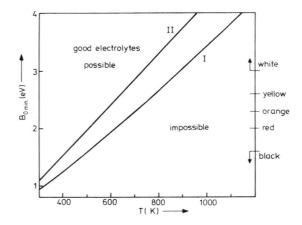

Fig. 11.54 Minimum value of band gap required for a good ionic conductor (Case I), Case II has a safety margin of 100 included.

As we have already discussed, the total conductivity of a material depends upon its resistance, which in the case of polycrystalline materials includes both the bulk resistance and the grain boundary resistance. The resistance of the material is very dependent upon a number of factors including the type of system, grain size, percentage of impurities present, porosity of the sintered material in the case of ceramic systems, state of stress, etc. There are a number of techniques available for measuring the value of the resistivity, and thus the value of conductivity. Some of the techniques can also be used to determine the separate contributions of various polarisation processes in the system, such as the kinetics of the electrode reactions in an electrolytic cell.

11.30 Methods of Measuring Ionic and Mixed Ionic/Electronic Conductors: DC Conductivity Measurements

11.30.1 *Two-probe technique*

The resistance of an electrolyte can be measured using a simple electrochemical cell consisting of the electrolyte material with a pair of electrodes, reversible to the charge carrying species, placed on either side. A steady-state potential difference can then be applied between the electrodes, and the current then measured. Figure 11.55 shows a typical equivalent circuit for this 2-probe dc technique. The conductivity of the material (σ) can be measured from the simple relationship given in equation 11.65.

$$\sigma = \left[\frac{IL}{VA} \right]$$ (11.65)

Where I is the current that flows through the cell, V is the voltage drop between the electrodes, A is the cross-sectional area of the electrolyte, and L is the length of the sample.

This technique can be used to obtain a reasonable value for the resistance of the material under scrutiny. However, there are a number of problems associated with this technique. The technique does not take into account any changes in the composition of the electrodes during the experiment, it is very difficult to maintain good contact between the probes and the electrodes, and it assumes that the electrode resistance is negligible compared to the electrolyte resistance.

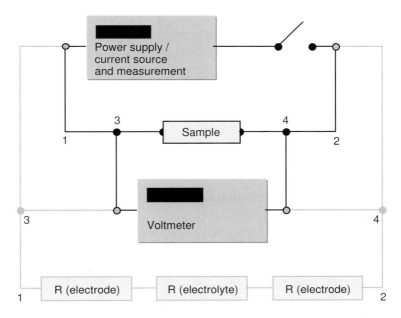

Fig. 11.55 The equivalent circuit for the two-probe technique (1 and 2 represent positions of the current probes, 3 and 4 represent positions of the potential probes, the equivalent circuit is given in light grey).

11.30.2 *Four-probe technique*

In order to eliminate the contributions due to the electrode/electrolyte interface, a four-probe technique was developed. In this method, current flows through two current probes, and the potential difference is measured between two other probes using a high impedance voltmeter. The set up of the four-probe technique is Fig. 11.56; in this case the value of L given in equation 11.65 is now the difference between the potential probes.

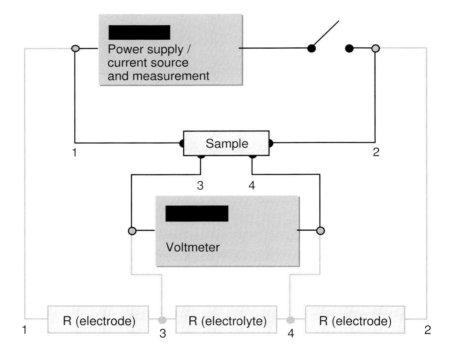

Fig. 11.56 The equivalent circuit for the four-probe technique (1 and 2 are positions of the current probes, 3 and 4 are positions of the potential probes; the equivalent circuit is given in light grey).

This technique is quite accurate and can be used to measure the conductivity of the material with varying temperature and partial pressure. The technique cannot, however, be used to differentiate between the grain boundary and bulk resistance contributions which are present in a polycrystalline material.

11.30.3 *Van der Pauw technique*

It is worth mentioning that the above methods are really only suitable if the material being measured is of a very well defined simple geometry (such as a bar). If the geometry of the material is much complicated, or of arbitrary shape, then the van der Pauw technique can be used to obtain the material's resistivity, and thus conductivity. This technique uses very fine contacts attached to the specimen, and looks at the resistance between a number of these contact points. The resistivity is then calculated as a function of these resistance values, and the thickness of the sample. It is not a commonly used technique in measuring the conductivity of ionic conductors, but is a good technique when the shape is too complicated to use any other method.

11.31 AC Techniques

Several AC techniques are used for measuring the conductivity or resistivity of ionic or mixed conductors, and include the following:

- Measuring the current/voltage response at constant frequency using a two or four probe technique as discussed above. The main problem with this technique is that as the temperature is increased, the time constants of the various resistance processes (electrolyte and electrode) changes, and the frequency response shifts to higher values.
- Impedance spectroscopy, where the response of the material (cell) is studied over a wide range of frequencies. It is the most common technique used today for measuring the conductivity of fast ion conductors, and thus we will examine this technique in more detail.

11.31.1 *Impedance spectroscopy*

This technique involves the application of an alternating current (AC) of varying frequencies across the material, and comparing the input and output signals to get information about the phase shift observed, and the impedance modulus.

Assume that an electric signal, U, is applied across the sample. If the sample is a solid electrolyte, then two electrodes (usually, but not always, Pt) are placed on either side (the sample is usually in the form of a pellet of known dimensions) to form a cell. The electric signal is given by:

$$U = U_o e^{i\omega t} \qquad (11.66)$$

where ω is the angular frequency, and t is the time constant; the resultant current passing through the cell is, thus,

$$I = I_o e^{i(\omega t + \phi)} \qquad (11.67)$$

where ϕ is the phase angle. The impedance of the cell (Z) is thus measured from

$$Z = \frac{U}{I} = \frac{U_o e^{-i\varphi}}{I_o} = Z_o e^{-i\varphi} \qquad (11.68)$$

Where Z_0, as shown in Fig. 11.57, is the modulus of the impedance of the sample. This can then be plotted onto a Nyquist diagram, with the opposite of the imaginary component, $Z = -Z_o \sin\phi$, on the y axis.

The admittance (1/Z) can also be plotted, and is sometimes preferable.

Now, each electrode or electrolyte process has a different time constant and therefore relaxes over a different frequency range. If the applied frequency range is large enough, then the contribution of each of the processes can be separated. In general, however, the time constant of each process decreases with increasing temperature and thus the response of the cell shifts to higher frequencies.

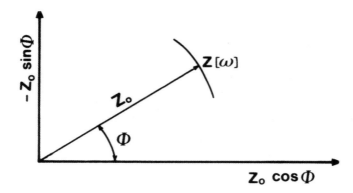

Fig. 11.57 The general principles of impedance spectroscopy.

In its simplest form, a solid electrolyte cell can be considered as the response from an RC circuit (ie: a combination of a resistance (R) and a capacitor (C) in parallel), as described in Fig. 11.58a. The response from the circuit is as shown in Fig. 11.58b, and is similar in nature to that observed for a solid electrolyte material. The semicircle is observed when a complex impedance plane is plotted, with the real impedance plotted on the x-axis, and the imaginary impedance plotted on the y-axis. From AC theory, the resistance of each of the components can be given by the intercepts on the x-axis. The value of the capacitor can be obtained from the apex frequency. Thus, in a real polycrystalline material, it is possible to separate out the bulk (intergrain) resistance from the grain boundary resistance, as shown in Fig. 11.58b. The electrode response is also separated out from the electrolyte response and, in fact, can be separated into three further semi-circles (corresponding to processes including charge transfer, diffusion, and adsorption/desorption).

Most semi-circle are not, however, so well defined, and the following can occur, which makes full interpretation difficult:

- Overlapping semicircles
- Semicircle depression
- Semicircle identification (shift to the left or right, or may disappear entirely with, for example, temperature).

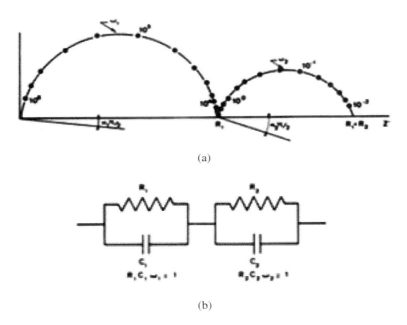

(a)

(b)

Fig. 11.58 (a) Typical impedance diagram and; (b) typical equivalent RC circuit obtained from an ideal system.

The set-up for impedance spectroscopy is quite straightforward, and usually includes a holder containing the sample which is then placed into a furnace. Leads, which are connected on either side of the sample and pressed against the electrodes, are attached to the outside of the furnace, and thus to the frequency response analyser (FRA). The FRA automatically analyses the input and output data, and, with the help of a computer, will plot the Nyquist diagram. A typical rig is shown in Fig. 11.59. The analysed data can give information pertaining to the conductivity of the bulk and grain boundary of the polycrystalline material. Any changes in the physical properties, affecting the electrical properties can be observed using this technique. Figure 11.60 (a-b), for example, shows the effect on the grain boundary and bulk conductivity for a coarse (11.60a) grained yttria stabilised zirconia (YSZ) sample, with a significant impurity phase present at the grain boundary, and a fine (11.60b) grained YSZ sample. It is quite apparent that the two plots are significantly different, due to the two different microstructures.

The cell can be set up to determine the effect of different gas partial pressures (for example oxygen partial pressures) on the resistivity or conductivity of the sample, and thus an idea of the electronic component can be realised (this depends on the response of the conductivity of the sample to the changing gas atmospheres).

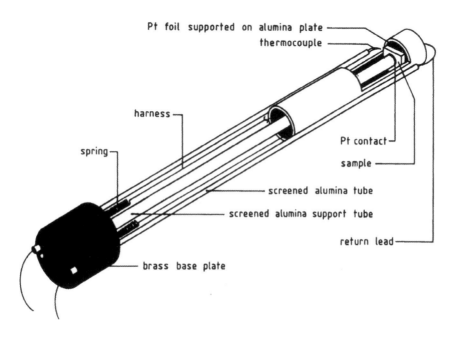

Pt foil supported on alumina plate

thermocouple

harness

spring

Pt contact

sample

screened alumina tube

screened alumina support tube

return lead

brass base plate

Fig. 11.59. Typical impedance spectroscopy rig to measure polycrystalline materials (after N. Bonanos, B.C.H. Steele, W.B. Johnson, W.L. Worrell, D.D. MacDonald, and M.C.H. McKubre, "Applications of Impedance Spectroscopy", Chapter Four in "Impedance Spectroscopy – Emphasizing Solid Materials and Systems", Edited by J. Ross MacDonald, Wiley Interscience, 1987).

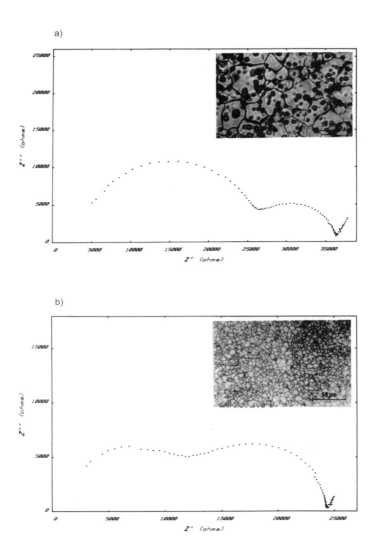

Fig. 11.60 (a) Complex impedance plot of YSZ at 425°C; (b) complex impedance plot of an alternative YSZ sample at 425°C (after B.C.H. Steele, J. Drennan, R.K. Slotwinski, N. Bonanos, and E.P. Butler, in Science and Technology of Zirconia, Advances in Ceramics, Volume 3, Editors, A.H. Heuer, L.W. Hobbs, The American Ceramics Society, Columbus, Ohio, 1981, p286).

11.32 Current Interruption Technique

This technique is a galvanostatic (constant current) technique, which involves passing a fixed current through a two, or more usually a three electrode cell set-up for a sufficiently long time so as to achieve a steady state potential. The applied current is then interrupted using a fast electronic switch, and the potential/time transient is measured, usually using a fast-storage oscilloscope, within a few milliseconds of the interruption. The voltage decays almost immediately, and the electrolyte resistance (IR) losses can be determined. The slow voltage decay is due to the electrolyte/electrode interface. Figure 11.61 shows the transient response of the solid electrolyte after the interruption of the steady state current. The technique is a very useful one, particularly when determining electrode over-potentials. Here, a fixed current would flow through an electrode (usually made deliberately dense) of known dimensions, and the transient response thus recorded.

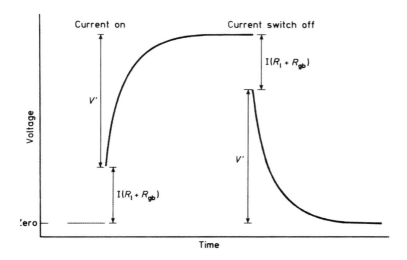

Fig. 11.61 Transient response of the electrolyte after the interruption of the current (after SPS Badwal, Chapter 11 in Materials Science and Technology, Volume 11, Structure and Properties of Ceramics, Ed M. Swain, VCH Publishers, (1994), Germany, p567–63 (Badwall 1994)).

11.33 Other Techniques

Other techniques commonly used include:

- *Potentiostatic techniques*, where a constant potential is applied between a working and a reference electrode, and the current flowing between the working and counter electrodes is measured.
- *Cyclic voltammetry, and linear sweep voltammetry*. In these techniques, a linear voltage ramp (either in one direction (linear sweep voltammetry) or where the sweep is frequently reversed and repeated many times (sweep voltammetry)) is imposed between a working and a reference electrode, and the current response is measured between the working and counter electrodes.
- *Chronopotentiometry and Chronoamperometry*, where the potential or current response is measured as a function of time.

11.34 Techniques for Measuring Partial (Mixed) Conductivity

11.34.1 *EMF technique*

There are a number of different methods for determining the transport number of a material. The most commonly used technique is known as the EMF method. In this technique, the material under investigation is placed between two reversible electrodes, which define the chemical potential for the mobile component of that material at each electrode/electrolyte interface. If each interface is at a different chemical potential, then an EMF will be present across the cell, which, for an oxygen concentration cell, is given by:

$$E_m = \frac{1}{nF} \int_{\ln \mu''_{O_2}}^{\ln \mu'_{O_2}} t_i \, d \ln \mu O_2 \qquad (11.69)$$

where $\mu O_2'$ and $\mu O_2''$ are the chemical potentials on either side of the electrolyte, F is the Faraday constant, and n is the number of electrons required in the chemical reaction at the electrodes.

   Now, the chemical potential can be given by:

$$\mu O_2 = \mu O_2{}^\circ + RT \; lnPO_2 \qquad (11.70)$$

Thus

$$E_m = \frac{RT}{nF} \int_{\ln PO_2''}^{\ln PO_2'} t_i \, d \ln PO_2 \qquad (11.71)$$

By measuring the EMF across the cell, the transport number can be determined from the approximation

$$E = \frac{RT}{4F} t_i \ln \left(\frac{PO_2{}'}{PO_2{}''} \right) \qquad (11.72)$$

This approximation can only be made if the sample is assumed to be constant within the material. However, in general, this is a reasonable assumption, provided that the chemical potential gradient is kept reasonably small. Thus, the ionic transport number is taken as the ratio of the measured EMF relative to the theoretical EMF.

Unfortunately, this technique cannot be used when the sample has predominantly electronic conductivity, because there is a possibility of internal short circuiting of the cell, causing a change in the oxygen concentration at the electrode/electrolyte interface, and the polarisation of both electrodes due to the passage of current. Thus, this technique is most suitable for measuring the ionic transport number of materials which are classed as ionic conductors, rather than mixed or electronic conductors.

11.34.2 *The polarized cell technique*

The polarised cell technique uses an ion-blocking electrode on one side of the cell, and a reference electrode on the other. A small DC potential is applied across the system to set up a chemical potential gradient between the blocking electrode and the reference electrode. The blocking electrode allows no ionic current to flow under steady state conditions. Thus, the conduction takes place only by the migration of electronic carriers. With the use of two additional probes positioned on the specimen, reversible only to electronic or ionic current, the conductivity of the material can be measured. A typical set-up is shown in Fig. 11.62.

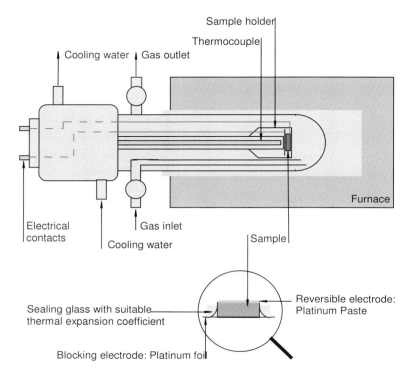

Fig. 11.62 Typical set-up for the polarised cell-technique (courtesy of M. Keppeler).

11.34.3 *Oxygen permeation*

By definition, the total electrical current in an open circuited cell must be zero. Now, we can consider a cell of the type O_2 (P'') /specimen/O_2 (P'), where P is the oxygen partial pressure, with $P' > P''$.

When the only charge carriers in the sample are oxygen ions, then the oxygen flux across the sample must remain zero. However, when electronic charge carriers exist, an oxygen ion flux can persist if an equal and opposite current flows due to the electronic charge carriers. Thus, when a fixed gradient of oxygen pressure is applied across the specimen, a steady state flux of neutral oxygen flows from the high oxygen partial pressure side (P´) to the low oxygen pressure side (P´´), given by the equation:

$$J_o = -D_a \nabla C_o \qquad (11.73)$$

Where C_o represents the concentration of oxygen, and D_a is the ambipolar diffusivity that is approximately given by:

$$D_a \approx t_i D_e + t_e D_I \qquad (11.74)$$

When the majority charge carrier in the material is the oxygen ion, then $t_I \approx 1$, and D_a is approximately equal to the diffusivity of the electronic charge carriers. It can further be shown that the permeation current density, j, can be expressed by:

$$\left(\frac{\partial j}{\partial \ln PO_2} \right)_{PO_2 = P''} = \left(\frac{\kappa T}{4qL} \right) \sigma_i t_i \qquad (11.75)$$

where L is the thickness of the sample.

Differentiation of the equation yields

$$j = \frac{\kappa T}{4qL} \sigma_i \int_{\ln P'}^{\ln P''} t_e \, d \ln PO_2 \qquad (11.76)$$

Thus, the transport number can be obtained by plotting the permeation current density, j, against $\ln P''$.

Important Concepts:

Point Defects: Lattice sites which contain an atom, ion or molecule not present in the stoichiometric material.

Line Defects: Usually considered to be dislocations. Edge, screw and mixed dislocations need to be considered.

Ionic Conductors: These are either anion or cation sites that are free to move throughout the structure. Conductivity occurs because of the high degree of disorder in the material.

Random Walk Theory: This is a mass transport process which occurs between well defined positions in the system. The diffusion is considered to be via hopping in a random way between these well defined positions.

Vacancy Mechanism: Diffusion which occurs between vacancies.

Interstitial Mechanism: Diffusion which occurs between interstitials.

Interstitialcy Mechanism: Diffusion which occurs when an interstitial atom displaces another atom from its original substitutional site.

Oxide Ion Conductivity: Ionic conductivity where the majority charge carrier is the oxygen ion (O^{2-})

Proton Conductivity: Ionic conductivity where the majority charge carrier is the proton (H^+)

Zirconium Dioxide: One of the most important ionic conductors, and the system which has been fabricated into many devices, such as solid oxide fuel cells, oxygen pumps, and oxygen gauges.

Kroger Vink (Brouwer) diagram: A plot of defect concentration (or partial conductivity) as a function of species activity (for example oxygen partial pressure in an oxide ion conductor).

Electrolytic domain boundary: The oxygen partial pressure at which the ionic conductivity is 100 times the electronic (or electron defect) conductivity.

Ionic domain: The oxygen partial pressure at which the ionic conductivity is equal to the electronic (or electron defect) conductivity.

Mixed conductor: A material which has both electronic (or electron hole) and ionic conductivity.

Patterson map: A plot of the conductivity of the ionic/electrolytic domain versus temperature.

Problems

1. Choose ONE ceramic material (metal oxide) that is an ionic/mixed/electronic conductor. Determine, what the diffusion and conduction mechanisms are for your material, and thus the defects, and how this is affected under the likely conditions found for your system (e.g. effect of partial pressure, temperature etc.). State the majority defect, and draw a simplified Brouwer diagram for your material. How would you measure the conductivity and diffusion coefficient of your particular material, and, if applicable, how would you determine the transport number.

2. Sketch a full Brouwer diagram (log defect concentration versus log PO_2) for an oxide MO_2 dominated by fully ionized oxygen and metal vacancies at under- and over-stoichiometry, respectively. Assume that intrinsic electronic equilibrium predominates close to the stoichiometric point. Use the rules listed in the lecture to solve this.

3. Sketch a Brouwer diagram for an oxygen deficient oxide (MO) doped with a substitutional higher valent dopant. Sketch a Brouwer diagram for an oxide $M_{1+x}O$ doped with a substitutional lower-valent dopant. Sketch another diagram, but this time log of defect concentrationa versus dopant concentration, for an oxide $M_{1+x}O$ doped with a substitutional lower-valent dopant.

4. The incorrect statement related to Schottky defecta is

    a. It is a stoichiometric point defect

    b. Equal number of cations and anions are missing from their lattice points.

    c. Shown by strongly ionic crystals with high co-ordination number.

    d. Density & covalent nature are increased.

5. Consider the following statements.

    a. Both Schottky and Frenkel defects are non stiochiometric defects

    b. Crystals with Schottky and Frenkel defects show little electrical conductivity

    c. Frenkel defects are shown by ionic compounds with high co-ordination numbers and with big sized cations.

    d. Crystals with Schottky and Frenkel defects are electrically neutral

The correct statements are

    a. a & b
    b. b & d
    c. c & d
    d. b, c & d

6.   Addition of little $SrCl_2$ to NaCl produces:

    a. Cation vacancies
    b. Anion vacancies
    c. Both cation & anion vacancies
    d. None

Reprint Permission

The American Ceramic Society, Ohio, USA: Figures 11.12, 11.13, 11.60

American Chemical Society, Washington, USA: Figures 11.33, 11.34

Chapman and Hall, London, UK: Figures 11.17, 11.32, 11.37(a), 11.41, 11.23(a), 11.24, 11.29

The Electrochemical Society Inc., Pennington, USA: Figures 11.21, 11.23(b), 11.37(a), 11.42

Elsevier Science Publishers, Amsterdam, The Netherlands: Figures 11.18, 11.19, 11.20, 11.35, 11.36, 11.37(b), 11.39, 11.43

Commission of the European Communities, Brussels, Belgium: Figure 11.45

Institute of Materials, London, UK: Figure 11.41

John Wiley and Sons, New York, USA: Figures 11.14, 11.19, 11.27, 11.31, 11.38, 11.59, 11.61, 11.42, 11.44

Kluwer Academic Publishers, Amsterdam, The Netherlands: Figure 11.25

Springer Verlag, Heidelberg, Germany: Figure 11.10

Chapter 12

Thin Film Electronic Materials

12.1 Introduction

A thin film can be defined as a low-dimensional material which is prepared by using elemental materials or inorganic/organic compounds. This microscopically thin layer of material is normally supported by a suitable substrate of metal, ceramic, semiconductor or polymer. Its thickness is typically less than several micrometers. However, in different research fields this thickness can be varied in a wide range from a few tens micrometers to a few nanometers.

Similar to bulk materials, thin film materials can be single crystalline, polycrystalline or amorphous, and their crystal size can be in micro- or nano-meter range. In comparison with bulk materials, materials in thin film forms have many fascinating characteristics, typically including large surface energy, abnormal (non-equilibrated) microstructure, non-stoichiometry, and size effects (e.g. quantum size effect), resulting from their unique atomic growth process.

As such thin film materials normally exhibit quite different chemical, electrical, mechanical, optical, and physical properties. Research on thin films therefore is very important and active branch of materials science and engineering. Thin films are also finding more and more applications in various electronic, optical, magnetic and mechanical devices, and other functional aspects such as for colour, decorative, instrumental, electromagnetic screening and protective purposes.

12.2 Techniques for Preparation of Thin Films

A variety of techniques have been developed to prepare thin films with varied composition, structure and performance. These methods could be categorised into three groups: (1) chemical/electrochemical deposition, (2) chemical vapour deposition (CVD) and (3) physical vapour deposition (PVD). The common features of these growth techniques are:

1. Generally has nucleation and growth stages;
2. Nucleation and growth processes are controlled by deposition conditions; and
3. Compositional, microstructural, and functional properties can be controlled during nucleation and growth stages.

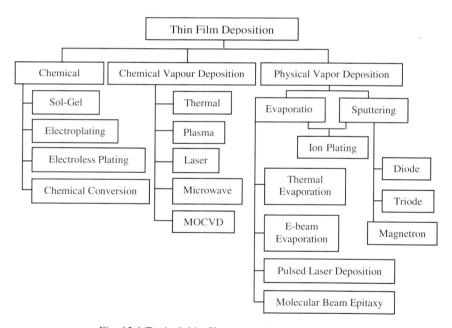

Fig. 12.1 Typical thin film preparation techniques.

In the following sections, some typical film growth techniques will be briefly introduced on their operation principle and application.

12.2.1 *Evaporation*

Growth of thin films using evaporation, in practice, could be done using two typical ways for generation of vapour from the selected source materials, namely, thermal evaporation and electron-beam evaporation. Atoms or molecules from the vaporization source reach the substrate situated in the deposition chamber with a relatively good vacuum ($< 10^{-4}$ Torr).

Thermal evaporation might be the oldest and the most accessible technique for growth of thin metallic films. The source materials are resistance-heated to an appropriate temperature at which there is an appreciable vapour pressure. For most common metals that vaporize below about 1500°C, evaporation can be achieved simply by contacting a hot surface that is resistively heated by passing a current through a material. Typical resistive heating elements are carbon (C), molybdenum (Mo), tantalum (Ta), tungsten (W) and BN/TiB_2 composite ceramics. The heated surface may have many configurations, such as basket, boat, crucible, and wire, to realise fast heating and uniform distribution of vapour. The major advantages of thermal evaporation are high film deposition rates and relative simple and cheap equipment. This technology however is limited to the evaporation of metals with relatively low melting points, such as aluminium (Al) and zinc (Zn).

Electron beam (E-beam) evaporation is another important evaporation technique. It uses high-energy electron beams (typically accelerated with a voltage of 5 to 20 kV) to "heat" the source materials that are placed in water-cooled copper hearth "pocket". This technique can vaporize most pure metals, including those with high melting points. It is particularly suitable for the evaporation of refractory materials, such as most ceramics (oxides and nitrides), glasses, carbon, and refractory metals. E-beam technology is probably the fastest deposition source available today. Using high-power E-beam sources, deposition rates as high as 50 microns per second have been attained. With adequate adjustment of the electron beam spot size, uniform films of high purity can be obtained.

Evaporation technology is commonly used for metallization of ceramic, semiconductor, and dielectric foils; preparation of functional coatings and thin films (optical, decorative, corrosion-resistant, or insulating); and fabrication of free-standing structures.

12.2.2 *Sputtering*

The sputtering process involves physical evaporation of atoms from a surface by momentum transfer from bombarding energetic atomic sized gas ions (such as argon ions) accelerated in an electric field.

$$Ar + e^- \rightarrow Ar^+ + 2e^-.$$

Sputtering deposition then is the deposition of species vaporized from the surface of a certain source material (i.e., target, can be element, alloy, compound, or their mixture) by the sputtering process. This can be performed in a vacuum using either low pressure plasma (<5 mTorr) or high pressure plasma (5–30 mTorr).

12.2.2.1 *DC diode sputtering*

The simplest configuration for sputtering deposition is a direct-current (DC) diode plasma device that consists of a cathode (i.e., target), an anode (on which the substrate is placed), a DC power source, and a vacuum chamber. Argon (Ar) is normally used to establish a discharge. The plasma can be established uniformly over a large area so that a solid large-area vaporization source can be used. Further this surface is not necessary to be planar but can be shaped so as to be conformal to a substrate surface. Typical plasma generation region is about one centimetre in width. The electrons in the plasma are hot, with a Maxwellian energy distribution and an equivalent thermal temperature of $1–5\times10^4$ K.

The particles vaporized from the target surface have an intrinsic high energy. As a result, thin films or coatings formed with DC diode sputtering generally show a superior adhesion to the substrate and good crystalline structures. However this technology exhibits relatively low deposition rates due to the low plasma density. In addition, the target must be electrically conductive since an insulating surface will develop a surface charge that will prevent ion bombardment of the surface. If the target is initially conducting but a non-conducting or poorly-conducting surface layer is built up during sputtering due to reactions with gases in the plasma, arcing on the target surface will happen and deteriorate the structural and functional performance of the growing film.

12.2.2.2 *Radio frequency sputtering*

Radio frequency (RF) sputtering is commonly used for deposition of electrically insulating materials, such as oxides and polymers. When a RF potential is capacitively coupled to a target, an alternating positive/negative potential appears on its surface. In a half cycle, positively charged ions are accelerated towards the surface with enough energy to cause sputtering while on alternate half cycle, electrons reach the surface to prevent any charge build-up. RF frequencies used for sputtering deposition are in the range of 0.5–30 MHz with 13.56 MHz being the most widely used. RF sputtering can be used for film deposition at low working gas pressures (< 1 mTorr).

12.2.2.3 *Magnetron sputtering*

In magnetron sputtering, a magnetic field is superposed to increase the plasma density and current density at the cathode (target), then to effectively increase the sputtering rate. The magnetic field is parallel to the cathode surface. The electrons, ejected from the cathode, are deflected to stay near the target surface.

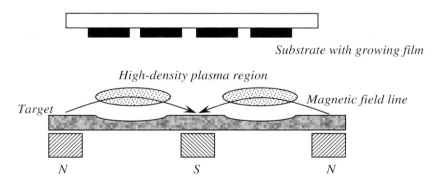

Fig. 12.2 A schematic showing magnetron sputtering configuration.

If the magnets behind the target are arranged properly, the electrons can circulate on a closed path on the target surface. This electron-trapping effect will increase the collision probability between the electrons and the sputtering gas molecules, creating a high density plasma. This enables the sputtering at low gas pressures with a high deposition rate. The most commonly used configuration of magnetron is the planar one where the erosion track is a closed or elongated circle.

12.2.3 *Chemical vapour deposition*

Chemical vapour deposition (CVD) involves the formation of a thin solid film on the heated area of a substrate situated in a reactor by chemical reactions of volatile precursors containing the desired species. The basic physicochemical steps in a CVD process may include the following key steps:

1. Evaporation and transportation of precursors into the reactor;
2. Chemical reactions of precursors in the reaction zone;
3. Mass transportation of reactive intermediates to the substrate surface;
4. Adsorption of the reactive intermediates on the substrate surface; and
5. Film formation through surface diffusion, nucleation, and surface chemical reactions.

CVD is an important technique for producing various functional thin films in semiconductor and electronic industry, including III-V and II-VI compound semiconductors, C, B, Si, borides, carbides, nitrides, oxides, silicides, sulfides, and also many metallic and non-metallic elements. These materials can be in forms of single-crystalline, polycrystalline, amorphous and epitaxial.

Typical reactions used in CVD process are shown below:

$$CH_4 \rightarrow C(\text{diamond}) + 2H_2,$$
$$MoCl_6 + 3H_2 \rightarrow Mo + 6HCl \ (400–1350°C),$$
$$Ni(CO)_4 \rightarrow Ni + 4CO \ (180–200°C),$$
$$WF_6 + 3H_2 \rightarrow W + 6HF \ (300–700°C),$$

$$2AlCl_3 + 3H_2 + 3CO_2 \rightarrow Al_2O_3 + 3CO + 6HCl \ (1050°C),$$
$$4BCl_3 + CH_4 + 4H_2 \rightarrow B_4C + 12HCl \ (1200-1400°C),$$
$$BCl_3 + NH_3 \rightarrow BN + 3HCl \ (>1300°C),$$
$$CH_3SiCl_3 \rightarrow SiC + 3HCl \ (900-1400°C),$$
$$3SiCl_4 + 4NH_3 \rightarrow Si_3N_4 + 12HCl \ (\sim850°C),$$
$$TiCl_4 + 2BCl_3 + 5H_2 \rightarrow TiB_2 + 10HCl \ (800-1100°C),$$
$$TiCl_4 + CH_4 \rightarrow TiC + 4HCl \ (850-1050°C),$$
$$TiCl_4 + NH_3 + H_2 \rightarrow TiN + 4HCl \ (>900°C).$$

Thermal energy required for CVD processes can be inputted through several methods such as direct resistance heating of the substrate or substrate holder, RF induction of the substrate holder, thermal radiation heating, or any suitable combination of these. Another way of inputting energy to a CVD process is to use high energy photons. The process of photo-assisted CVD involves interaction of light radiation with precursor molecules either in the gas phase or on the growth surface.

One way of reducing growth temperatures is to use plasma-assisted or plasma-enhanced CVD (PECVD). With this technique deposition can occur at a low temperature, even close to ambient, since electrical energy rather than thermal energy is used to initiate homogeneous reactions for the production of chemically active ions and radicals that can participate in heterogeneous reactions.

Metal-organic chemical vapour deposition (MOCVD) is a type of CVD that utilizes metal-organic precursors. Originally, metal-organic (or organometallic) precursors were those compounds containing a direct metal-carbon bond (e.g., metal alkyls or metal carbonyls), but they have been broadened to include precursors containing metal-oxygen bonds (e.g. metal-alkoxides, metal-β-diketonates), metal-nitrogen bonds (e.g. metal alkylamides), and even metal hydrides (e.g. trimethylamine alane).

12.2.4 *MBE*

Molecular beam epitaxy (MBE) is commonly used to develop epitaxial films with atomic thicknesses on to single crystal substrates using atomic and molecular beams produced from Knudsen cells under high or ultrahigh vacuum conditions (better than 10^{-8} Pa). This technique was originally developed in the mid-1960s, when homoepitaxial films of silicon were grown from molecular beams of monosilane (SiH_4) by Joyce and Bradley. The tremendous works of A.Y. Cho and J.R. Arthur at Bell Laboratories triggered the incredible expansion of MBE.

The MBE process is essentially a refined form of vacuum evaporation. Ultra-pure elements are heated in Knudsen cells (or effusion cells) and start to sublimate. This provide an angular distribution of atoms or molecules in a beam. Evaporated atoms or molecules (beams) are then directed to the substrate surface where they may react chemically with each other or other gaseous species introduced into the vacuum chamber in purpose and condense onto the heated (continuously rotated, when needed to improve homogeneity) single crystal substrate. If the substrate is of the same nature as the growing film, the term of homoepitaxy is often used; otherwise the process is called heteroepitaxy.

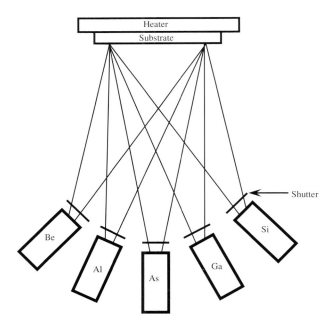

Fig. 12.3 A schematic of molecular beam epitaxy system.

Figure 3 is a schematic of MBE system. The mechanical shutters in front of each cell are under well control, enabling precise control of the composition, doping, microstructure and thickness of the growing layer at the monolayer level. The operation time of a shutter can be around 0.1 s, much shorter than the time needed to grow one monolayer (typically 1–5 s). Reflection High Energy Electron Diffraction (RHEED) is often used for monitoring the growth process. The oscillation of the RHEED signal exactly corresponds to the time needed to grow a monolayer and the diffraction pattern on the RHEED window gives direct indication over the state of the surface.

A number of modifications have been introduced to the basic MBE growth technology. Some of the solid-state elemental sources were replaced by various gas and/or organo-metallic sources. For example, for the growth of III-V semiconductor compounds, elemental As and P had been substituted by arsine and phosphine. Flux modulation techniques was also introduced to MBE as a major modification, this led to the so-called migration enhanced epitaxy (MEE). In MEE process, both cation and anion fluxes are modulated during growth, which leads to enhanced the cation migration distances at fairly low temperatures when the anion surface adatom population is low, thus alleviating the surface roughness of the growing surfaces.

12.2.5 *PLD*

Pulsed laser deposition (PLD) uses a focused high-power pulsed laser beam to ablate a target surface of the desired composition to generate some volatile phases. When the laser pulse is absorbed by the target, its energy is converted to electronic excitation and then thermal, chemical and mechanical energy, resulting in evaporation, ablation, plasma formation and even exfoliation. The ejected species expand into the surrounding vacuum in the form of a plume containing many energetic species including atoms, molecules, electrons, ions, clusters, particulates and molten globules. These materials are then condensed onto a suitable substrate as a thin film. There are three possible growth modes in PLD:

1. Step-flow growth: this mode is often observed during deposition on a high miscut substrate or at elevated temperatures. Atoms landed on the substrate surface diffuse to atomic step edges and form into surface islands. The growing surface is viewed as steps travelling across the surface;
2. Layer-by-layer growth: islands continue to nucleate on the surface until a critical island density is reached. As more material is added, the islands continue to grow until coalescence between them appears, resulting in a high density of pits on the surface. When additional atoms are added to the surface they diffuse into these pits to complete the layer. This process is repeated for each subsequent layer; and
3. 3D growth: this mode is similar to the layer-by-layer growth. The difference is that once an island is formed an additional island will nucleate on top of the previous island. Continuing growth in one layer will not persist, leading to a roughened surface when additional materials are added.

PLD is often carried out in a high or ultra-high vacuum chamber. Reactive gaseous species, such as oxygen, can be introduced to initiate reactive deposition of oxide or more complicated materials.

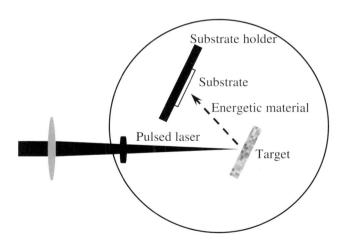

Fig. 12.4 A schematic of pulsed laser deposition system.

PLD is the primary option for growing various complex materials such as transition metal oxides, nitrides, high-temperature superconducting thin films (e.g. $YBa_2Cu_3O_7$), high-quality multi-layer thin films and superlattices. The development of new laser technology, such as lasers with high repetition rate and short pulse durations, made PLD a very competitive tool for the growth of thin, well defined films with complex stoichiometry.

12.2.6 *Sol-gel*

The sol-gel process is a wet-chemical technique based on the evolution of a colloidal system through the formation of an inorganic or hybrid sol followed by its gelation to form a continuous polymer network (gel). Metal alkoxide precursors can be used as precursors for classical synthesis of such sols. Alkoxysilanes, such as tetraethoxysilane (TEOS) and tetramethoxysilane (TMOS), and alkoxides of zirconium, titanium, cerium, tin and aluminium are commonly used in the sol-gel route.

The common reactions that take place during the sol-gel process are:

1. Replacement of the alkoxide groups (–OR) with hydroxyl groups (–OH) during the hydrolysis stage due to the interaction of alkoxide molecules with water;

$$Si\text{-}OR + H_2O \rightarrow Si\text{-}OH + ROH.$$

2. Condensation of two –OH groups or –OH group with –OR group, which produces M–O–M bonds and water or alcohol. Usually, the condensation process starts before the end of the hydrolysis step.

$$Si\text{-}OH + RO\text{-}Si \rightarrow Si\text{-}O\text{-}Si + ROH.$$

The overall reaction can be written as:

$$Si(OR)_4 + 2H_2O \rightarrow SiO_2 + 4ROH$$

The sol can be deposited on a substrate to form a film using either dip coating or spin coating. The first sol-gel process was described by Ebelmen in 1844 for synthesis of silica and the commercial production of sol-gel coatings onto flat glass appeared in the early 1960s. Sol-gel processes are now playing a central role in synthesis of multi-component ceramics, polymers, porous solids, and other nanostructured materials.

12.2.7 *Electrodeposition*

Electrodeposition refers to the deposition of a thin layer of pure metals or alloys from solutions by the passage of an electric current to reduce cations of a desired material from a solution. Electrolytes are usually aqueous, although many materials have been deposited from non-aqueous solutions. The part to be covered by the deposit is the cathode of the circuit. The anode is normally the metal to be plated onto the cathode though non-consumable anode may be used as well. The compositional, structural, and functional properties of an electrodeposit are determined by many external factors including the electrolyte composition, pH, temperature and agitation, the potential applied between the electrodes, and the current density. The resulting films can be crystalline or amorphous, metallic or non-metallic.

12.3 Thin Film Conducting Materials

Thin film conductors are widely used as interconnectors, electrodes for thin film resistors and capacitors, and welding regions in various electrical and electronic devices. The resistivity of thin film conducting materials is in general higher than that of the bulk material with the same composition, mainly due to the presence of surface reflective effects, high concentration of impurities and defects. The basic requirements for thin film conducting materials may include:

1. High conductivity;
2. Good adhesion to the underlying substrate;
3. Good compatibility with other materials; and
4. Compatible to photolithographic and other processes.

Thin film conductors could be roughly categorized into two main groups: elemental thin film and multi-layered thin films. The former is mainly fabricated using a single metallic element, such as Al. While the latter may have two, three, or more metallic constitutes, such as Cr-Au, Ti-Pd-Au, Ti-Cu-Ni-Au, etc.

12.3.1 *Cu*

Copper has good electrical conductivity, thermal conductivity, high strength, high resistance to fatigue, and excellent solderability. Thin films of Cu are commonly prepared by using chemical vapour deposition, thermal evaporation, sputtering, or electrochemical plating. A common problem associated with Cu films is that they have relatively poor adhesion to most dielectric substrates. Inter-layers composed of Al, Cr, Nb, or Ti had been used to enhance the adhesion in order to achieve desired mechanical bonding. Another technique is to add small amounts of alloying elements, such as Cr and Ti, to Cu to improve the interfacial bonding, especially after heat treatment or annealing.

12.3.2 *Al*

Aluminium has a lower conductivity than Cu, but Al and most of its alloys have advantages of acceptable degree of adhesion to various substrate materials including polymeric dielectrics, good corrosion resistance in natural atmosphere and in many chemicals and solutions. Thin films of Al are normally produced by vacuum evaporation.

12.3.3 *Au*

Gold has high electrical conductivity and excellent corrosion resistance and environmental stability in a wide range of temperature. Its thin films can be prepared through evaporation or sputtering. However Au does not adhere well to substrate materials so that underlayers composed of Ti or Cr might be needed. As such Au thin films are more often used as a top layer to protect other metallic layers.

12.3.4 *Ag*

Silver has excellent conductivity. Its thin films, prepared by electrochemical deposition or evaporation, are mainly used for contacts.

12.3.5 *Cr-Au and NiCr-Au*

Cr-Au and NiCr-Au systems are the most widely used multi-layered conducting thin films. Cr and NiCr films are used to improve the interfacial adhesion while Au film is the conducting film. A problem with these multi-layered films of different components is that at elevated temperatures, inter-diffusion between these two sub-layers may occur, leading to the formation of undesired compounds or structural defects, and then resulting in increased electrical resistance, poor solderability, and decreased surface property, reliability, and stability. In comparison with Cr-Au system, NiCr-Au system has a lower inter-diffusion probability.

12.3.6 *Ti-Pd(Pt)-Au*

This system has better resistance stability under humid conditions compared with NiCr-Au system. It also has excellent corrosion resistance since Ti has superior self-passivation ability under most atmospheric conditions. The Pd or Pt layer sandwiched between the bottom Ti layer (to improve interfacial bonding) and the top Au layer is mainly used as a barrier layer that can effectively inhibit the inter-diffusion between these two layers, therefore resulting in a higher thermal stability. Ti, Pd, Pt, and Au layers can be prepared by evaporation or sputtering techniques.

12.3.7 *FeCrAl-Cu-Au*

This system is widely used as end contacts for thin film resistors, interconnectors, and welding joints. When compared with NiCr-Au thin films, FeCrAl-Cu-Au films have better interfacial bonding strength, solderability, and stability. However inter-diffusion between FeCrAl and Cu layers could still happen when the temperature of heat treatment is high.

12.4 Thin Film Resistors

Approximately 40-50% of components in an electronic circuit are resistors. Thin film resistors widely used in integrated circuits are fundamental to the desired functions of the devices, including biasing of active devices, serving as voltage dividers, and assisting in impedance matching. Research and development of thin film resistors mainly focus on finding, understanding, and improving materials with very low temperature coefficient of resistance (TCR), low voltage coefficient of resistance (VCR), and sufficient stability.

Typically there are three major design principles of discrete thin film resistors, including: CHIP-type, MELF-type, and Leaded type. The heart of a thin-film resistor consists of a thin layer of resistor material deposited onto a suitable substrate (e.g., silicon, GaAs, or Al_2O_3). The deposition was normally achieved by using sputtering. The contacts or interconnections on the ends were then connected to circuit components.

The resistance value of a thin film resistor is given by the relation:

$$R = R_s \times \frac{l}{w}, \qquad (12.1)$$

where R_s is the sheet resistance of the film, l is the length and w is the width of the resistor.

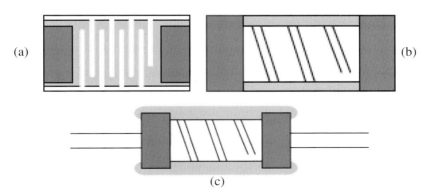

Fig. 12.5 Different types of thin film resistors. (a) CHIP-type, (b) MELF-type and (c) Leaded type. (after R.W. Kuehl, Microelectronics Reliability, **49**, 2009, pp51-58).

A precision resistor is characterized by $|\text{TCR}| < 25$ ppm/K. Thin film resistor materials can comprise a variety of materials, including metals, alloys, nitrides or cermets.

12.4.1 *Ni-Cr*

Ni-Cr or Ni-Cr-Al alloys are the classical materials system of thin film resistors that are extensively used as discrete loads or potentiometers in hybrid circuits. A typical alloy composition of Ni-Cr is 80:20 (wt.%), however the ratio of Ni:Cr can vary from 40:60 to 80:20. The resistivity is typically ranging from 110 to 300 $\mu\Omega\cdot$cm. The TCR is more negative for films with higher Cr content. TCR control of Ni-Cr thin film resistors can be achieved by using Au, Al and O_2 as dopants.

Ni-Cr resistors are unstable under high humidity; therefore a coating, such as silicon monoxide and polyimide, is often employed to increase the stability in a wide variety of atmospheric conditions. Ni-Cr films were commonly prepared by DC or RF magnetron sputtering or by vacuum evaporation.

12.4.2 *Cu-Ni*

Cu-Ni alloys have resistivities below 50 $\mu\Omega\cdot$cm, making them suitable for low-ohmic applications. In the Cu-Ni alloy system, two compositions, i.e., 34 and 56 at.% Ni, show zero TCR. Thin films of Cu-Ni were generally sputtering deposited onto suitable substrates with a thickness typically smaller than 500 nm.

12.4.3 *Cr*

Chromium films are attractive for resistive films because they have very good adhesive strength to substrate. Cr thin films have higher resistivities than Ni-Cr films; and their TCRs are not as good as Ni-Cr. Cr films can be deposited by either sputtering or evaporation.

12.4.4 *Ta*

Tantalum films were widely used as resistive coatings since 1960s. These films have characteristics of negative temperature coefficient of resistance, wide range of resistivity at low temperatures (10-1000 $\mu\Omega\cdot$cm at 0-30% oxygen contents), high stability with respect to aging and temperature, high dielectric constant (20-40), transparency in the visible and IR range of 400-800 nm, high durability, superior corrosion resistance, and excellent mechanical protection to the underlying layers.

During the 1990s, Ta films were successfully applied in microelectronics applications. Ta thin films were commonly deposited onto insulating substrate by sputtering. Special attention is required during deposition since the sputtered films are highly prone to contamination.

12.4.5 *Ta-Al*

Amorphous Ta-Al film resistors have been achieved at a certain composition ratio of Ta to Al for application to thermal bubble inkjet printheads where high corrosion resistance at high temperature is required. When deposited by sputtering using Ta and Al targets, the Ta-Al films with a composition ratio of about 1:1, exhibiting amorphous-like microstructure with nanocrystalline grains embedded in an amorphous matrix. These as-deposited amorphous-like Ta-Al films have the merits of high resistivity and good thermal stability at annealing temperatures of up to 550°C.

12.4.6 *Tantalum nitride*

Tantalum nitride (Ta_2N or TaN) thin films have some attractive physical properties such as high resistivity and low temperature coefficient of resistance (TCR). These films also have higher stability under various environmental conditions than Ta films. Resistors composed of Ta_2N are more stable than those formed with TaN.

Tantalum nitride films were usually prepared by sputtering during which nitrogen was introduced into the working chamber to achieve Ta-N combination. The deposition conditions, such as powder density forwarded to the target, flow and partial pressure of nitrogen, substrate temperature, and bias voltage, influence the electrical properties of the as-deposited films. If oxygen was also introduced into the sputtering chamber, tantalum oxynitride films can be obtained. These films have higher resistivity, and their TCR values could be tuned by varying the partial pressures of oxygen and nitrogen.

12.4.7 *Cermet*

Cermets are an important class of ceramic-metal composites for thin film resistors. Cr-Si-O is a typical composition. Their thin films were generally deposited by sputtering. Cr-Si-O cermet is comprised of a continuous insulating SiO_2 matrix in which Cr and its silicides-monoxides serve as the conductors and semiconductors, respectively. As-deposited cermets of 80Cr and 20SiO have a resistivity of 550 $\mu\Omega\cdot$cm. A post-annealing at 400°C would decrease the resistance to ~400 $\mu\Omega\cdot$cm. Higher resistivities of 10^3 to 10^5 $\mu\Omega\cdot$cm can be achieved by varying the percentage of SiO and Cr. The thermal stability of the Cr-Si-O cermet material is a concern in some applications.

12.4.8 *RuO₂*

RuO_2 thin films can be deposited using various techniques, such as chemical vapor deposition (CVD), and sputtering (DC or RF). The resistance of the films is temperature dependent. The films show positive TCR in a wide temperature region from 4.2 to 300 K. Thin films of RuO_2 on SiO_2/Si or ceramic alumina substrates with negative TCRs had also been reported by controlling the conditions of sputtering and *in-situ* annealing. The realization of both negative and positive TCRs of the film made it feasible for zero TCR resistor applications.

12.5 Transparent and Conductive Thin Films

A combination of optical transparency and electrical conduction can be achieved in different types of materials. Extremely thin films of metals such as Ag, Au or Cu can be transparent. Their luminous transmittance can be up to ~50% for a single film. Glass with Ag coatings are in widespread use in modern fenestration technology for providing thermal insulation and, in climates requiring space cooling, for diminishing throughputs of near-infrared solar radiation.

A second materials group is found among the wide-bandgap oxide semiconductors, i.e., transparent conducting oxides (TCOs). The first observation appears to have been in Cd oxide (CdO). Electrical conductivity and optical transparency were found to co-exist when a thin film of sputtering deposited Cd metal after incomplete thermal oxidation and post-deposition heat treatment in air. Though the electrical conductivity of this material changed with time, its oxide was indeed a representative of conductor. The reason is that resident oxygen deficiency lends free charge carriers to associated metal defect energy levels near the bottom of the metal-like conduction band of the oxide.

Since the early discovery, appreciable values of electrical conductivity have been reported in many single, binary, ternary, and quaternary metal oxide systems, with indium oxide (In_2O_3) as an excellent example. Intensive research efforts led to films of doped SnO_2 and In_2O_3:Sn (commonly known as indium tin oxide or ITO). A large number of alternative oxides have also been explored, including Cd_2SnO_4, $CdSnO_3$, $CdSb_2O_6$, $CdIn_2O_4$, ZnO, $ZnSnO_3$, Zn_2SnO_4, $Zn_2In_2O_5$, $MgIn_2O_4$, $GaInO_3$ and $In_4Sn_3O_{12}$. While transmission through these materials in the visible region of the spectrum can be quite good, electrical conductivities still remain considerably below those of metals.

12.5.1 *Figure of merit for TCOs*

The measure of the performance of TCOs (or so-called figure of merit) is the ratio of the electrical conductivity σ (measured per ohm per cm) to the visible absorption coefficient α (measured per cm),

$$\sigma/\alpha = -\{R_s \times ln(T + R)\}^{-1}, \qquad (12.2)$$

$$R_s = \rho/t, \qquad (12.3)$$

where R_s is the sheet resistance, ρ is the resistivity in $\Omega\cdot$cm, and t is the TCO film thickness in cm.

The sheet resistance has units of ohms, but is conventionally specified in units of ohms per square (Ω/D). T is the total visible transmission and R is the total visible reflectance. The larger is the value of σ/α the better is the performance of the TCO films.

12.5.2 Tin oxide (SnO₂)

SnO_2 doped with F or Sb is widely used for energy-efficient windows in building applications. $F:SnO_2$ deposited from $SnCl_2$ precursors typically has the cassiterite structure (similar to rutile) with a direct bandgap energy of 4.0 eV and an indirect bandgap of 2.6 eV. Films show a resistivity on the order of 6×10^{-4} $\Omega \cdot$cm, mobility of 20 cm^2/V$\cdot$s, and carrier concentration of 5–8$\times 10^{20}$ cm^{-3}. Sb-doped films show similar properties.

Although the electrical conductivity of SnO_2-based materials is typically not as good as that of ITO films, SnO_2 is inexpensive both in terms of raw materials and processing. It can be easily deposited using chemical methods such as spray pyrolysis from the chlorides or from organometallic precursors. SnO_2 is receiving great attention for photovoltaic applications, especially for heterojunctions with intrinsic thin film cells and related cells such as amorphous or microcrystalline Si.

12.5.3 Indium tin oxide (ITO)

Crystalline In_2O_3 exhibits a bixbyite structure with a unit cell containing 40 atoms and two non-equivalent cation sites. Crystalline ITO (*c*-ITO), amorphous ITO (*a*-ITO), and amorphous IZO (*a*-IZO) are the most important three indium oxide based TCO materials. The most commonly used technique for the deposition of these TCOs is DC or RF magnetron sputtering deposition. The targets are sintered ceramic In_2O_3 containing 3–10 wt.% SnO_2 and 7–10 wt.% ZnO for ITO and IZO, respectively. Crystalline ITO deposited onto heated substrates (250–350 °C) offers relatively low resistivities (1–3$\times 10^{-4}$ $\Omega \cdot$cm). Both *a*-ITO and *a*-IZO films have slightly inferior electrical transport properties compared with *c*-ITO.

If a lower resistivity is demanded, an extremely thin metallic film could be embedded between two layers of oxide to form a sandwiched structure. A natural choice is the multilayered system of ITO/Ag/ITO that combines the electrical conductivity of ITO and Ag with the optical transparency of ITO, with a refractive index of ~2, to boost the transmittance of the Ag. Studies of ITO (40 nm)/Ag (15 nm)/ITO (40 nm) found that the Ag film lowered the overall film resistivity by a factor of ~20, while the optical transmittance exceeded 80% in the full luminous wavelength range.

12.5.4 *Zinc oxide (ZnO)*

In the undoped state, ZnO is highly resistive because its native point defects are not efficient donors. Substitutional doping with Al, In or Ga is commonly used to increase its conductivity. Al-doped ZnO (AZO) has been a research focus. However Al dopant requires a high degree of control over the oxygen potential in the sputtering gas atmosphere due to its high affinity with oxygen. Ga is less reactive and has a higher equilibrium oxidation potential, which makes it a better choice for ZnO doping applications. Furthermore, the slightly smaller bond length of Ga-O (0.192 nm) compared with Zn-O (0.197 nm) minimize ZnO lattice deformation at high substitutional concentrations.

The electrical properties of doped ZnO films strongly depend on the deposition methods and conditions. Al and Ga-doped ZnO (AZO or GZO) films with a resistivity on the order of 10^{-4} Ω·cm have been prepared by vacuum arc plasma evaporation (VAPE), metal organic molecular beam deposition (MOMBD), metal organic chemical vapour deposition (MOCVD), and magnetron sputtering. AZO films with a lower resistivity of the order of 10^{-5} Ω·cm have been prepared by PLD.

At present, AZO and GZO films with a resistivity of $2-3\times10^{-4}$ Ω·cm, a refractive index of approximately 2.0, and an average transmittance above 85% in the visible range can be obtained on large area substrates with a high deposition rate at a temperature above ~200°C. These films have transparent electrode properties comparable to those of ITO films. They also offer significant benefits of low cost relative to In-based systems and high chemical and thermal stability. However Zn is more chemically active in an oxidizing atmosphere than Sn and In, the activity and amount of oxygen must be precisely controlled during the deposition.

12.5.5 *Application*

As far as applications are concerned, TCOs are being used extensively in the window layers of solar cells, as front electrodes in flat panel displays (FPD), low emissivity windows, electromagnetic shielding of cathode ray tubes in video display terminals, as electrochromic (EC) materials in rear-view mirrors of automobiles, EC-windows for privacy (so-called smart windows), oven windows, touch-sensitive control panels, defrosting windows in refrigerators and airplanes, invisible security circuits, gas sensors, biosensors, organic light emitting diodes (OLED), polymer light emitting diodes (PLED), antistatic coatings, cold heat mirrors, etc. Some new applications of TCOs have been proposed recently such as holographic recording media, high-refractive index waveguide overlays for sensors and telecommunication applications, write-once read-many-times memory chips (WORM), electronic ink and etc.

12.6 Thin Film Magnetic Materials

12.6.1 *Soft magnetic thin films*

Magnetic recording heads are the key parts for achieving high-density magnetic recording. Soft magnetic thin films are commonly used for write head core materials. The basic requirements for a soft magnetic material suitable for application in thin film recording heads include:

1. High magnetic saturation ($B_S \gg 1\text{T}$);
2. Low coercivity for low hysteresis loss ($H_c < 80$ A/m);
3. Optimal anisotropy field (H_k) for high permeability (μ);
4. Zero or near zero magnetostriction (λ) for reduced domain noise and head instability;
5. High-electrical resistivity (ρ) for low eddy current loss;
6. Well-defined uniaxial anisotropy for domain structure control; and
7. Good thermal stability, excellent corrosion and wear resistance.

Soft magnetic thin film materials can be roughly divided into three main categories: crystalline, micro-/nano-crystalline and amorphous.

12.6.1.1 *Crystalline soft magnetic films*

Ni-Fe alloy (Permalloy) is one of the typical crystalline alloys. Ni80Fe20 alloy has been used for thin film inductive head since 1979. However, as the coercivity (H_c) of magnetic recording media greatly increased, soft magnetic thin films with high-saturation magnetic flux density (B_s) were desired for writing magnetic signals on the recording media with high H_c. Therefore, various soft magnetic films with high B_s were investigated. Fe-Al-Si (Sendust) and Fe-Ga-Si-Ru alloy films exhibit higher saturation flux densities around 1.3 T, and are applied for metal-in-gap (MIG) type VCR heads. In these alloy films, good soft magnetic properties are realized by annealing at a temperature above 500°C.

12.6.1.2 *Micro-/nano-crystalline soft magnetic films*

Micro- or nano-crystallization could be used to improve the saturation flux density of soft magnetic films. In these films, magnetocrystalline anisotropy is reduced by controlling the microscopic crystalline structure. Various films prepared by electrodeposition in solutions had been reported to possess high B_s values, such as CoFe with B_s = 1.8–1.9 T, CoNiFe with B_s = 1.6–1.8 T. Sputtering-deposited nano-crystalline films, such as FeN-based films and NiFe/CoFeN/NiFe multi-layered films exhibited very high values of B_s at around 1.4-1.9 T and 2.4 T, respectively.

Electrochemical deposition is a preferred technique for the deposition of CoNiFe soft magnetic films with a nano-crystalline structure due to its advantages of simplicity and cost-effectiveness. Electrodeposition with conventional chemical baths could provide films possessing B_s greater than 1.6 T and H_c larger than 400 A/m. Co65Ni12Fe23 is a typical example of soft magnetic films successfully obtained with B_s as high as 2 T. This thin film is composed of very fine crystal grains with diameters typically ranging from 10 to 20 nm. Applying this material to the write core of the magnetic recording head in hard disk drives, a superior write performance was obtained compared with normal magnetic recording heads with a conventional NiFe soft magnetic film.

A distinct disadvantage associated with electrodeposited films is that they are more susceptible to corrosion than the films prepared by physical vapour deposition (such as sputtering) due to the presence of organic additives (e.g. saccharin or thiourea) in the chemical bath. The films are anodically active; as such the formation of a passive film on the surface is relatively difficult.

Nano-crystalline films of soft magnetic iron-based alloys are normally design with the composition of Fe-M-X, where M is a transition metal from group III-V of the periodic table (typically Zr, Hf, Nb and Al) and X is C, N or O. These soft magnetic films were prepared by using (reactive) sputtering method. The fine nanostructure can be obtained by crystallization of amorphous as-deposited alloy films or by *in-situ* direct growth in sputtering process. These materials normally possess high thermal stability, improved corrosion resistance and high resistivity and have been proven to satisfy various requirements as a potential candidate for a thin film head material. In addition, they can also be used as thin film inductors and flux gate magnetic sensors.

12.6.1.3 *Amorphous soft magnetic films*

Soft magnetic amorphous thin films with excellent magnetic, magnetotransport and magnetoelastic properties can find wide applications in sensors and transducers. Co-based binary and/or ternary (e.g. Co-Fe-B, Co-Ta-Zr and Co-Nb-Zr) amorphous films, prepared by sputtering or laser ablation, have better soft magnetic properties than Permalloy and Sendust based films. Anisotropy control can be achieved by annealing in a magnetic field. One disadvantage of Co-based amorphous alloys however is the reduction of the crystallization temperature in the high Co content region. Amorphous alloys with compositions modulated by nitride layers and non-nitride layers show remarkable thermal stability. Their soft magnetic properties are kept after an annealing at a temperature above 700°C. This thermal stability improvement is believed to be caused by the phase transformation to the micro-crystalline state.

12.6.2 *Hard magnetic thin films*

Hard magnetic films have a broad application area ranging from magnetic recording media to magnetic micro-electromechanical systems (MEMS) including magnetic sensors and actuators. Promising candidates in these fields are rare-earth transition metal compounds such as NdFeB and SmCo. A large variety of techniques have been developed to prepare hard magnetic thin films, such as evaporation, sputtering, MBE, CVD, electroless deposition, and electrodeposition.

NdFeB combines a high saturation magnetisation (~1.6 T) with an anisotropy field of ~6.8 T at moderate material cost. On heated substrates NdFeB can be grown with a c-axis texture, resulting in a perpendicular magnetic anisotropy with respect to the film plane. Using sputtering deposition, textured films with coercivities up to 0.7 T and isotropic films with coercivity up to 1.6 T had been reported. By choosing an appropriate buffer layer, substrate temperature, and film thickness in PLD process, $Nd_2Fe_{14}B$ films with a very fine grain size, a pronounced perpendicular magnetic anisotropy, and a coercivity up to 1.5 T could be prepared. A further modification of the film morphology could increase the coercivity to ~2 T.

SmCo-based films are the choice when large in-plane coercivity and remanence are required, due to the large in-plane macroscopic anisotropies when the films are deposited at temperatures of ~400°C. The highest anisotropy field known is for $SmCo_5$ ($\mu_0H_A \approx 35$ T, $\mu_0M_s = 1.14$ T); the highest saturation magnetisation is found for Sm_2Co_{17} ($\mu_0H_A \approx 10$ T, $\mu_0M_s = 1.25$ T). Thick SmCo films with a Sm content ranging from 11–19 at.% had been found to exhibit excellent magnetic properties. When the film is sputter deposited onto a heated substrate, the hard magnetic phase can form directly upon deposition; and the films possess a magnetic in-plane texture. The c-axis is hereby randomly oriented in the film plane. For thick films (1–2 μm), this texture is observed on several different substrates; and the degree of texture improves when Sm content increases. Highly in-plane textured SmCo films have been prepared by magnetron sputtering and PLD.

Co-rich alloys containing As, Bi, Cr, Cu, Fe, Mn, Mo, Ni, P, Pd, Pt, Sb and W are also promising candidates of hard magnetic materials. Their thin films have been deposited electrochemically. The alloying elements tend to concentrate at the grain boundaries, leading to the formation of a structure consisting of isolated magnetic Co grains surrounded by non-magnetic or weakly magnetic boundaries. This microstructural feature provides an increased energy barrier for magnetic re-alignment of the domains, and thereby increases the overall coercivity (H_c) of the films, making them magnetically hard.

Promising hard magnetic thin film materials also include CoPt and FePt due to their high magnetocrystalline anisotropy and magnetic saturation. Specifically, $L1_0$ ordered phase materials (Co50Pt50 and Fe50Pt50) show extremely high coercivity. Most CoPt and FePt magnetic thin films were prepared using sputtering or MBE. They were deposited as multilayered structures and then annealed to produce ordered phases.

12.6.3 *Giant magnetoresistive film*

Giant magnetoresistance (GMR) is based on spin-dependent scattering at both the interfaces and within the individual layers. A number of factors are crucial in metallic ferromagnets; but all of them derive from the spin-split band structure in a ferromagnetic material. GMR effect was discovered in the exchange coupled in Fe/Cr/Fe trilayers by Grünberg and in Fe/Cr multilayers by Fert in 1988. The GMR multilayer shows a high resistance at the remanent state due to the anti-parallel alignment of the magnetization of the adjacent layers. When a sufficiently strong magnetic field is applied in the plane of the magnetic layer, the magnetization of the multilayer will be aligned parallel to the external field, leading to a lower magnetic scattering and then a low resistance state. In general, the electrical resistance decrease can be typically around 10–80%.

12.6.3.1 *Granular GMR*

GMR effect had been observed in thin films composed of solid precipitates of a magnetic material in a non-magnetic matrix. Such a material system requires that the matrix metal and the precipitate metal are immiscible. Typical examples are Fe-Ag, Co-Cu, and Co-Ag in which Cu or Ag matrices are containing Fe or Co granules. Granular films can be prepared by evaporation, sputtering, or ion implantation. The GMR effect is varying with the concentration, size, and morphology of magnetic granules which further depend on the deposition and post-annealing conditions.

12.6.3.2 *Multilayer GMR*

GMR has been observed in many multilayered structures of the form FM/NM. FM represents a transition-metal ferromagnetic layer (Fe, Co, Ni or their alloys) and NM is a non ferromagnetic transition metal (Cr, Cu, Ag, Au etc.). Two or more ferromagnetic layers are separated by a very thin (~1 nm) non-ferromagnetic spacer. The simplest system for examining GMR may consist of a thin layer of nonmagnetic Cr sandwiched between two layers of magnetic Fe, i.e., Fe/Cr/Fe. Multilayer stacks of 10 or more layers may also be established.

To observe GMR effect in multilayered system, some requirements need to be met:

1. The magnetic moments of the adjacent magnetic layers can be re-oriented upon the application of magnetic field;
2. The thickness of a single layer must be much smaller than that of the average free path of electrons; and
3. The scattering rates of the up-spin and down-spin electrons in ferromagnetic metals are sufficiently different.

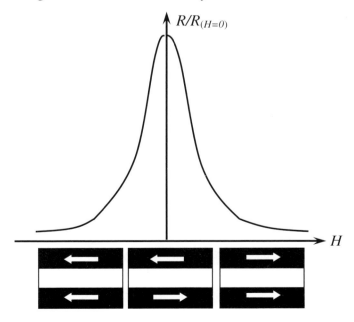

Fig. 12.6 Schematic of GRM effect in multilayer.

GMR multilayers have been mainly produced by physical vapour deposition techniques, such as electron-beam evaporation, laser ablation, and sputtering.

12.6.3.3 *Application*

GMR can be used in the areas of sensing where size, speed and sensitivity are important parameters. Typical examples may include nanoscale arrays of GMR sensors for 100 µm scale spatially resolved eddy current detection, biological sensors for molecule tagging, galvanic isolators, magnetic sensors for traffic control and engine management systems, magnetic separation and electronic compasses. GMR is utilized in hard drive read heads through the integration of a spin valve. There has been considerable effort for developing a magnetic random access memory (MRAM) which would have the advantages of non-volatility, radiation hardness and low energy consumption.

Summary:

Thin film electronic materials with tailored composition, structure, property and performance are the platform for the development of microelectronic, optoelectronic, magnetic, mechanical, sensing and/or photovoltaic devices. Many techniques have been developed to prepare thin films with unique functions and applications. Obviously, thin film technology and thin film material have become to be an international research frontier with the focus on: (1) advanced thin film materials (including conducting, semiconducting and superconducting films, magnetic (particularly giant magnetoresistive) films, and light emitting films); (2) mechanisms concerning nucleation and growth of thin films; (3) advanced technologies for thin film preparation; (4) new techniques for analysis, evaluation and monitoring of thin films.

In this chapter, an overview of the typical techniques for the formation of thin films was given. Then the characteristics and applications of metal and/or oxide based conducting and magnetic thin film systems were briefly introduced.

Important Concepts:

Thin film: a low-dimensional material which is prepared by using elemental materials or inorganic/organic compounds. This layer is normally supported by a suitable substrate of metal, ceramic, semiconductor or polymer. Its thickness is typically less than several micrometers.

Chemical vapour deposition (CVD): a process involves the formation of a thin solid film on the heated area of a substrate situated in a reactor by chemical reactions of volatile precursors containing the desired species

Physical vapour deposition (PVD): a process involves creation of vapours (using evaporation, sputtering or laser ablation) and their subsequent condensation onto a substrate to form a film.

Sputtering deposition: this is the deposition of species vaporized from the surface of a certain source material (i.e., target, can be element, alloy, compound, or their mixture) by the sputtering process.

Molecular beam epitaxy (MBE): a process commonly used to develop epitaxial films with atomic thicknesses using atomic and molecular beams produced from Knudsen cells under high or ultrahigh vacuum conditions.

Pulsed laser deposition (PLD): a process uses a focused high-power pulsed laser beam to generate a plume containing many energetic species (including atoms, molecules, electrons, ions, clusters, particulates and molten globules) that condense onto a suitable substrate as a thin film.

Transparent and conductive films: films that combine optical transparency and electrical conduction.

Giant magnetoresistance effect: this effect is observed in artificial multilayered structures composed of alternating ferromagnetic and nonmagnetic thin films. In the presence of an external magnetic field, the electrical resistance of the structure significantly decreases (typically 10–80%).

Questions:

1. Compare the major differences between chemical vapour deposition and physical vapour deposition techniques in terms of operation principle, processing condition, system, precursor, cost, etc.
2. Discuss the advantage and disadvantage of using an additional magnetic field in sputtering deposition.
3. RF sputtering is normally used for deposition of semiconducting or insulating films. However this method has relatively low deposition rate, please discuss the potential ways that can be used to enhance the deposition rate of high quality films.
4. What are the fundamental requirements for a thin film conducting material? Please list the most commonly used thin film conducting materials.
5. Discuss the characteristics of major thin film resistors.
6. How to evaluate the performance a transparent and conducting thin film?

Chapter 13

Organic Electronic Materials

13.1 Introduction

Organic electronics has been an active field of study for recent decades. The research activity is extremely broad and the topics span many areas, including materials synthesis and characterization, deposition technology, device design and construction, performance evaluation and optimization, and computational modeling. Many discoveries and accomplishments have been reported in the last few years. The impetus behind this is that it is possible to enable new applications by circumventing some of the limitations of conventional inorganic electronic materials, and to achieve overall comparable device performance but at considerably reduced cost.

The electronic devices and products fabricated with organic materials (insulating, conducting, semiconducting or superconducting) cover a large area including thin film transistors, memories, photovoltaics, light emitting diodes, electrophoretic display pixels, micro-fluidic channels, etc. Some of them had been brought into the market by big companies such as Motorola, Philips and Pioneer. Integrated circuits for smart identification tags, cards, highly-valued electronic papers and active-matrix display are focuses of current research.

13.2 Conducting Polymers

13.2.1 *Introduction*

It is generally accepted that organic compounds and polymers will not conduct electricity well. And the truth is that most of them are applied as excellent electrical insulators. Discovery of conducting polymers obviously established a new field of polymer research; and conducting polymers as functionalized materials hold a special and an important position in materials science.

As the name indicated, conducting organic materials are organic compounds that can conduct electricity. These conducting materials can be metallic conductors or semiconductors and their typical electrical conductivity can range from 10^{-16} to 10^5 S/cm. The electrical conduction can be achieved through different mechanisms. Conducting polymers, thus, could be categorized into two large groups according to the conducting mechanisms: (1) conducting polymeric composites and (2) intrinsically conducting polymers. Conducting polymer composites utilize insulating polymers as the matrix in which conducting materials (particles, fibers or whiskers) are added in to achieve electrical conduction. The most commonly used conducting materials are metallic powder, carbon black, and carbide (such as nickel carbide and tungsten carbide).

On the other hand, the electrical conduction in conducting polymers is linked to the chemical structure of the polymer. The conductivity is highly flexible and can be tuned from insulating state to superconducting state by chemical modification or by changing the degree and nature of doping. Polymers also offer the advantages of light-weight, flexibility, high corrosion resistance, chemical inertness and the ease of processing. Conducting polymers therefore are promising candidates for various applications such as sensors, battery electrodes, screening materials, displays and solar cells.

13.2.2 *Discovery of conducting polymers*

Research on conducting polymers intensified after the discovery of poly (sulphur nitride) $[(SN)_x]$ in 1975, which possesses metallic conductivity and becomes superconductor at 0.29 K. Conducting polymer complexes, in the form of tetracyano and tetraoxalato-platinates and the Krogman salts charge transfer complexes, had been known earlier. However, the idea of using polymers for their electrical conducting properties actually emerged in 1977 when Heeger, MacDiarmid and Shirakawa carried out their prize winning work with π-conjugated polyacetylene (PA), a polymer that exists as a black powder. They enhanced the electrical conductivity of PA by several orders of magnitude by doping with oxidizing agents, e.g. I_2, AsF_5 and $NOPF_6$ (*p*-doping) or reducing agents (*n*-doping), e.g., sodium napthalide. They were awarded the Nobel Prize in Chemistry for 2000 jointly "for the discovery and development of conductive polymers".

Since discovery of conductive PA, other π-conjugated polymers, such as polypyrrole (PPy), polyaniline (PANI), polythiophenes (PTH), poly(p-phenylene) (PPP), poly(p-phenylenevinylene) (PPV), and poly(2,5-thienylenevinylene) (PTV) have been reported as conducting polymers.

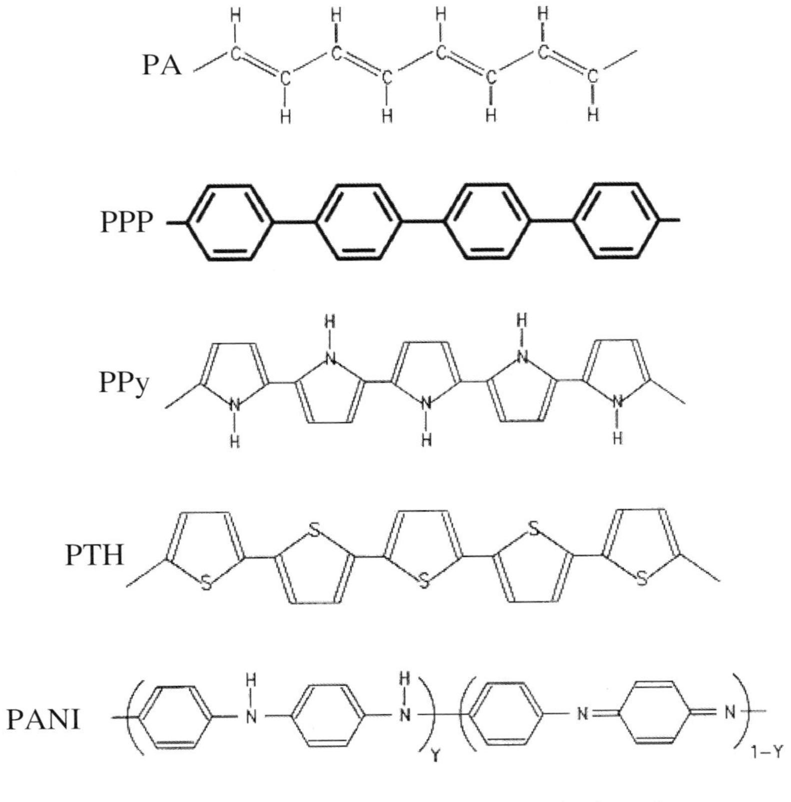

Fig. 13.1 Chemical structures of some typical conducting polymers.

13.2.3 *Conducting mechanism of conjugated polymers*

The mechanism of conduction in polymers is complex and not well understood at this moment. However, it was believed that the electronic properties are resulted from various types of charge carriers, which are entirely different from that of metallic conductors and inorganic semiconductors. Conjugated polymers have σ-bonds and π-bonds. Usually, the σ-bonds are fixed and immobile due to the covalent bonds between the carbon atoms. On the other hand, the π-electrons are relatively localized and responsible for the unusual electronic properties. This extended π-conjugated system of the conducting polymers have single- and double-bonded sp^2 hybridized atoms along the polymer chain. The planar conformation of the alternating double-bond system, which maximizes sideway overlap between the π molecular orbitals, is critical for conductivity.

The concepts of solitons, polarons, and bipolarons have been used to explain the electronic phenomena in conjugated polymer systems. Upon doping (oxidizing or reducing a neutral polymer and providing a counter anion or cation), charge carriers can be introduced into the polymer. The polymer chain is distorted when electrons or holes are injected at the electrodes, and the charge carriers couple with the polymer chain distortion to form a mobile polaron (radical ions) or soliton. When the density of injected electrons or holes increases, the coupling of charge carriers with the polymer chain gives rise to the formation of bipolaron (dications or dianions). Transportation of charge may involve scattering and trapping processes, recombination, tunneling, and hopping.

Conductivity can be augmented generally by increasing the doping percentage and changing the dopant. In essence, conductivity in conducting polymers is governed by a variety of factors including polaron length, the conjugation length, the overall chain length and by the charge transfer to adjacent molecules. These are explained by a large number of models based on intersoliton hopping, hopping between localized states assisted by lattice vibrations, intra-chain hopping of bipolarons, variable range hopping in three dimensions, small polaron transport, solitonic transport, and charging energy limited tunneling between conducting domains.

13.2.4 *Synthesis of conducting polymers*

Conducting polymers can be prepared by several techniques based on chemical and electrochemical principles. Typical chemical synthesis includes addition and condensation polymerization. These chemical approaches can be used to produce powders or thick films of ordered structures with high yield.

Electrochemical synthesis can only be used for those systems in which the monomer can be oxidized in the presence of a potential to form reactive radical ion intermediates for polymerization. Potential control is critical for the production of high-quality materials. Electrochemical polymerization of conducting polymers is generally achieved by using: (1) constant current or galvanostatic; (2) constant potential or potentiostatic; and (3) potential scanning/cycling methods in a standard three-electrode cell. The most commonly used anodes are gold (Au), chromium (Cr), nickel (Ni), palladium (Pd), platinum (Pt), titanium (Ti) and conducting (indium tin oxide coated) glass slides. Semiconductors such as *n*-type silicon, gallium arsenide (GaAs), cadmium sulphide (CdS), and graphite have also been used.

Electrochemical technique can easily change the oxidation state of the polymer by potential cycling between the oxidized, conducting state and the neutral, insulating state, or by using suitable redox compounds. It can also be used to prepare free standing, homogeneous and self-doped films. The surface area and film thickness of a conducting polymer can be more easily controlled through electrochemical synthesis. Actually it is the preferable synthetic approach for fabrication of thin and/or ultrathin functional films that can be used as polymeric electrodes and sensors. It is also capable of producing copolymers and graft copolymers. Polyazulene, polythiophene, polyaniline, and polycarbazole have been synthesized using this approach. Moreover, electrochemical synthesis has the obvious advantages of simplicity, reproducibility, applicable to a variety of substrates, and ease of processing control. Nowadays, electrochemical methods are favored for the synthesis of conducting polymers.

Photo-polymerization and plasma polymerization have also been developed. For example, polypyrrole (PPy) has been synthesized from pyrrole using tris (2,2'-bipyridyl ruthenium) complex as a photosynthesizer. This technique can control the polymerization reaction on any desired surfaces. Plasma-assisted polymerization makes use of molecules occurring in various plasma environments. This technique can be used to prepare very thin but uniform polymeric films that are highly cross-linked and strongly adhere to a desired substrate.

13.2.5 *Characterization of conducting polymers*

A variety of techniques are available for compositional, optical, structural and functional characterization of conducting polymers.

13.2.5.1 *Charge transfer mechanism*

Cyclic voltammetry (CV) provides basic information on the oxidation potential of the monomers, film growth, the redox behavior of the polymer and the surface concentration. Cyclic voltammograms are very useful for analysis of charge transfer rate, charge transport process and for unveiling the complex interactions that occur within polymer segments, at specific sites, and between ions and solvent molecules.

Chronoamperometry is commonly used to determine the charge transport diffusion coefficient, and also to study phase formation, phase transitions and relaxation. It can determine the total charge consumed. Electrochemical impedance spectroscopy (EIS) represents another powerful tool for investigating charge transfer rate and charge transport process in films and membranes.

13.2.5.2 *Chemical species analysis*

Spectroscopic techniques, including UV-Vis, Fourier Transform Infrared (FTIR) and Raman spectroscopy, have been routinely combined with electrochemical methods to monitor the chemical changes occurring in the surface film. Electron spin resonance spectroscopy (ESR) has been widely used to investigate the nature of charged defects formed upon doping of conducting polymers. It is particularly capable of characterizing radicals and radical cations produced by oxidation or reduction during electropolymerization. ESR parameters, including resonance intensity, asymmetry, line-width, spin-lattice relaxation time and g-value, give important information about doping level and spin nature of defects. Investigation using ESR together with other electrochemical techniques would be much helpful for elucidating structure features and reaction mechanisms.

13.2.5.3 *Surface morphology*

Scanning electron microscopy (SEM) is the most frequently used technique for acquisition of high resolution images of the surface morphology of polymers. Meanwhile, three-dimensional, atomic-scale images can be obtained using scanning probe microscopy (scanning tunneling microscope (STM) and/or atomic force microscope (AFM)). STM can be applied even under wet conditions, i.e., both the probe tip and the sample are immersed in the electrolyte.

13.2.5.4 *Composition*

Non-destructive X-ray photoelectron spectroscopy (XPS) is frequently used to obtain atomic information on the surface composition of conducting polymers.

13.2.5.5 *Crystal structure*

The crystallinity of polymers can be investigated by X-ray diffraction (XRD), X-ray absorption near edge structure (XANES), and extended X-ray absorption fine structure (EXAFS) techniques.

13.3 Semiconducting Organic Materials

13.3.1 *Energy band model*

The molecules of an organic semiconductor have many discrete energy levels (i.e., molecular orbitals) which are formed as a result of the combination of a large number of atomic orbitals. Some of the molecular orbitals are filled with electrons while others are empty. The splitting of interacting molecular orbitals leads to the formation of a spatially delocalized band-like electronic structure. This creates two states: the highest occupied molecular orbital (HOMO) and the lowest unoccupied molecular orbital (LUMO). These terms thus replace the concepts of valence and conduction bands in inorganic semiconductors. There is also a finite energy gap between the HOMO and LUMO states. It is this gap that makes most organic materials insulating or semiconducting. In conjugated polymers system, the HOMO-LUMO bandgap is mainly determined by the degree of dimerization and the recombination length of carriers.

13.3.2 *Charge carriers*

In inorganic crystalline semiconductors, *n*-type and *p*-type refer to the type of dopant, and therefore the majority carrier, in a semiconductor. Both holes and electrons can usually be transported reasonably well. Charge carriers in organic semiconductors are more complex. A number of charge carriers can be found in organic semiconductors, typically including exciton, negative/positive polaron, negative/positive bipolaron and negative/positive soliton.

Charge sustained on the unsatisfied bonds can be moved on the molecule by an electric field. This phenomenon is responsible for the generation of charge states in organic semiconductors. A similar situation can be generated by doping processes which might oxidize or reduce one of the bonds along the chain, creating electron donor or acceptor. Addition of charge creates new and unfilled electronic energy states that lie within the original HOMO-LUMO energy gap. The charges can be transported via polarons (a species comprises of both the electron or hole and the structural deformation of the lattice around it). Polarons can become combined in more complicated structures with greater charge or carriers such as excitons (where positively and negatively charged polarons attract each other and form an electrically neutral energy carrier) and bipolarons (a stable combination of two nearby polarons with a lower energy state).

In general, organic semiconductors support either positive or negative charge carriers. It is thus common in the literature to refer to hole transporting organic semiconductors as *p*-type and electron transporting materials as *n*-type. Recent research however revealed that certain organic semiconductors do not have any particular properties favoring electrons or holes.

13.3.3 *n-type organic semiconductors*

To carry electrons easily the conjugated carbon backbone needs to be electron deficient, i.e., has a high electron affinity. Typical examples of *n*-type organic semiconducting materials in early research are naphthalene tetracarboxylic dianhydride (NTCDA), perylene and its derivatives such as perylene tetracarboxylic dianhydride (PTCDA), perylene tetracarboxylic diimide (PTCDI) and derivatives. In general, these materials exhibit relatively low mobility.

Recently it had been reported that an *n*-type organic material, 3',4'-dibutyl-5, 5''-bis(dicyanomethylene) -5, 5''-dihydro-2,2':5', 2''-terthiophene (DCMT), has a mobility as high as 0.2 cm^2/V·s. Polyacetylene (PA) has been deliberately *n*-doped via ion implantation of Li$^+$ at low levels (10^{17}/cm^3). Problem associated with this material is that these doping ions are mobile when an external field is applied, so the doped organic material is not compatible with device operation.

By adding strong electron-withdrawing groups such as -F, -CN and -Cl to the outer rings of molecules, good candidates for *n*-type semiconductors may be created. Such an example was shown as F$_{16}$CuPc (copper hexadecafluorophthalocyanine). The hexadecahalogenated metallophthalocyanines were found to function as *n*-type semiconductors with an electron mobility of 0.03 cm^2/V·s and excellent air-stability. Aromatic perfluorocarbons such as perfluoro-*p*-sexiphenyl (C$_{36}$F$_{26}$) and perfluoropentacene (C$_{22}$F$_{14}$) can also be efficient *n*-type semiconductors.

In general, the development of high performance *n*-type organic semiconductors is relatively slow, because of (1) they have low stability and high sensitivity to oxygen and moisture due to the organic anions, in particular carbanions, which react with oxygen and water under operating conditions; and (2) their rapid degradation in mobility and on/off ratio, although improved oxidative stability and mounting mobilities had been reported. Stable *n*-type organic semiconductors with high carrier mobility are still a very important challenge for materials and chemistry scientists.

13.3.4 *p-type organic semiconductor*

Most organic semiconductors tend to transport holes better than electrons. The *p*-type organic semiconductors mainly consist of oligomers, pentacene and phthalocyanine.

13.3.4.1 *Pentacene*

Pentacene is a polycyclic aromatic compound with five condensed benzene rings and has been widely studied as a *p*-type semiconductor for organic field-effect transistors (OFETs) primarily due to routinely obtainable high thin film transistor hole mobility (>1 cm^2/V·s). Vapor deposited pentacene TFTs claimed exceptional mobilities ranging from 2–5 cm^2/V·s, through the use of special polymeric surface treatment. The high mobility of pentacene is a direct result of significant orbital overlap from edge-to-face interactions among the molecules in their crystal lattice.

Major problems for processing of pentacene are its instability in air and low solubility in organic solvents. Due to its insolubility, pentacene can only be deposited using vacuum processes. Oxidation disrupts transport and crystallization in devices. In addition, pentacene can condense into two closely related but not perfectly matched crystal phases, leading to polymorphic crystal growth and then mismatched grains and decreased device performance.

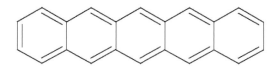

Fig. 13.2 The structure of pentacene.

Much work was then focused on the development of soluble pentacene precursors in which pentacene can be generated *in-situ*. The precursor was cast into thin films from a solution of methylene chloride or toluene. The resulting film can be converted into well packed pentacene films on exposure to heat (~140°C) through the retro Diels

Alder reaction. The conversion process can also be engineered to be photo-patternable. It appeared that the resultant TFT mobility is typically around 0.1–0.2 cm²/V·s, lower than through vapor deposition routes. However, enhanced mobility could be achieved by chemical modification of substrate and by optimization of the processing and conversion conditions of the precursor.

Another strategy is to develop functionalized pentacene. One approach is to synthesize heterocyclic analogs fused with thiophene rings. The thiophene rings allow greater intra- and inter-molecular overlap between the π-conjugated units, and also provide sites for attachment of solubilizing alkyl chains. Attachment of bulky groups to pentacene passivates the most reactive sites of the molecule, therefore constraining the crystallization into a single highly favorable phase, and imparting solubility. One such example is the development of functionalized pentacene, such as TIPS pentacene, which is light enough to be purified and deposited using vacuum sublimation.

13.3.4.2 *Oligomers*

Oligomers consisting of conjugated oligothiophene and polymers are promising charge transport semiconductors. Conjugated oligothiophenes have good solubility and can be easily purified. They form polycrystalline thin films with some *p*-orbital overlap parallel to the substrate. Several end group substitutions, such as long alkyl *n*-hexyl chains, have been developed to improve the film self-organization (ordering and stacking).

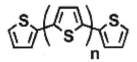

Fig. 13.3 The structure of oligothiophene.

A carrier mobility of 1.1 cm²/V·s had been reported for alkyl-substituted oligothiophene. The charge carrier mobility can be improved by using high temperatures for film deposition.

13.3.4.3 *Phthalocyanines*

The semiconducting properties of phthalocyanines (Pcs) were first observed in 1948. Cu-phthalocyanine was reported to have a mobility of 0.02 cm^2/V·s in a *p*-channel FET. Some fused aromatic compounds, such as dihydrodiazapentacene, bisdithienothiophene, diphenylbenzo-dichalcogenophenes and dibenzothienobisbenzodithiophene, have been successfully synthesized with high mobilities of 0.006, 0.05, 0.17 and 0.2 cm^2/V·s, respectively.

Phthalocyanines exhibit excellent thermal and chemical stability; and devices derived based on them are stable in air for months. They also show excellent photoelectric characteristics, therefore have been widely used as solar cells, optical limiters and photoconductors.

13.3.5 *Ambipolar organic semiconductor*

Some organic semiconductors can be both *p*-type and *n*-type and do not have any particular properties favoring electrons or holes. This is known as ambipolar behavior. When incorporated in a transistor design, the charge carrier type will be determined at the interface between the semiconductor and the metal electrode injecting the charge.

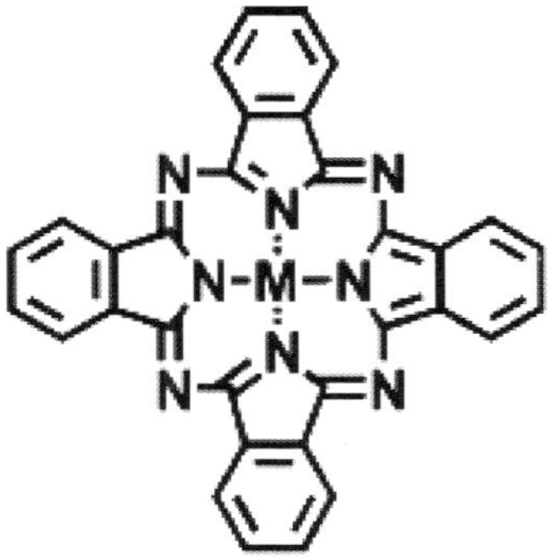

Fig. 13.4 The structure of metal phthalocyanines.

For example, aluminum chlorophthalocyanine (AlPcCl) is usually regarded as a *p*-type semiconductor. However, after AlPcCl deposited on a gate insulator with thin gold electrodes was subject to annealing at 150 °C, a layer of electric dipoles was formed at the interface between the gold and AlPcCl. This changed the alignment of electronic states so that the sample switched to *n*-type behavior. Pristine pentacene could also have ambipolar behavior.

It should however be indicated that this concept is somewhat controversial. Some researchers believe that an ambipolar semiconductor that conducts holes for negative gate voltages and electrons for positive gate voltages will only turn off for a very limited voltage range, or for no voltage at all. This seems detrimental for any practical complementary circuits. While other researchers accept that simultaneous injection of electrons and holes is an advantage of ambipolar organic semiconductors. A change in carrier type achieved by interface modification in a single semiconductor will be a useful technology.

13.3.6 *Applications*

13.3.6.1 *Diodes*

Rectifying junctions based on organic semiconductors are using a heterojunction between an inorganic semiconductor and an organic semiconductor or a Schottky junction between organic semiconductor and metal. *p*-type Si, *n*-type Si and organic semiconductors have been used to achieve stable junctions. Schottky junctions have been fabricated using PPy, PAni, polyacetylene, poly(alkylthiophene) and poly(3-alkylthiophene) (P3AT).

13.3.6.2 *Field effect transistors*

A typical organic field effect transistor (FET) consists of an organic semiconductor deposited onto a dielectric (e.g. SiO_2) and use an underlying gate electrode for conductivity control (Fig. 5). This device operates in an enhancement mode, i.e., the conductivity is increased by the applied bias in the device.

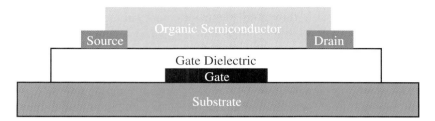

Fig. 13.5 Schematic of organic thin film field effect transistor.

One of the important parameters determining the performance of these devices is the mobility. The charge carrier mobilities in these FETs are found to be lower than that of the inorganic semiconductor devices. The primary reason for the underperformance of the polymers is the lack of well-organized crystals. Consequently, carrier mobilities remain very low and limited by inter-chain hopping rather than intra-chain transport. To achieve high carrier mobilities, highly organized molecular or polymeric layers are necessary as chain-to-chain or molecule-to-molecule charge transfer generally limits the mobility. Higher levels of crystallinity assist such transfers. Carrier mobilities in organic TFTs exceeding 2 cm^2/V·s with on/off current ratios in excess of 10^8 have been achieved.

Semiconducting polymer based FETs are much cheaper than Si based devices; and the possibility of making flexible and flat panels open a new area of large-area cost-effective plastic electronics if the slow response and limited lifetime can be significantly improved.

13.3.6.3 *Light emitting diodes*

Electroluminescence (EL) in organic semiconductor was discovered in 1990 in PPV, which has a bright yellow colour owing to the onset of absorption around 517 nm, corresponding to a HOMO-LUMO energy gap of 2.4 eV. Electroluminescence has also been reported from poly(pphenylene), polyfluorene (PFO, blue light: 380–420 nm) and P3AT.

The emission wavelength of organic semiconductors can be tuned in whole visible region through chemical modification. For example, copolymer engineering can be used to tune the emission wavelength in PFO-related materials. Copolymerization with PPV-like or benzothiadiazole (BT) units has led to red and green emissions, respectively. This makes them excellent candidates as luminescent materials for replacing inorganic semiconductors. Other advantages may include their low operating voltage, low materials cost, high flexibility, easy processing, and possibility of fabricating large area display devices.

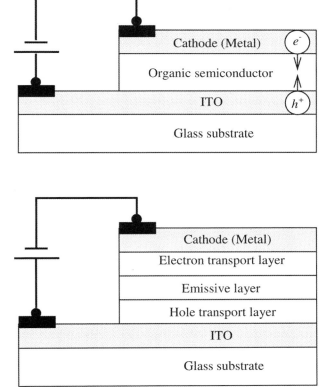

Fig. 13.6 Schematics diagram of organic light-emitting devices.

Organic light emitting diodes (OLEDs) are thin film multilayer devices in which active charge transport (and/or light emitting) materials are sandwiched between two thin film electrodes, of which at least one must be transparent to light. Generally high work function (~4.8 eV), low sheet resistant (~20 Ω/□), and optically transparent indium tin oxide (ITO) is used as an anode, while the cathode is a low work function metal such as Al, Ca, Mg or their alloys. The active layers consist of one or more films. If a single layer of film is used, the film must be able to transport both holes and electrons, accept both efficiently from the contacts, and emit light by carrier recombination. This has been proven to be difficult for a single material.

When a bias is applied, the electric field across the polymer layer(s) is increased. Oppositely charged carriers (electrons and holes) are injected by the anode and cathode, and are swept through the device driven by the electric field. Some of the electrons and holes combine within the device to form singlet and triplet excitons. The radiative recombination of the excitons in the polymeric layer gives light.

The external efficiency of OLED is determined by several factors, and can be described as:

$$\eta = \gamma \times r_{st} \times q \times \eta_{Coupling} , \qquad (13.1)$$

where γ, r_{st}, q and $\eta_{Coupling}$ are charge balance, ratio of singlets to triplets, photoluminescent quantum efficiency and optical output coupling efficiency, respectively. Thus the ways can be used to improve the device efficiency include:

1. The number of electrons and holes injected into the device should be balanced and their recombination probability optimized;
2. The triplet states in OLED devices are non-radiative. Development of new phosphorescent materials has enabled the triplet lifetime to be reduced significantly;
3. The photoluminescent quantum efficiency, the ratio between radiative and non-radiative decay processes should be maximized by controlling metal diffusion and exciton-plasmons coupling at the cathode-polymer interface; and

4. The output coupling should be increased by increasing the surface roughness of the substrate-air interface; by optimizing geometry of the surface to reduce the total internal reflection and to enhance surface emission; by modifying the distribution of the different types of optical modes; by two-dimensional patterning and formation of photonic crystals.

The performance of OLEDs has been continuously and rapidly improved. However, there are still some technical issues need to be addressed. The emission intensity of OLEDs deteriorates with time. The absolute efficiency of the device (optical power out per Watt of input power) also varies significantly from device to device. Furthermore, lifetimes also vary depending on the injection current and the wavelength of operation and the contacts and other materials used. Obviously major improvements in efficiency, stability and full color tuning are necessary to enable OLED-based products to reach the market and compete with the existing technologies.

13.3.6.4 *Sensors*

Gas and solution sensors based on organic semiconductors have been demonstrated successfully for a range of polymers. For example, PPys and PTs show conductivity changes upon exposure to both oxidizing gases (e.g. NO_2) and reducing ones (e.g. NH_3).

The major approach to make sensors is to use the operating principle of FET. This is based on the fact that these organic materials can form weak chemical interactions with a variety of species; and the semiconducting behavior of these organic materials permits the signals related to the chemical information can be amplified to the level at which they can be conveniently manipulated by solid state electronics.

The chemical interactions between the species to be detected and the semiconductor result in a marked change in the drain current. Such changes in current are reversible and the original current can be restored by reverse biasing the transistor. This repeatability is an important requirement for sensing technology. The availability of a large number of organic semiconductors will permit systematic identification through 'fingerprinting', in which a particular species produces a unique pattern of responses with different semiconductors.

Organic transistor sensors applicable for sensing water-based species open up a new area of opportunity for detection of various biochemicals and biological materials such as lactic acid and glucose. In a typical design, the source and drain contacts were covered with a hydrophobic insulator and this is to protect the high electric field regions from contact with the water. A microfluidic channel running along the transistor electrical channel was integrated to facilitate analyte delivery.

13.4 Organic Superconductors

In 1964, W.A. Little of Stanford University explored the potential means of increasing the temperature at which certain materials can transfer into a superconducting state and suggested that "it is possible to synthesize organic materials that, like certain metals at low temperatures, conduct electricity without resistance". He also established a model describing the chemical structure of the proposed superconducting organic polymer. This model was based on the use of a long conjugated polymer such as a polyacetylene molecule grafted by polarizable side groups. Though this idea had not been fully realized it did launch and turn out to be a very strong stimulant for the development of organic superconductors. Since then, several different types of organic superconductors have been identified such as Bechgaard salts, fullerenes, acenes, polythiophene, and oligomers of poly-phenylene-vinylene.

13.4.1 *One-dimensional superconductor*

Following Little's concept, a new molecule, tetrathiafulvalene (TTF), was synthesized by Wudl. This molecule contains four sulphur heteroatoms in the fulvalene skeleton that can easily donate electrons when it is combined to electron-accepting molecules. This accomplishment had an extraordinary influence on the further development of organic conductors and allowed the successful synthesis of the first stable organic metal, i.e., the charge transfer complex with metal-like characteristics, tetrathiafulvalene-7,7,8,8-tetracyano-p-quinodimethane (TTF-TCNQ), in 1973. The system is made up of two kinds of flat molecules each forming segregated parallel conducting stacks. TTF serves as the electron donor while TCNQ is the electron acceptor.

This compound can be recognized as an organic (metal) conductor as the orbitals involved in the conduction are associated with the molecule as a whole rather than with a particular atom with carriers in each stacks provided by an inter-stack charge transfer at variance with other organic conductors such as the doped conjugated polymers. A giant conductivity peak of the order of 10^5 $(\Omega \cdot cm)^{-1}$ at 60 K was attributed to the precursor signs of an incipient superconductor though it is not a real superconductor. The synthesis of TTF-TCNQ, from the practical point of view, found a way to suppress the metal-insulator transition and was recognized as a very important step toward the achievement of organic superconductor.

Bechgaard salt, $(TMTSF)_2PF_6$ (TMTSF: tetramethyl-tetraselenafulvalene), could be regarded as the first organic superconductor. It has a superconducting transition temperature of 0.9 K at 1.2 GPa. This system is named after Klaus Bechgaard of the University of Copenhagen. He and other three French colleagues established the theory of organic superconductivity in 1979. In this system, TMTSF served as the electron donor and PF_6^- as the electron accepting anion. The general formula of Bechgaard salt is $(TMTSF)_2X$, where X can be either an octahedral or tetrahedral anion such as AsF_6^-, SbF_6^-, TaF_6^-, NbF_6^-, ReO_4^-, and ClO_4^-, in addition to PF_6^-. This material system is non-ideal Type-II superconductor.

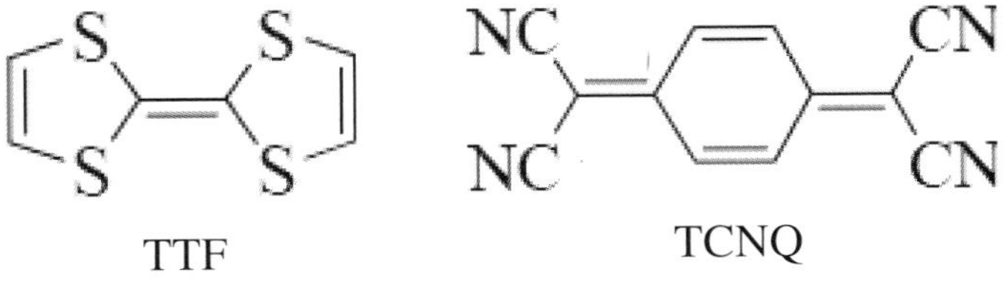

Fig. 13.7 Chemical structures of TTF and TCNQ.

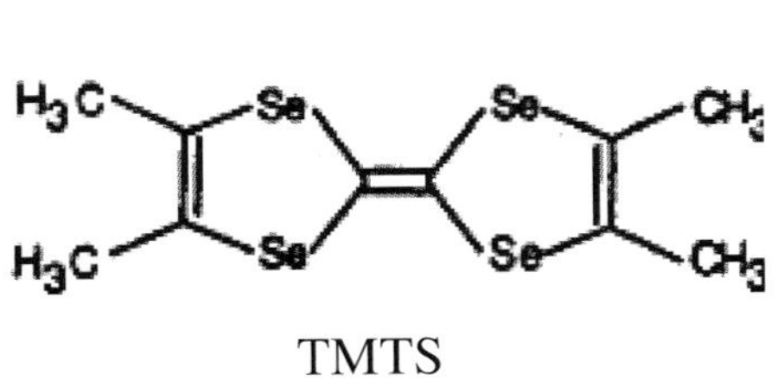

TMTS

Fig. 13.8 Chemical structure of TMTSF.

Bechgaard salts are grown as very thin crystals and usually described as quasi-one dimensional superconductor. Their conductivity is extremely anisotropic due to the stacking of segregated sheets of electron acceptors and donors. Conductivity is the highest along the axis upon which the donor molecules are stacked. Most of them also require considerable pressure to enter a superconducting state. Use of smaller tetrahedral anion could mimic the effect of pressure through closer packing. For example, perchlorate, ClO_4^- was utilized as the electron acceptor in Bechgaard salt to display superconductivity at an ambient level of pressure.

13.4.2 *Two-dimensional superconductor*

Superconductivity exhibited in two dimensions was first realized in charge transfer salts based on bis(ethylenedithio)tetrathiafulvalene (BEDT-TTF), a quasi-two dimensional electron donor molecule. In this system, half of eight sulfur atoms are located at the periphery of the donor molecule, leading to a better orbital overlap. This overlapping of molecular orbitals leads to the formation of broad electron energy bands in the crystal.

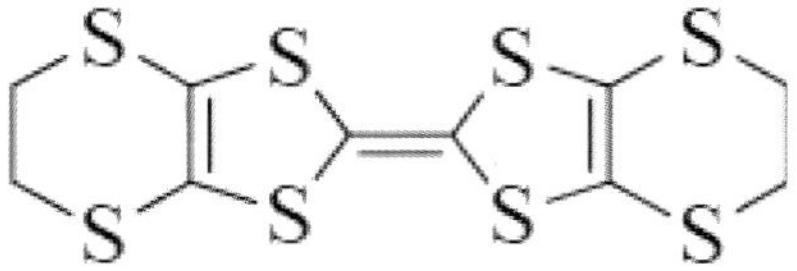

BEDT-TTF

Fig. 13.9 Chemical structure of BEDT-TTF.

A variety of electron acceptors, including linear, tetrahedral and polymeric anions, have been used to prepare two-dimensional conductors with remarkably different compositions. However, superconductivity only occurs when certain anions have been employed, e.g., centrosymmetric anions. In addition to BEDT-TTF, other donors have also been used, such as BEDO-TTF [(bisethylenedioxy)tetrathiafulvalene] and DODHT [(1,4-dioxane-2,3-diyldithio)dihydrotetrathiafulvalene].

The critical temperatures of BEDT-TTF based charge transfer salts are 13.1 K at 0.03 GPa and 12.3 K at ambient pressure.

13.4.3 *Other superconductors*

Superconductivity has been observed in doped fullerene (fulleride) with critical transition temperatures (T_c) up to about 40 K, a value much higher than those yet obtained with quasi-one and two dimensional organic systems. The fullerene consists of 60 carbon atoms that join in a closed sphere, and is often referred to as a buckyball. The C_{60} superconductors can be prepared by:

1.   doping with alkali metal (A) or AM (M = Hg, Bi);
2.   liquid-phase reaction of C_{60} powder with A in toluene under inert atmosphere; or
3.   reaction with alkali and alkali earth azides.

The common superconducting phase is the *fcc* A_3C_{60}. The Cs-doped C_{60} (Cs_3C_{60}) has been reported to have a T_c of 40 K under high pressure (1.5 GPa) and Cs_2RbC_{60} has a T_c of 33 K at ambient pressure.

Hole doped fullerenes normally exhibit higher critical temperatures than electron doped fullerenes mainly due to the higher density of states in the conduction band. The highest critical transition temperature for any organic superconductor, 117 K, was achieved with a buckyball doped with holes and intercalated with $CHBr_3$.

Recently, superconductivity has also been discovered in pure carbon fullerenes. Single-walled and multi-walled carbon nanotubes were reported to have a T_c of 15 K and 12 K, respectively.

13.5 Organic Piezoelectric Materials

Piezoelectric effect has long been reported in biological materials, such as timber, cellulose, wool, bone, tendon and blood vessel of animals. In 1960, this effect was further discovered in polymers. Exceptional piezoelectricity was found in electrically poled polyvinylidene fluoride (PVDF) by H. Kawai in 1969, research has then really risen.

The piezoelectric behaviour and performance of PVDF has been well investigated. PVDF is formally known as 1,1 difluoro-ethylene and has a chemical composition of $(CH_2\text{-}CF_2)_n$. The chain structure of PVDF is schematically shown in Fig. 13.10. It is a semi crystalline polymer with a crystal volume fraction of 35–50%. The crystal structure after extrusion is a classical α phase. This spherulitic non-polar phase has a helical (TGTG') configuration. Mechanical rolling or stretching is then required to induce solid state phase transition to form a highly polar β phase with a planar zig zag, all trans (TTTT) chain configuration. The resulting film has its z-axis perpendicular to the surface, the x-axis in the direction of elongation, and y-axis in the plane of the surface of the film. The succeeding poling produces the preferential orientation of the CF_2 dipoles in PVDF in a direction vertical to the plane of the film. After this biaxial orientation, a spontaneous polarization due to dipole orientation is produced in the film. The variation of this residual polarization under the external stress will cause anisotropic piezoelectricity.

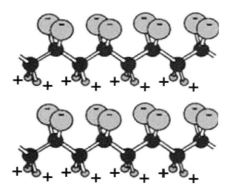

Fig. 13.10 Chain structure of PVDF (M. Lindner, H. Hoislbauer, R. Schwodiauer, S. Bauer-Gogonea and S. Bauer, Charged cellular polymers with "ferroelectretic" behaviour, IEEE Transactions on Dielectrics and Electrical Insulation, **11**, 2004, pp255-263).

The piezoelectric tensor can be described by five piezoelectric coefficients: longitudinal piezoelectric coefficients d_{31} and d_{32}, the transverse piezoelectric coefficient d_{33} and the shear coefficients d_{15} and

d_{24}. Typical values of the piezoelectric coefficients of PVDF are d_{33} = -20 – -23 pC/N, d_{31} = 16 pC/N and d_{32} = 3 pC/N for uniaxially stretched, and d_{33} = -20 – -25 pC/N, d_{31} = d_{32} = 5 pC/N for biaxially stretched PVDF, with a maximum temperature of use around 75–80°C.

Piezoelectric behaviour has then been reported in a number of other polymers including polyvinylidene cyanide and its copolymers, aromatic and aliphatic polyureas, polyvinyl chloride (PVC), aromatic polyamides, PVDF copolymers with trifluoroethylene (P[VDF-TrFE]), tetrafluoroethylene (P[VDF-TFE]), and hexafluoropropylene (P[VDF-HFP]), PVDF blends with polymethyl methacrylate (PMMA), polyvinyl fluoride (PVF), polyvinyl acetate, and Nylon 11. Copolymers depict slightly larger values of piezoelectric coefficients and an extended temperature range of use up to 110°C.

In comparison with inorganic piezoelectric material, organic piezoelectric materials have the advantages of smaller acoustical impedance, low Young's modulus, large mechanical deflection, flexibility, ease of thin film fabrication, and nearly perfect match with water, other liquids, and human body; thus show a broad prospect for hydrophone, sensor, actuator, cable, stress gauge, transducer, and various equipment parts for medical applications.

Summary:

Organic electronic materials mainly include organic conductors, semiconductors, superconductors, piezoelectric and thermoelectric materials. The most exciting development of organic electronic materials in optoelectronics is that devices based on them can have comparable or even better performance than those based on conventional inorganic materials. Particularly, these devices of specific functions can be fabricated using materials with tailored composition, structure and property. Obviously, organic synthesis has advantages of lower-technology processing and higher flexibility. Consequently, organic electronic materials are rapidly finding more and more applications in biology, chemistry, electronics, physics and information technology.

This chapter introduces the concept, synthesis, characterization and application of typical organic conducting, semiconducting, superconducting and piezoelectric materials.

Important Concepts:

Conducting organic materials: these are organic compounds that can conduct electricity. These materials have an electrical conductivity typically ranging from 10^{-16} to 10^5 S/cm. They can be categorized into two groups according to the conducting mechanisms: (1) conducting polymeric composites and (2) intrinsically conducting polymers.

Highest occupied molecular orbital (HOMO): this is the orbital of highest energy that could act as an electron donor in organic semiconductors and is similar to the valence band of inorganic semiconductors.

Lowest unoccupied molecular orbital (LUMO): this is the orbital of lowest energy that could act as the electron acceptor in organic semiconductors and is similar to the conduction band of inorganic semiconductors.

Ambipolar organic semiconductor: organic semiconductors that can be both *p*-type and *n*-type and do not have any particular properties favouring electrons or holes.

Questions:

1. Compare the conducting mechanisms of conducting polymers and conventional inorganic conductors/semiconductors.
2. Discuss the potentials ways that can be used to improve the efficiency and performance of organic light emitting diodes.
3. Discuss the potentials ways that can be used to enhance the carrier mobility of organic semiconductors.

4. For an ambipolar organic semiconductor, how can its carrier type be controlled?

5. Discuss the structural feature and materials development of organic superconductors.

6. Discuss the structural characterisitcs and origin of piezoelectric property in PVDF.

Chapter 14

Nanomaterials for Electronic Device Applications

14.1 Introduction

Nanomaterials and nanotechnology have attracted enormous attention in the recent decade from both scientific and industrial communities. The definition of nanotechnology from National Nanotechnology Initiative (NNI) of USA is: (1) research and technology development at the atomic, molecular or macromolecular levels, in the length scale of approximately 1–100 nm range; (2) creating and using structures, devices and systems that have novel properties and functions because of their small and/or intermediate size; and (3) ability to control or manipulate on the atomic scale. The definition from The Royal Society UK is "the design, characterisation, production, and application of structures, devices, and systems by controlling shape and size at the nm scale."

From the above definitions, it can be seen that nanomaterials and nanotechnology cover an extremely large area. This chapter is focused on nanomaterials and their applications as electronic devices.

Nanomaterials may be defined as the materials whose structural elements, crystallites, molecules or clusters have at least one dimension smaller than 100 nm. Using a relatively broad and simple approach, nanostructured materials can be classified into the four main groups:

1. Zero-dimensional (quantum dots or nanoclusters);
2. One-dimensional (nanorods, nanowires, or nanotubes);
3. Two-dimensional (nanobelts); and
4. Three-dimensional (nanosprings).

A wealth of interesting and new phenomena are associated with nanostructured materials, with the best established examples including size-dependent excitation or emission, quantized conductance, single electron tunnelling (SET), and metal-insulator transition. Obviously, quantum confinement of electrons by the potential wells of nano-sized structures provides the most powerful method to control these fundamental chemical, electrical, optical, and magnetic properties, and so to create a variety of materials, structures, and devices with unprecedented property, performance, and function. The most successful example is provided by the evolution of integrated circuits in microelectronics, where a smaller size has meant more components per chip, faster operation speed, lower cost, less power consumption, and in short, a greater performance.

14.2 Techniques for Preparation of Nanomaterials

One goal of nanotechnology is to make nanostructured materials with tailored properties compared to those of large-sized counterparts. In recent years, a number of chemical, mechanical and physical methods have been developed for the preparation of nanostructured materials with well-controlled composition, dimension, shape, structure and property. The mechanisms concerning crystal nucleation, continuing growth, structural/morphological evolution and property modulation, were intensively investigated. Results obtained enrich the knowledge base of materials science and nanotechnology. These nanostructured materials are currently the candidates for building of various micro- and nano-scale devices through top-down or bottom-up paradigms.

14.2.1 *Quantum dots*

Quantum dots (QDs), also known as nanocrystals, are nano-scale particles that exhibit a phenomenon of quantum size effect. For semiconductors, when the size of particles decreases, their energy gap between conduction and valence bands increases, leading to a blue shift of the emission wavelength, i.e., size-dependent excitation or emission.

Semiconductor QDs have been prepared by a variety of physical and chemical methods, such as vapour-phase deposition and colloidal chemistry approaches.

14.2.1.1 *Vapour-phase approach*

QDs can be prepared by physical vapour deposition techniques, such as MBE or MOCVD, onto appropriate substrates. There are two growth modes. The first one is Stranski-Krastinow (S-K) growth. In this growth mode, nano-sized islands are formed when several monolayers (about 3–10) of atoms or molecules are deposited onto the substrate surface. The deposited and substrate materials have a large lattice mismatch. Epitaxial growth initiates in a layer-by-layer fashion and transforms to 3D island growth to minimize the strain energy contained in the film. The islands then grow coherently on the substrate without generation of misfit dislocations until a certain critical strain energy density, corresponding to a critical size, is exceeded. Beyond this critical size, the strain of the film-substrate system is partially relieved by the formation of dislocations near the edges of the islands. This type of growth has been demonstrated for Ge/Si, InGaAs/GaAs, InP/GaInP and InP/AlGaAs.

The second mode of growth first produces a near-surface quantum well (QW) and then deposits coherent S-K islands on top of the outer barrier layer of the QW that have a large lattice mismatch with the barrier that subsequently produces a compressive strain in the island. The large resultant strain field can extend down into the QW structure to produce a quantum dot with three-dimensional confinement. Such types of growth have been reported for InGaAs and GaAs QDs on a GaAs/InGaAs/GaAs QW and on an AlGaAs/GaAs/AlGaAs QW.

14.2.1.2 *Chemical synthesis*

14.2.1.2.1 Synthesis in organic solvent

Soft chemical synthesis using organic solvents has been widely used to prepare nanocrystal quantum dots of various semiconductors. The synthesis of monodisperse colloids via homogeneous nucleation requires a temporal separation of nucleation and growth of the seeds. This is achieved typically through two experimental techniques.

In the first method, reagents are rapidly injected into hot coordinating solvent so that the precursor concentration in the reaction solution is raised above the nucleation threshold. This leads to an instantaneous nucleation, which is quickly quenched by the fast cooling of the reaction mixture and by the decreased supersaturation after the nucleation burst. Another approach is to achieve the degree of supersaturation necessary for homogeneous nucleation by supplying thermal energy, i.e., by heating-up the reaction solution. At the critical temperature, precursors will decompose sufficiently and quickly to result in supersaturation. Growth is maintained by the addition of monomer from the solution to the nanocrystal quantum dot nuclei. Once the concentration of monomer is sufficiently depleted, continuous growth can be sustained by Ostwald ripening mechanism, i.e., growth of larger particles is at the expense of sacrificial dissolution of smaller particles with higher surface energy.

Temperature, reaction time and precursor concentration are the most important processing variables for control of average size and size distribution of QDs produced using this approach. It was observed that a relatively longer reaction time intends to increase the average size. Lower nucleation temperatures support lower monomer concentrations and yield nuclei of a larger size, whereas higher growth temperatures favour the formation of large-sized QDs since the rate of monomer addition to existing particles is increased and Ostwald ripening occurs more readily as well. Precursor concentration can influence both the nucleation and the growth processes. If all other processing conditions are kept at the same, a higher precursor concentration would tend to promote the formation of fewer and larger nuclei, thus, larger QD size.

This synthesis method, initially developed for CdSe QDs, has been adapted to II–VI and III–V semiconductors including CdS, CdTe, Cu- and Mn-doped ZnSe, InP, InAs, InN, GaP, GaN and AlN.

14.2.1.2.2 Aqueous synthesis

Synthesis of water insoluble II–VI and IV–VI semiconductor QDs can be realized by using chemical precipitation reaction in aqueous media. The pioneering work on aqueous alternative to the organometallically synthesized CdSe-based QDs was demonstrated by the development of thiol-capped CdTe QDs.

In a typical procedure, $Cd(ClO_4)_2 \cdot 6H_2O$ (or other soluble Cd salt) is dissolved in water and an appropriate amount of thiol is added. The use of different thiols is to control reaction kinetics, to passivate surface dangling bonds and to provide stability, solubility and surface functionality of QDs. After pH adjustment, H_2Te gas is passed through the solution. NaHTe solution can be taken out and injected in the reaction flask. CdTe precursors are formed at the stage of Te precursor addition; formation and growth of QDs proceed upon refluxing at 100°C. The reactions can be expressed below:

$$Cd^{2+} + H_2Te \rightarrow Cd\text{-}(SR)_xTe_y + 2H^+, \tag{14.1}$$

$$Cd^{2+} + NaHTe \rightarrow Cd\text{-}(SR)_xTe_y + H^+ + Na^+, \tag{14.2}$$

$$Cd\text{-}(SR)_xTe_y \rightarrow CdTe \ (QDs). \tag{14.3}$$

A major technical problem associated with aqueous synthesis of QDs is that their size distribution in water is relatively broad. One of the most popular approaches is the size-selective precipitation from solvent – non-solvent mixtures. The method exploits the difference in solubility of smaller and larger particles. As-prepared nanoparticles with a broad size distribution are dispersed in a solvent and a non-solvent is added. The largest nanoparticles, exhibiting the greatest attractive van der Waals forces, aggregate before the smaller particles. The aggregates primarily consisting of the largest nanoparticles can be separated by centrifugation or filtration and re-dissolved. The next portion of non-solvent is added to the supernatant to isolate the second size-selected fraction, and so on. Each size-selected fraction can be subjected again to size selection to further narrow the size distribution.

Aqueous synthesis approach has been successfully used to prepare QDs of CdS, CdSe, CdTe, $Cd_xHg_{1-x}Te$, HgTe and ZnSe.

14.2.2 *One-dimensional nanostructures*

One-dimensional (1D) nanomaterials have attracted much attention in recent years. It is envisaged that the next generation nanocircuits, nanotools, nanowire lasers, photon tunnelling devices, and near field photo-waveguide devices, will be built using various functional 1D semiconductor nanomaterials. As such, lots of efforts have been made to synthesize and characterize the compositional, structural, and functional properties of 1D nanostructured materials and further to assemble these materials using novel approaches for micro- and nano-devices. These are the frontier of research in nanotechnology.

14.2.2.1 *Homogeneous nanorods and nanowires*

14.2.2.1.1 Vapour-Liquid-Solid growth

Vapour-Liquid-Solid (VLS) growth was originally developed by Wagner & Ellis to produce micrometer-sized single-crystalline whiskers in 1960s. In a typical process, species from the vapour phase condense onto a miscible liquid, followed by supersaturation of a species and its subsequent precipitation, forming the solid phase. The anisotropic crystal growth is promoted by the presence of the liquid alloy/solid interface.

VLS mechanism is very successful for fabrication of single crystalline nanowires in large quantities. The growth process starts with the dissolution of gaseous reactants into nano-sized liquid droplets of a metal catalyst, followed by nucleation and growth of single-crystalline rods and/or wires. The 1D growth is induced and dictated by the liquid droplets, whose sizes remain essentially unchanged during the entire growth process. Because each liquid droplet serves as a virtual template to strictly limit the lateral growth of an individual nanostructure, the size of the catalyst particle largely determines the diameter of each growing rod/wire. Smaller catalyst islands yield thinner rods/wires.

Vapour phase reactant species required for nanostructure growth can be generated using various methods of chemical and/or physical nature, such as chemical vapour deposition, laser ablation and thermal evaporation. Metal catalyst particles can be deposited onto substrates using evaporation, laser ablation, sputtering of a source target, decomposition of flowing gaseous precursors, or metallorganic chemical vapour deposition (MOCVD). The metal catalyst should be able to form a miscible liquid phase with the nanostructure component to be grown, and the metal in the solid phase cannot be more thermodynamically stable than the nanostructure material.

The VLS process is widely used for the synthesis of a rich variety of one-dimensional nanostructural materials, typically including elemental semiconductors (Si, Ge), III–V semiconductors (GaN, GaAs, GaP, InP, InAs), II–VI semiconductors (ZnS, ZnSe, CdS, CdSe), oxides (ITO, ZnO, MgO, SiO_2, CdO), carbides (SiC, B_4C), and nitrides (Si_3N_4).

14.2.2.1.2 Vapour-Solid growth

Vapour-Solid (VS) growth involves the transportation and condensation of a flowing vapour-phase precursor produced by chemical reduction, evaporation, laser ablation, MOCVD or sublimation. Epitaxial growth of 1D nanostructure, such as nanowires and nanorods, of a number of metal oxides can be achieved without metal catalyst particles on the substrate, based on this growth mechanism.

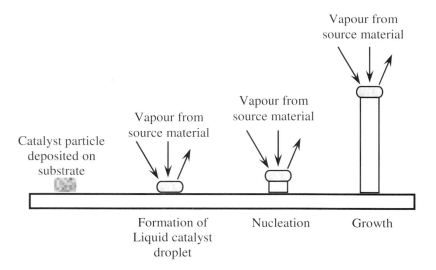

Fig. 14.1 Synthesis of 1D nanostructures through VLS mechanism.

The mechanisms responsible for VS growth are not completely elucidated. Based on thermodynamic and kinetic considerations, the growth of nanorods and nanowires could be possibly through (1) an anisotropic growth mechanism, (2) Frank's screw dislocation mechanism, (3) a different defect-induced growth model, or (4) self-catalytic VLS. In an anisotropic growth mechanism, 1D growth is accomplished by the preferential reactivity and binding of gaseous reactants along specific crystal facets and also the tendency for a system to minimize its surface energy. In the dislocation and defect-induced growth models, structural defects are known to have larger sticking coefficients for gas phase species, thus allowing enhanced reactivity and deposition of gas phase reactants at these locations.

The VS method has been used to create nanowires, nanoribbons and a variety of more complex morphologies of Al_2O_3, CdO, CuO, Ga_2O_3, In_2O_3, SeO_2, SiO_2, SnO_2, ZnO and β-SiC.

14.2.2.1.3 Carbothermal reactions

Synthesis of 1D nanostructures using carbothermal reactions is a relatively simple process. Carbon, such as activated carbon and carbon nanotubes, is mixed with an oxide at elevated temperatures to generate sub-oxidic vapour species which can further react with C, O_2, N_2 or NH_3 to produce nanorods or nanowires of carbides, oxides or nitrides (e.g. SiC, B_4C, Si_3N_4, Al_2O_3, MgO, ZnO).

The most commonly used reactions can be depicted as the follows:

$$\text{metal oxide} + C \rightarrow \text{metal suboxide} + CO,$$

$$\text{metal suboxide} + O_2 \rightarrow \text{metal oxide nanowires},$$

$$\text{metal suboxide} + NH_3 \rightarrow \text{metal nitride nanowires} + CO + H_2,$$

$$\text{metal suboxide} + N_2 \rightarrow \text{metal nitride nanowires} + CO,$$

$$\text{metal suboxide} + C \rightarrow \text{metal carbide nanowires} + CO.$$

A popular example of carbothermal reaction for low-dimensional nanomaterials synthesis is heating a mixture of Ga_2O_3 and carbon in N_2 or NH_3 to produces GaN nanowires.

14.2.2.1.4 Template-assisted growth

In this technique, the template with special structural features serves as a scaffold and the growing material is in-situ shaped into a nanostructure having a dimension and morphology complementary to that of the template.

The most widely used template is the porous alumina produced by anodization, i.e., the so-called anodic alumina membranes (AAMs) containing a hexagonally packed 2D array of cylindrical pores with a uniform size. Other popular templates include mesoporous materials with nano-scale channels and polycarbonate membranes. These nanoscale channels are filled using solution, sol-gel, or electrochemical methods to produce nanorods or nanowires. Free-standing nanostructures are then released from the templates by removal of the host matrix.

A variety of materials have been fabricated into nanowires using AAMs. These inorganic or organic materials typically include Ag, Au, Pt, In_2O_3, MnO_2, SnO_2, TiO_2, ZnO, CdS, CdSe, CdTe, conducting polymers (polypyrole, poly(3-methylthiophene), polyaniline) and carbon nanotubes (CNTs).

Filling of single-walled carbon nanotubes (SWNTs) can also be used to produce a variety of metal nanowires of 1–1.4 nm diameter. Nanowires of Ag, Au, Pd and Pt have been synthesized by employing sealed tube reactions, as well as solution methods.

Mesophase structures that are self-assembled from surfactants can be used as templates for synthesis of versatile low-D nanostructures. A typical example of this approach is the utilisation of anisotropic structures which are formed by spontaneous organization of surfactant molecules at critical micellar concentration (CMC). The rod-shaped micelles can be used as soft templates to promote the formation of nanorods when coupled with appropriate chemical or electrochemical reactions. Nanowires of CuS, CuSe, CdS, CdSe, ZnS and ZnSe have been grown by using Na-AOT and Triton X as the surfactants.

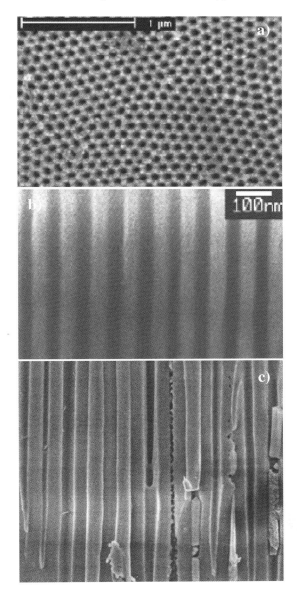

Fig. 14.2 Top-view (a) and cross-sectional (b) images of anodic aluminium oxide films. (c) copper nanowires deposited inside the nanochannels of the AAO film. (R. Bertholdo, M.C. Assis, P. Hammer, S.H. Pulcinelli and C.V. Santilli, Journal of the European Ceramic Society, **30**, 2010, pp181-186).

14.2.2.1.5 Solution-Liquid-Solid growth

The operation principle of Solution-Liquid-Solid growth (SLS) is similar to that of VLS growth. In this method, metallorganic liquid-phase precursors are decomposed in solution at low temperatures. Low melting-point metal nanoclusters (such as Bi, In and Sn) are used as the catalysts. This strategy had been successfully used to synthesize nanorods, nanowires and nanowhiskers of III–V or II–VI semiconductors, including GaAs, InAs, InN, InP, GaP, $Al_xGa_{1-x}As$, CdSe, CdTe and ZnTe in hydrocarbon solvents at relatively low temperatures (<300°C).

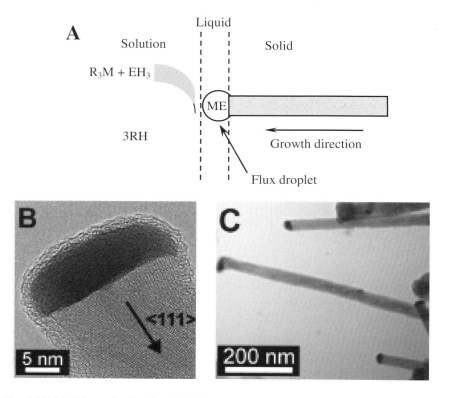

Fig. 14.3 (a) Schematic showing the Solution-Liquid-Solid growth mechanism (redrawn from T.J. Trentler, K.M. Hickman, S.C. Geol, A.M. Viano, P.C. Gibbons and W.E. Buhro, *Science*, **270**, 1995, pp1791-1794); (b) Si nanowire with a Au seed at the tip; and (c) Si nanowires grown using Bi nanocrystals as seeds (A.T. Heitsch, D.D. Fanfair, H.Y. Tuan, and B.A. Korgel, *Journal of the American Chemical Society*, **130**, 2008, pp5436-5437).

A variation of this method is the so-called supercritical fluid-liquid solid (SFLS) method. This solution phase self-assembly approach had been used to produce high-quality, defect-free, single-crystalline silicon (Si) and germanium (Ge) nanowires. These Si nanowires with a diameter of 4-5 nm were seeded and directed with monodisperse Au nanoclusters with a diameter of 2.5 nm in supercritical hexane at 500 °C and 200–270 bar. Most recently, Si nanowire growth by the SLS mechanism at atmospheric pressure using trisilane (Si_3H_8) as a reactant in octacosane ($C_{28}H_{58}$) or squalane ($C_{30}H_{62}$) was reported. In this study, gold (Au) or bismuth (Bi) nanocrystals were used as seeds.

14.2.2.1.6 Solvothermal synthesis

Solvothermal method is extensively used as a solution route to produce semiconductor nanowires and nanorods. In this process, a solvent is mixed with metal precursors and crystal growth regulating or templating agents, such as amines. This mixed solution is then transferred into an autoclave maintained at relatively high temperatures and pressures to carry out the crystal growth and assembly process.

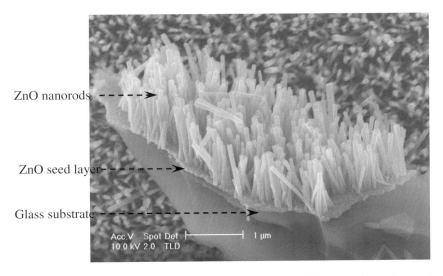

Fig. 14.4 Aligned ZnO nanorod arrays grown from a ZnO seed layer on glass substrate in a solution of $Zn(NO_3)_2$ and HMT.

A variation of this growth technique is the homoepitaxy with a textured thin film deposited on a non-epitaxial substrate (such as silicon or glass) to act as a nanowire nucleation/seeding layer. This approach has been used for the growth of high-quality vertically-aligned ZnO nanowire/nanorod arrays. The seed layer is a very thin ZnO film which can be prepared by vapour-phase based technique or other simple methods such as decomposition or hydrolysis of zinc salts in aqueous solutions. The seeded substrate is then immersed into aqueous solution containing zinc precursor, e.g., $ZnCl_2$ or $Zn(NO_3)_2$, and HMT, ammonia, or NaOH, and heated to a temperature typically ranging from 60 to 95°C for a certain period.

14.2.2.2 *Axial heterogeneous one-dimensional nanostructures*

In catalyst assisted growth, one-dimensional axial heterostructures can be created by changing the flow of vapor-phase semiconductor precursor. A requirement for this is that the nanocluster catalyst should be suitable for the growth of different 1D nanostructures under similar growth conditions. In principle, the vapor-phase reactants can be generated by laser ablation of solid targets or CVD.

By using Au as the catalyst, p-Si/n-Si, Si/Ge, InAs/InP and GaAs/GaP junctions and superlattices had been successfully fabricated by dopant or source material modulation. These novel structures provide versatile building blocks for nanoscale electronic and photonic devices.

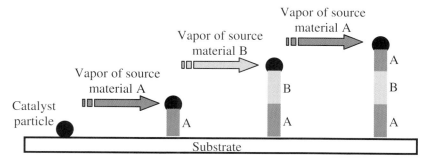

Fig. 14.5 Schematics showing the growth of axial heterogeneous one-dimensional nanostructures.

14.2.2.3 *Core-Shell nanostructures*

One-dimensional core-shell structures can be achieved by coating a core nanowire/nanorod to produce a shell composed of a different material. The growth for the core and the shell can be the same and can occur in the same reaction chamber or can be done under completely different conditions.

14.2.2.3.1 Multi-step growth

In this process, the core (nanorod or nanowires array) is synthesized from one material first by VLS process. Radial growth of a second material is then achieved by depositing the second material on the surface of the core to form the shell. In general, the shell is produced by atomic layer deposition (ALD), pulsed laser deposition (PLD), sputter deposition or pyrolysis. Particularly, the axial and radial growth of core and shell can be realized by changing the reaction kinetic. For example, intrinsic silicon (19 nm in diameter) core and p-type Si shell (p-Si) can be fabricated by altering growth modes during CVD.

This procedure has been reported to successfully prepare core-shell nanostructures of Ge/Si, GaP/GaN, n-GaN/InGaN/p-GaN, GaN/MnGaN, GaN/AlN/AlGaN, GaAs/AlInP, GaAs/Ga$_x$In$_{1-x}$P, Ga$_2$O$_3$/TiO$_2$, Ga$_2$O$_3$/ZnO, ZnO/Er$_2$O$_3$, ZnO/TiO$_2$, MgO/LaCaMnO$_3$, MgO/LaSrMnO$_3$, MgO/La$_{0.67}$Ca$_{0.33}$MnO$_3$, MgO/PbZr$_{0.58}$Ti$_{0.42}$O$_3$, MgO/YBa$_2$Cu$_3$O$_{6.66}$ and Ge/Si/Al$_2$O$_3$/Al.

14.2.2.3.2 One-step growth

In the one-step growth mode, all precursors for the core and the shell are introduced to the growth chamber at the same time. This occurs due to differences in the reactivity of the core and shell materials.

An excellent example is the fabrication of CdS/ZnS core-shell heterojunctions. Precursors for CdS (Cd33) and ZnS (Zn44) were fed into the system simultaneously. Cd33 has a higher reactivity than Zn44 so it decomposes at the catalyst first, leading to the growth of CdS core. The formed CdS core subsequently serves as a catalyst for thermal decomposition of Zn44, resulting in the deposition of ZnS layer on the core surface. Ge/SiO$_x$, GaN/BN, GaP/C, and GaP/SiO$_x$/C core/shell nanocables and In/GaN, ZnS/SiO_2, $ZnO/ZnGa_2O_4$ and ZnO/(Mg,Zn)O radial nanowires/nanorods can be prepared using this approach.

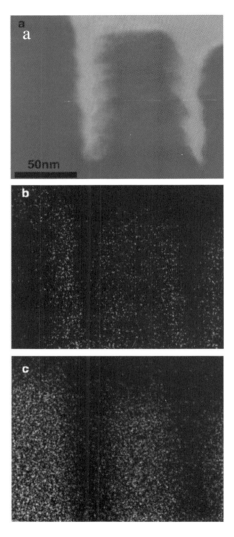

Fig. 14.6 (a) The cross-sectional image of the $ZnO/ZnGa_2O_4$ core-shell nanorods fabricated by reactive evaporation; (b) and (c) Corresponding Ga and Zn EDX spectroscopic mapping images. (C.L. Hsu, Y.R. Lin, S.J. Chang, T.S. Lin, S.Y. Tsai and I.C. Chen, Chemical Physics Letters, **411**, 2005, pp221–224).

14.2.2.4 *Nanotubes*

14.2.2.4.1 Carbon nanotubes (CNTs)

CNTs were first produced by arc discharge. In this process, a DC arc voltage (12–25 V and 50–120 A) is applied between two graphite rods in the presence of catalytic particles installed in a vacuum chamber back filled with He or Ar. The anode evaporates to form fullerenes while, a small part of the evaporated anode is deposited on the cathode, which includes CNTs.

There are several variations that can affect the arc discharge process. The presence of specific metal catalytic particles determines whether the method produces single-walled carbon nanotubes (SWNTs) or multi-walled carbon nanotubes (MWNTs). The use of Ni-Co, Co-Y and Ni-Y catalysts increases the amount of SWNTs produced with pure metals. In contrast, adding W to the Co decreases the yield of SWNTs. By changing the ratio of Ar to He, the diameter of the SWNTs can be controlled, with a larger percentage of Ar yielding smaller diameters. The overall gas pressure has been shown to affect the weight percent yield of SWNTs. Both SWNTs and MWNTs made from this process are now commercially available though extensive purification is needed before use.

The laser vaporization method was applied for production of CNTs in 1996, especially SWNTs in large-scale (gram quantities). The system typically includes a furnace, a quartz tube with a window, a target which is carbon composite doped with catalytic metals (Co or Ni), and flow control systems for the buffer gas (Ar or He). A laser beam (typically a YAG or CO_2 laser) is introduced through the window and focused onto the target. When the target is heated with the laser, carbon phases containing metal catalysts were ablated. The catalysts were then cooled and crystallized into nanoparticles from which the SWNTs grow.

CVD is another popular method for CNTs fabrication. In this process, a hydrocarbon vapour is thermally decomposed in the presence of a metal catalyst at a sufficiently high temperature (600–1200 °C). This hydrocarbon source can be solid, liquid, or gas, such as acetone, acetylene, alcohol, benzene, camphor, ethylene, methane, naphthalene, toluene, and xylene. When liquid is employed, it is first heated in a flask to generate vapour that is carried into the reaction furnace by an inert gas flow. When solid hydrocarbon sources are employed, they can be vaporized in another furnace at low temperature.

A wide range of transition and rare earth metals (Fe, Co, Ni, V, Mo) have been investigated for the synthesis of CNTs by CVD. The catalysts can also be solid, liquid or gas, and can be placed inside the furnace or fed in from outside. Pyrolysis of the catalyst vapour at a suitable temperature liberates metal nano-particles *in situ*. Alternatively, substrates coated with nano-particles of catalyst can be placed in the hot zone of the furnace to catalyze CNT growth. Catalytically decomposed carbon species of the hydrocarbon dissolve in the metal nano-particles and, after reaching supersaturation, precipitate out in the form of a fullerene dome extending into a carbon cylinder. When the substrate-catalyst interaction is strong, a CNT grows up with the catalyst particle rooted at its base, i.e., attached to the surface (base growth model). When the substrate-catalyst interaction is weak, the catalyst particle is lifted up by the growing CNT and continues to promote CNT growth at its tip (tip growth model).

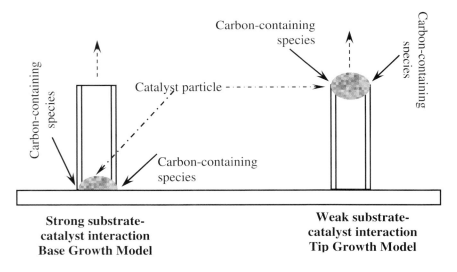

Fig. 14.7 Growth models of CNTs.

These two mechanisms have been indirectly observed for growth of MWNTs and SWNTs, depending on the catalyst type, hydrocarbon source, and growth temperature. In general, tip growth is considered to be the dominant mechanism for MWNTs, while base growth is dominant for SWCNT growth.

Formation of SWNTs or MWNTs is governed by the size of the catalyst particle. In general, when the particle size is a few nanometers, SWNTs form, whereas particles a few tens of nanometers wide favour MWNT formation. Temperature also affects. Low-temperature CVD (600–900 °C) yields MWNTs, whereas a higher temperature (900–1200 °C) reaction facilitates the growth of SWNTs.

CVD seems to offer the best chance of developing a controllable process for the selective production of carbon nanotubes with well-defined properties. Decomposition of C_2H_2 over SiO_2-supported Fe catalysts had been reported to produce very long CNTs (~2 mm). These nanotubes were produced at very high purity, in well-aligned arrays, and in uniform lengths. Successful development of plasma-enhanced hot filament CVD (PE-HF-CVD) also allow CNTs growth at lower temperatures.

14.2.2.4.2 Other inorganic nanotubes

A strategy, called Epitaxial Casting had been developed for the synthesis of various inorganic nanotubes. The principle for the success of this approach is that the core and sheath materials exist in epitaxial registry but possess different chemical stabilities. GaN nanotubes with inner diameters of 30–200 nm and wall thicknesses of 5–50 nm had been prepared using this approach. Hexagonal ZnO nanowires were fully covered by thin GaN layers which were established by epitaxial growth in a MOCVD system. The cores of ZnO nanowires were then removed preferentially by thermal reduction and evaporation in a NH_3/H_2 atmosphere, leading to the formation of ordered arrays of GaN nanotubes.

Similarly, amorphous SiO_2 nanotubes were synthesized using Si nanowires as templates. In this process, thermal oxidation was employed to convert the outer layer of the Si nanowires into a uniform oxide sheath. The remaining inner Si nanowires core was then etched away with XeF_2.

Fig. 14.8 ZnO nanotube obtained by etching of ZnO nanorod in KOH (0.05 M) solution at 95°C for 4 hrs.

AlN, BN, $H_3Ti_2O_7$, In_2O_3 MoS_2, V_2O_5 and WS_2 have been derived using the above-mentioned techniques which have quite different capability of controlling over nanotube positioning and shell thickness.

Chemical etching in aqueous solution have been recently reported for the fabrication of free-standing or aligned nanotubes of ZnO. The first step of this process is to grow ZnO nanorods, mainly using aqueous solution growth and electrodeposition techniques. In the second step, the nanorods are immersed into an etching solution, containing acetic acid, ethylenediamine (EDA), or potassium hydroxide of very low concentration. The solution is then heated to a low temperature and maintained for a certain time period. Selective dissolution leads to the formation of tubular structure.

14.3 Micro-/Nano-devices Using Nanostructured Materials

Low-dimensional nanostructural materials are believed to play an important role in future and emerging nanoscale electronic, optical, and magnetic devices through the bottom-up paradigm. In particular, carbon nanotubes, quantum dots, semiconductor nanocrystals, semiconductor nanowires, nanoscale thin films and organic molecules, may find more and more applications in the near future.

14.3.1 *Carbon nanotube transistor*

The structure of carbon nanotube transistors is very similar to that of the conventional field-effect transistors (FET). The most obvious structural difference is that a single carbon nanotube is serving as the channel. This concept was first demonstrated at Delft University of Technology and IBM in 1998. The first examples were made by dispersing carbon nanotubes on SiO_2-covered Si wafers with prefabricated noble metal electrodes. The Si substrate is used as the bottom gate electrode to modulate the channel conduction and FET action. FETs fabricated in this fashion showed high resistances in the ON state, in excess of 1 MΩ.

The second generation FET has a separate top gate over each individual nanotube. The top-gated CNT FETs allow local gate biasing at low voltage, high speed switching, and high density of integration. In comparing carbon nanotube transistors with Si-based FETs, the figure of merit (normalized the mutual conductance to the channel width) is about 20 times better than existing complementary metal oxide semiconductor (CMOS) devices. Thus, CNT transistors have extremely good limiting performance over Si, and can be used to form simple circuits as demonstrated by IBM.

Although the device structures/configurations of the Si-based and CNT-based transistors are similar, their scaling rules are different. As for their small diameters, CNTs are already ultra-thin-body FETs. Si-based MOSFETs cannot be expected to provide this kind of confinement. Semiconducting carbon nanotubes are ballistic at low bias and length scales below hundreds of nanometers. A decrease in the channel length below this value will not result in any change in resistance. However, the device will still switch faster due to time-of-flight considerations. On the other hand, carbon nanotube transistors with channel lengths above 10 μm will behave more like bulk-switching MOSFETs, showing current saturation.

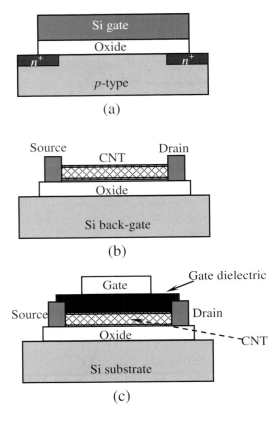

Fig. 14.9 Schematic of a *n*-channel Si MOSFET (a), carbon nanotube transistor with bottom gate (b), and carbon nanotube transistor with top gate (c).

Due to the cylindrical geometry of carbon nanotubes, the gate capacitance in long-channel devices scales with the inverse logarithm of the oxide thickness rather than inversely, as in the planar MOSFET geometry.

In a carbon nanotube transistor, the subthreshold slop, defined as

$$S = [d(log\ I)/dVG]^{-1}. \qquad (14.4)$$

S is dependent on the oxide thickness, while according to MOSFET modeling, this value should be a constant 60 mV/decade at room temperature. Finally, the band gap of carbon nanotubes can in principle be "tuned" by synthesis of nanotubes with the appropriate diameter.

For Si-based MOSFETs, SiO_2 is so far the only choice. The relatively low dielectric constant of SiO_2 ($k = 3.9$) however limits its use in transistors as gate lengths scale down to tens of nanometers. MOSFETs also rely on the nearly perfect interface that forms between silicon and its oxide. Most high-k dielectrics are accompanied by unacceptable mobility degradation in silicon MOSFETs. The gate stack of carbon nanotubes transistors can be optimized to improve their performance. In this respect, carbon nanotubes are much more versatile than silicon, because the carbon nanotube surface has no dangling bonds. Surface roughness scattering is therefore not an issue. Several high-k dielectrics ($k = 20$–30) have already proven successful in carbon nanotube transistors: Al_2O_3, HfO_2, ZrO_2, $SrTiO_3$ and Si_3N_4.

14.3.2 *Single electron transistor*

As the semiconductor feature size enters the sub-50 nm range, the so-called single electron effect will appear in addition to the well-known quantum effect. This effect is related to the quantized nature of the electronic charge: Charging each electron to a small confined region requires a certain amount of energy in order to overcome the Coulomb repulsion; if this charging energy is greater than the thermal energy, $k_B T$ (k_B Boltzman constant and T temperature), a single electron added to the region could have a significant effect on other electrons entering the confined region.

The single electron transistor (SET) is a typical single electron effect device. In conventional transistors, on/off switching is achieved via the motion of large numbers of electrons through the channel between the source and drain, while in single electron devices, this function is operated via the movement of single electrons into and out of a quantum-confined island (or referred to as quantum dot). Such devices exploit an effect called Coulomb blockade (CB) that is the increased resistance at small bias voltages of an electronic device comprising at least one low-capacitance tunnel junction.

A simple SET consists of two tunnel junctions (barriers) sharing one common electrode with a low self-capacitance, i.e., the island or quantum dot. The island is connected to the source and drain through two barriers. In such a way the straight channel in the conventional field effect transistor is replaced by a quantum dot channel. The electrical potential of the island can be tuned by the gate, capacitively coupled to the island. Discrete energy levels are formed inside the quantum dot.

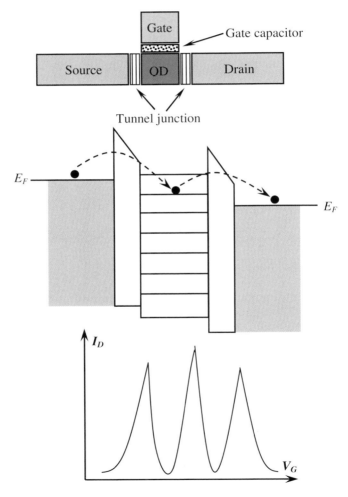

Fig. 14.10 Schematic of structure (a), band diagram (b), and *I-V* characterisitcs (c) of single electron transistor.

How does a SET work? The key point is that electrons pass through the island in quantized units. Electrons can only flow from the source to the drain when an energy level inside the quantum dot is aligned with the source-drain Fermi level. This energy alignment is regulated by the gate voltage. In the blocking state, no accessible energy levels are within tunneling range of the electron on the source. All energy levels on the island with lower energies are fully occupied. When a positive voltage is applied to the gate the energy levels of the island are lowered. The electron can tunnel onto the island, occupying a vacant energy level. From there it can tunnel onto the drain electrode where it inelastically scatters and reaches the drain Fermi level. Thus, instead of a linear relation between the drain current and the gate voltage as in a conventional transistor, a periodic oscillation would occur with each conductance peak representing the addition of a single electron to the quantum dot. This oscillating *I-V* characteristic is called Coulomb oscillation.

Single electron transistors, operating by means of one-by-one electron manipulation utilizing the Coulomb blockade effect has the advantages of faster operating speed, very low power consumption and are very promising as new nanoscale electronic devices. If the single electron transistors are used as ultra-large-scale integrated circuits (ULSIs) elements, the ULSI will have the attributes of extremely high integration and extremely low power consumption.

14.3.3 *Semiconductor nanowire/nanobelt transistor*

One-dimensional nanostructures, such as nanorods, nanowires and nanobelts, represent key building blocks for a variety of future electronic and optoelectronic devices. A typical example of such a device with broad potential for applications is the nanowire field effect transistor (NWFET), in which the nanowires or nanobelts of semiconductors are used as the conducting channel. NWFETs fabricated from group IV, III–V and II–VI semiconductors have demonstrated promising FET characteristics in top-gate, back-gate and surround-gate FET geometries.

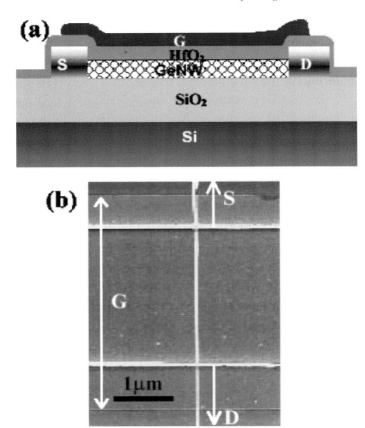

Fig. 14.11 Top-gated Ge nanowire FETs, (a) schematic side view and (b) SEM top view. (D.W. Wang, Q. Wang, A. Javey, R. Tu, H.J. Dai, H.S. Kim, P.C. McIntyre, T. Krishnamohan and K.C. Saraswat, Applied Physics Letters, **83**, 2003, pp2432-2434).

Silicon nanowires (SiNWs) have been extensively studied partially due to the dominance of Si devices in the semiconductor industry and partially due to the well-developed techniques for growth of SiNWs with well-defined compositional, structural, and functional properties. In a typical device geometry, SiNWs are first transferred into liquid suspension after growth via gentle sonication, then dispersed onto a degenerately doped Si substrate with a SiO_2 insulating layer. Source and drain electrodes are normally formed by E-beam or photo-lithography, with the degenerately doped Si substrate serving as the back gate.

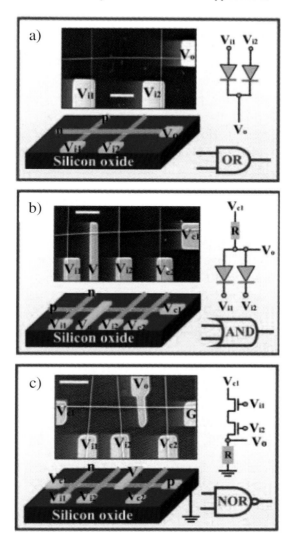

Fig. 14.12 Schematic and SEM of nanowire nano-logic gates. (a) logic OR gate constructed from a 2 by 1 crossed NW *p-n* junction; (b) logic AND gate constructed from a 1 by 3 crossed NW junction array; (c) logic NOR gate constructed from a 1 by 3 crossed NW junction array. Scale bar in SEM represents 1 μm. (Y. Huang, X. Duan, Y. Cui, L. Lauhon, K. Kim and C.M. Lieber, Science, **294**, 2001, pp1313-1317).

Depending on the type of charge carrier of the semiconductor nanowire, the NWFETs can be classified into two major groups, i.e., *n*-channel and *p*-channel nanowire transistors, which are fabricated using *n*-type and *p*-type nanowire, respectively. These transistors will respond in opposite ways to the potential applied to the gate (V_G). When a positive V_G is applied, the holes in the *p*-type nanowire deplete and the conductivity decreases. However, this will leads to an accumulation of electrons and an increase of conductivity in the *n*-type nanowire channel. Conversely, a negative V_G will increase the conductivity of *p*-type nanowire and decrease the conductivity of the *n*-type nanowire.

NWFETs can exhibit performance comparable to the best reported for planar devices made from the same materials. The carrier mobility has been shown to match or exceed those of planar Si; and the devices can achieve gain. Studies have also demonstrated the high electron mobility of epitaxial InAs NWFETs with a wrap-around gate structure. Furthermore, *p*- and *n*-type nanowires can be assembled into crisscross arrays where the junctions of the crossed wires serve as on-off switches. AND, OR, NOR and XOR logic gates have been demonstrated from arrays of Si and GaN nanowires FETs.

NWFETs have also emerged as extremely powerful sensors for ultrasensitive, direct and label-free detection of biological and chemical species. Binding to the surface of an NWFET is analogous to applying a gate voltage, which leads to the depletion or accumulation of carriers and subsequent changes in the NW conductance. The small diameters and high performance of NWFETs yield high sensitivity.

14.3.4 *Si nanocrystal memory*

To increase the storage density and speed and to lower the power consumption of semiconductor memories, the size of each memory cell must be reduced dramatically. This is the impetus for the development of nanoscale semiconductor memories.

A floating-gate Si nanocrystal memory utilizes quantum dot(s) as charge-storage cells which are embedded in the oxide layer between the gate and the channel of a field-effect transistor. When the gate energy is lowered with regard to that of the source and the drain, electrons are injected to the floating gate storage nodes from the channel. Electrons stored on the island screen the charge in the channel and hence lead to less channel charge for the same applied gate-to-channel potential. This is effectively a change in the threshold voltage. These electrons trapped on the islands that influence the conduction of a channel underneath them, and thus the conduction of the channel is a measure of the storage of the electrons. Barriers, used for storage of the electrons, are thus important to the write, erase and refresh conditions. For the floating gate memory, the tunnel oxide provides a very good energy barrier for the charge carriers. So it would be difficult for the charges to leak out from the floating gate, and hence the memory can be nonvolatile. But read of the device, and the amount of signal delivered by the device, are related to the field-effect.

In a single-electron memory, the number of electrons that can be injected into the floating gate during each writing process can be exactly controlled by means of the energy constraint. For example, if the energy increase due to the storing of a single electron (charging energy plus quantization energy) on the floating gate is large compared with the potential energy difference between the channel and the floating gate, transfer of other electrons will be blocked. This makes it possible to store one bit of information by a single electron, and the information can be read by sensing the threshold difference between the two states.

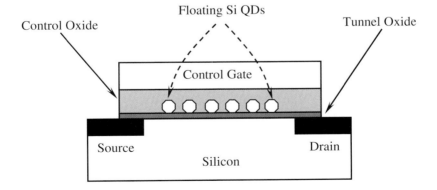

Fig. 14.13 Schematic of memory using Si QDs (nanocrystals/NCs).

To establish such a single-electron MOS memory, two basic requirements must be met:

1. the width of silicon MOSFET channel is less than the Debye screening length, and
2. the floating gate is a nanoscale square (dot).

The narrow channel ensures that storing a single electron on the floating gate is sufficient to screen the entire channel width from the potential on the control gate, leading to a significant threshold voltage shift. A small floating gate is used to significantly increase the quantum confinement energy and the electron charging energy, suppressing the charge fluctuation on the floating gate.

Nonvolatile memories using Si quantum dots (nanocrystals) are being actively developed as a faster, cheaper, higher density alternative to today's flash memory technology, which is widely used in cell phones, digital cameras, PDAs, and other consumer electronic products. To bypass certain scaling limitations of conventional flash memory devices, engineers have replaced the continuous floating gate layer with discrete Si nanocrystals. This allows for a reduction of the thickness of the oxide layer beneath the nanocrystals without sacrificing the reliability of the device. Nanoscale particles can store charge with minimal information loss since any charge leakage that occurs is limited to individual clusters. Low power operation and a simple fabrication process are other benefits of this nanocrystal structure. IBM, Micron Technology, Motorola and Lucent Technologies have made significant advances in this area.

Summary:

The nano-scale is not just another step toward miniaturization, but is a qualitatively new scale. The physical and chemical properties of nanomaterials are significantly different from those of bulk materials with the same chemical composition. The uniqueness of the structural characteristics, electronic structure and chemistry of nanomaterials constitutes the basis of nanoscience and nanotechnology. Low-dimensional semiconductor nanostructures with fascinating size-dependent characteristics (multifunctionality) therefore are very attractive for building up of future high-performance electronic, magnetic and photonic nanodevices using bottom-up paradigms.

In this chapter, typical techniques capable of producing nanostructured materials with well-defined composition, structure and property were first introduced. This was followed by a section discussing how these materials can be used to fabricate electronic devices.

Important Concepts:

Quantum dots (QDs): they are nano-scale semiconductor particles (2–10 nm in diameter) that exhibit a phenomenon of quantum size effect.

Single electron transistor: a device in which the effect of Coulomb blockade can be observed. It turns on and off again every time one electron is added to it. The behavior of the device is entirely quantum mechanical.

Coulomb blockade (CB): an effect describes the increased resistance at small bias voltages of a semiconductor device comprising at least one low-capacitance tunnel junction.

Questions:

1. Make a comparison between vapour-liquid-solid and solution-liquid-solid mechanisms for growth of low-dimensional nanomaterials.
2. One-step growth approach has been used to prepare core-shell low-dimensional nanomaterials (nanorods or nanowires). What is the major requirement for the occurrence of this type of growth mode?
3. Carbon nanotubes can be synthesized by chemical vapour deposition techniques. Please discuss the two main mechanisms behind the growth of CNTs.
4. Compare the fundamental differences between CNTs and conventional field effect transistors.
5. Discuss the configuration and operation principle of the simplest single electron transistor.

Bibliography

Book references:

Appleby, A. J. and Foulkes, F. R. (1989). *Fuel Cell Handbook*, (Van Nostrand Reinhold, New York).

Askeland, D. (1990). *The Science and Engineering of Materials*, (Chapman & Hall).

ASM Metals Handbook, Vol.2, 9th edn., 1979.

Badwal, S. P. S. (1994). *Ceramic Superionic Conductors*, Chapter 11 in *Materials Science and Technology, A Comprehensive Treatment, Volume 11*, Edited by R. W. Cahn, P. Haasen, and E. J. Kramer, Structure and properties of Ceramics, Volume editor M. Swain, (VCH Publishers).

Barsoum, M. (1997). *Fundamentals of Ceramics*, (McGraw Hill Series in Materials science and Engineering, New York).

Barsoum, M. (1997). *Fundamentals of Ceramics*, (McGraw Hill).

Blackwell, G. R. (2000). *The Electronic Packaging Handbook*, (CRC Press, Boca Raton FL).

Borg, R. G. and Dienes, G. J. (1977). *An Introduction to Solid State Diffusion*, (Academic Press, Inc.).

Braithwate, N. and Weaver, G. (1990), *Electronic Materials*, (Butterworths).

Braunovic, M., Konchits, V. V. and Myshkin, N. K. (2006). *Electrical Contacts: Fundamentals, applications and technology*, (CRC Press, Boca Raton, FL).

Bruce, P. J. (1997). *Solid State Electrochemistry (Chemistry of Solid State Materials, No. 5)*, (Cambridge University Press).

Cardwell, D. A. and Ginley, D. S. (2003). *Handbook of Superconducting Materials, Volume* I: *Superconductivity, Materials and Processes*, (Institute of Physics Publishing, Bristol and Philadelphia).

Chin, G. Y., Wernic, J. H. (1981). *Magnetic Materials*, (Wiley).

Cullity, B. D. (1972), *Introduction to Magnetic Materials*, (Addison-Wesley).

Datta, M., Osaka, T. and Schultze, J. W. (2005). *Microelectronic Packaging*, (CRC Press, Boca Raton London New York Washington, D.C.).

Doering, R. and Nishi, Y. (2007). *Handbook of Semiconductor Manufacturing Technology*, 2nd Ed. (CRC Press, Boca Raton, FL).

Faraday, M. (1936). *Experimental Researches in Electricity*, (Taylor and Francis, London), Art., 1339.

Geller, S. (1977). *Solid Electrolytes, Topics in Applied Physics, Volume 21*, (Springer-Verlag, Berlin, Heidelberg).

Geller, S. (1984). *Solid Electrolytes, Topics in Applied Physics, Volume 21*, (Springer-Verlag Publishers).

Gellings, P. J. and Boumeester, H. J. M. (1997). *The CRC Handbook of Solid State Electrochemistry*, (CRC Press).

Grosvenor, C.R.M. (1988), *Materials for Semiconductor Devices*, (IOM).

Harper, C. A. (2003). *Electronic Materials and Processes Handbook*, 3rd Ed. (McGraw-Hill, NY, USA).

Hull, D. (1965). *Introduction to Dislocations*, (Pergamon Press, New York)

It is quite hard to find any books that relate solely to the measurement of ionic and mixed ionic/electronic properties in materials. In general, a lot of the text books listed above contain chapters, or parts of chapters, dealing with this issue. The only book worthwhile mentioning is that edited by Macdonald on impedance spectroscopy.

Jackson, K. A. and Schröter, W. (2000). *Handbook of Semiconductor Technology*, (Wiley-VCH, Weinheim, Germany).

Jiles, D. (1994). *Electronic Properties of Materials*, (Chapman & Hall).

Kofstad, P. (1972). *Nonstoichiometry, Diffusion and Electrical Conductivity in Binary Metal Oxides*, (Wiley, New York).

Kroger, F. A. (1964). *The Chemistry of Imperfect Crystals*, (North Holland, Amsterdam).

Kroger, F. A. and Vink, H. J. (1956). *Solid State Physics*, Volume 3, Editors F. Seitz, and D. Turnbull, (Academic Press), Chapter 5.

Kymissis, I. (2009) *Organic Field Effect Transistors: Theory, fabrication and characterization*, (Springer, NY, USA).

Lebed, A. (2008) *The Physics of Organic Superconductors and Conductors*, (Springer Berlin Heidelberg, NY, USA).

Mattox, D. M. (1998). *Handbook of Physical Vapor Deposition (PVD) Processing: Film Formation, Adhesion, Surface Preparation and Contamination Control*, (William Andrew Publishing/Noyes, USA).

Minh, N. Q. and Takahashi, T. (1995). *Science and Technology of Ceramic Fuel Cells*, (Elsevier).

Nalwa, H. S. (2002). *Handbook of Thin Film Materials*, (Academic Press, USA).

National Research Council (1989). *Materials Science and Engineering for the 1990s*, (National Academy Press).

O'Connell, M. J. (2006). *Carbon Nanotubes: Properties and Applications*, (CRC Press, Boca Raton, FL).

Oda, S. and Ferry, D. (2006). *Silicon Nanoelectronics*, (CRC Press, Boca Raton, FL).

Pierson H. O. (1999). *Handbook of Chemical Vapor Deposition: Principles, Technology and Applications*, 2nd Ed. (Noyes Publications, Norwich, New York, USA).

Proceedings of the First and Second European Seminars on Solid Oxide Fuel Cells.

Proceedings of the International Conferences on Solid Oxide Fuel Cells, Volumes 1 to 5.

Richerson, D. (1992). *Modern Ceramic Engineering*, (Marcel Dekker Inc.).

Rickert, H. (1982). *Electrochemistry of Solids*, (Springer Verlag, Heidelberg).

Robins, D. (1989). *Introduction to Superconductivity*, (IBC Tech. Serv. Ltd.).

Rockett, A. (2008). *The Materials Science of Semiconductors*, (Springer, NY, USA).

Rogach, A. L. (2008). *Semiconductor Nanocrystal Quantum Dots: Synthesis, Assembly, Spectroscopy and Applications*, (Springer/Wien, Austria).

Rose, Shepard and Wulff, (1966). *Electronic Properties*, (John Wiley & Sons).

Ross Macdonald, J., Ed. (1987). *Impedance Spectroscopy: Emphasizing Solid Materials and Systems*, (John Wiley and Sons).

Science and Technology of Zirconia, Volumes 1 to 5 are worth reading.

Sheahen, T. (1994). *Introduction to High-Temperature Superconductivity*, (Plenum Press).

Smith, W. F. (1993). *Foundations of Materials Science and Engineering*, (McGraw-Hill).

Speyer, R. (1994). *Thermal Analysis of Materials*, (Marcel Dekker Inc.)

Steele, B. C. H. (1996). *Ceramic Ion Conductors and their Technological applications*, The Institute of Materials, British Ceramic Proceedings, No 56.

Subbarao, E. C. (1980). *Solid Electrolytes and Their Applications*, (Plenum Press, New York).

Swalin, R. A. (1972). *Thermodynamics of Solids*, Second Edition, (Wiley Interscience, New York).

Takahashi, T., Ed. (1989). *Ionic Conductors, Recent Trends and Applications*, (World Scientific Publishing Company, Singapore).

Tubandt, C. (1932). *Handbuch der Experimenatlphysik*, edited by W. Wien, and F. Harms, Akademische Verlagsgesellschaft, Leipzig, 12, Part 1, 303.

Wan, M. X. (2008). *Conducting Polymers with Micro or Nanometer Structure*, (Springer Berlin Heidelberg, NY, USA).

Warnes, L. A. A. (1990). *Electronic Materials*, (McMillan Education Ltd.).

West, A. R. (1984). *Solid State Chemistry and Its Applications*, (John Wiley and Sons).

Zwikker, C. (1954). *Physical Properties of Solid Materials*, (Pergamon).

Journal references:

Baraton, M. I. (2009). Transparent conductive oxide materials: Financial stakes and technological challenges, *International Journal of Nanotechnology*, 6, pp. 776–784.

Bertholdo, R., Assis, M. C., Hammer, P., Pulcinelli, S. H. and Santilli, C. V. (2010). Controlled growth of anodic aluminium oxide films with hexagonal array of nanometer-sized pores filled with textured copper nanowires, *Journal of the European Ceramic Society*, 30, pp. 181–186.

Canfield, P. C. and Crabtree, G. W. (2003). Magnesium diboride: Better late than never, *Physics Today*, March, pp. 34–40.

Chung, C. K., Chang, Y. L., Chen, T. S. and Su, P. J. (2006). Annealing effects on microstructure and properties of Ta-Al thin film resistors, *Surface & Coatings Technology*, 201, pp. 4195–4200.

Du, J. G., Ko, W. H. and Young, D. J. (2004). Single crystal silicon MEMS fabrication based on smart-cut technique, *Sensors and Actuators A*, 112, pp. 116–121.

Exarhos, G. J. and Zhou, X. D. (2007). Discovery-based design of transparent conducting oxide films, *Thin Solid Films*, 515, pp. 7025–7052.

Fortunato, E., Ginley, D., Hosono, H. and Paine, D. C. (2007). Transparent conducting oxides for photovoltaics, *MRS Bulletin*, 32, pp. 242–247.

Gerard, M., Chaubey, A. and Malhotra, B. D. (2002). Application of conducting polymers to biosensors, *Biosensors & Bioelectronics*, 17, pp. 345–359.

Gudiksen, M. S., Lauhon, L. J., Wang, J. F., Smith, D. C. and Lieber, C. M. (2002). Growth of nanowire superlattice structures for nanoscale photonics and electronics, *Nature*, 415, pp. 617–620.

Heitsch, A. T., Fanfair, D. D., Tuan, H. Y. and Korgel, B. A. (2008). Solution-liquid-solid (SLS) growth of silicon nanowires, *Journal of the American Chemical Society*, 130, pp. 5436–5437.

Hsu, C. L., Lin, Y. R., Chang, S. J., Lin, T. S., Tsai, S. Y. and Chen, I. C. (2005). Vertical $ZnO/ZnGa_2O_4$ core-shell nanorods grown on ZnO/glass templates by reactive evaporation, *Chemical Physics Letters*, 411, pp. 221–224.

Huang, Y., Duan, X. F., Cui, Y., Lauhon, L. J., Kim, K. H. and Lieber, C. M. (2001). Logic gates and computation from assembled nanowire building blocks, *Science*, 294, pp. 1313–1317.

Jerome, D. (2008). The development of organic conductors: Organic superconductors, *Solid State Sciences*, 10, pp. 1692–1700.

Kaiser, A. B. (1991). Metallic behaviour in highly conducting polymers, *Synthetic Metals*, 45, pp. 183–196.

Kelley, T. W., Baude, P. F., Gerlach, C., Ender, D. E., Muyres, D., Haase, M. A., Vogel, D. E. and Theiss, S. D. (2004). Recent progress in organic electronics: Materials, devices, and processes, *Chemistry of Materials*, 16, pp. 4413–4422.

Kiukkola, K. and Wagner, C. *J. Electrochem. Soc.* 10 (1957) 379.

Kuehl, R. W. (2009). Stability of thin film resistors – Prediction and differences base on time-dependent Arrhenius law, *Microelectronics Reliability*, 49, pp. 51–58.

Lacobs, I. S. (1979). *J. Appl. Phys.,* 50 pp. 7294.

Li, S. D., Yu, Z. and Burke, P. J. (2004). Silicon nitride gate dielectric for top-gated carbon nanotube field effect transistors, *The Journal of Vacuum Science and Technology B*, 22, pp. 3112–3114.

Lindner, M., Hoislbauer, H., Schwodiauer, R., Bauer-Gogonea, S. and Bauer, S. (2004). Charged cellular polymers with "ferroelectretic" behavior, *IEEE Transactions on Dielectrics and Electrical Insulation*, 11, pp. 255–263.

Lu, X. M., Fanfair, D. D., Johnston, K. P. and Korgel, B. A. (2005). High yield solution-liquid-solid synthesis of germanium nanowires, *Journal of the American Chemical Society*, 127, pp. 15718–15719.

Myung, N. V., Park, D. Y., Yoo, B. Y. and Sumodjo, P. T. A. (2003). Development of electroplated magnetic materials for MEMS, *Journal of Magnetism and Magnetic Materials*, 265, pp. 189–198.

Nernst, W. and Elektrochem, Z. 6 (1900) 41.

Osaka, T., Asahi, T., Kawaji, J. and Yokoshima, T. (2005). Development of High-performance magnetic thin film for high-density magnetic recording, *Electrochimica Acta*, 50, pp. 4576–4585.

Patel, N. K., Cinà, S. and Burroughes, J. H. (2002). High-efficiency organic light-emitting diodes, *IEEE Journal on Selected Topics in Quantum Electronics*, 8, pp. 346–361.

Pattanaik, G., Kirkwood, D. M., Xu, X. L. and Zangari, G. (2007). Electrodeposition of hard magnetic films and microstructures, *Electrochimica Acta*, 52, pp. 2755–2764.

Saxena, V. and Malhotra, B. D. (2003). Prospects of conducting polymers in molecular electronics, *Current Applied Physics*, 3, pp. 293–305.

Sinnott, S. B. and Andrews, R. (2001). Carbon nanotubes: Synthesis, properties, and applications, *Critical Reviews in Solid State and Materials Sciences*, 26, pp. 145–249.

Sun, Y. M., Liu, Y. Q. and Zhu, D. B. (2005). Advances in organic field-effect transistors, *Journal of Materials Chemistry*, 15, pp. 53–65.

Talin, A. A., Dean, K. A. and Jaskie, J. E. (2001). Field emission displays: A critical review, *Solid-State Electronics*, 45, pp. 963–976.

Tiwari, S. Rana, F. Hanafi, H. Hartstein, A. Crabbe, E. F. and Chan, K. (1996). A silicon nanocrystals based memory, *Applied Physics Letters*, 68, pp. 1377–1379.

Trentler, T. J., Hickman, K. M., Geol, S. C., Viano, A. M., Gibbons, P. C. and Buhro, W. E. (1995). Solution-liquid-solid growth of crystalline III-V semiconductors: An analogy to vapor-liquid-solid growth, *Science*, 270, pp. 1791–1794.

Van Den Broek, J. J., Donkers, J .J. T. M., Van Der Rijt, R. A. F. and Janssen, J. T. M. (1998). Metal film precision resistors: Resistive metal films and a new resistor concept, *Philips Journal of Research*, 51, pp. 429–447.

Wang, D. W., Wang, Q., Javey, A., Tu, R., Dai, H. J., Kim, H., McIntyre, P. C., Krishnamohan, T. and Saraswat, K. C. (2003). Germanium nanowire field-effect transistors with SiO_2 and high-*k* HfO_2 gate dielectrics, *Applied Physics Letters*, 83, pp. 2432–2434.

Wang, F. D., Dong, A. G., Sun, J. W., Tang, R., Yu, H. and Buhro, W. E. (2006). Solution-liquid-solid growth of semiconductor nanowires, *Inorganic Chemistry*, 45, pp. 7511–7521.

Weber, J., Singhal, R., Zekri, S. and Kumar, A. (2008). One-dimensional nanostructures: Fabrication, characterisation and applications, *International Materials Reviews*, 53, pp. 235–255.

Xu, N. S. and Ejaz Huq, S. (2005). Novel cold cathode materials and applications, *Materials Science and Engineering R*, 48, pp. 47–189.

Yang, C., Zhong Z. H. and Lieber, C. M. (2005). Encoding electronic properties by synthesis of axial modulation-doped silicon nanowires, *Science*, 310, pp. 1304–1307.

Index